Signals and Communication Technology

This series is devoted to fundamentals and applications of modern methods of signal processing and cutting-edge communication technologies. The main topics are information and signal theory, acoustical signal processing, image processing and multimedia systems, mobile and wireless communications, and computer and communication networks. Volumes in the series address researchers in academia and industrial R&D departments. The series is application-oriented. The level of presentation of each individual volume, however, depends on the subject and can range from practical to scientific.

Indexing: All books in "Signals and Communication Technology" are indexed by Scopus and zbMATH

For general information about this book series, comments or suggestions, please contact Mary James at mary.james@springer.com or Ramesh Nath Premnath at ramesh.premnath@springer.com.

Ha Nguyen • Long Le • Pradeepa Yahampath •
Ebrahim Bedeer Mohamed
Editors

30th Biennial Symposium on Communications 2021

Springer

Editors
Ha Nguyen *(deceased)*
Saskatoon, SK, Canada

Long Le
University of Quebec
Montreal, QC, Canada

Pradeepa Yahampath
University of Manitoba
Winnipeg, MB, Canada

Ebrahim Bedeer Mohamed
University of Saskatchewan
Saskatoon, SK, Canada

ISSN 1860-4862 ISSN 1860-4870 (electronic)
Signals and Communication Technology
ISBN 978-3-031-06946-8 ISBN 978-3-031-06947-5 (eBook)
https://doi.org/10.1007/978-3-031-06947-5

This Springer imprint is published by the registered company Springer Nature Switzerland AG
The registered company address is: Gewerbestrasse 11, 6330 Cham, Switzerland

To late Professor Ha Nguyen without whose dedication and hard work this book would not have been possible

Contents

PERIDOT Codes: Replacing Identifiers, Sequence Numbers, and Nonces with Permutations

Florian Euchner and Christian Senger

1 Introduction

Low-power wide-area network (LPWAN) protocols such as LoRaWAN, Sigfox, MIOTY, and others are commonly used to wirelessly transmit packets with very small application payloads, sometimes as small as a single bit. Devices connected to such networks are often uplink-only, and their transmissions are powered by small batteries or by energy harvesting.

Nevertheless, since LPWAN devices may transmit sensitive information, the various protocols define authentication schemes based on message authentication codes (MACs) and symmetric encryption schemes. These authentication and encryption methods need nonces, i.e., a number has to be transmitted alongside the application payload, and this number should be reused as rarely as possible, preferably only once. In practice, the sequence number (SN) of packets, a counter that is traditionally used to detect packet loss, gets a dual use as nonce. Protocol designers have to find a trade-off between longer SNs (improved security due to longer intervals between nonce reuses) and low energy consumption: The transmission of every single bit consumes energy, and the device's overall energy consumption is almost proportional to the number of bits transmitted since most other tasks (e.g., reading sensor values) are negligible in their energy consumption.

For instance, in the case of Sigfox, an uplink packet with a single-bit application payload has a total length of 112 bit including the preamble, 12 bit of which are used as SN and nonce [1]. This obviously makes replay attacks feasible since SNs have to be reused every $2^{12} = 4096$ packets. LoRaWAN, on the other hand, employs a 32-bit SN internally, and only the 16 least significant bits of which are transmitted—

F. Euchner (✉) · C. Senger
Institute of Telecommunications, University of Stuttgart, Stuttgart, Germany
e-mail: euchner@inue.uni-stuttgart.de; senger@inue.uni-stuttgart.de

H. Nguyen et al. (eds.), *30th Biennial Symposium on Communications 2021*,
Signals and Communication Technology, https://doi.org/10.1007/978-3-031-06947-5_1

Fig. 1 Channel model: a single Packet Erasure Channel

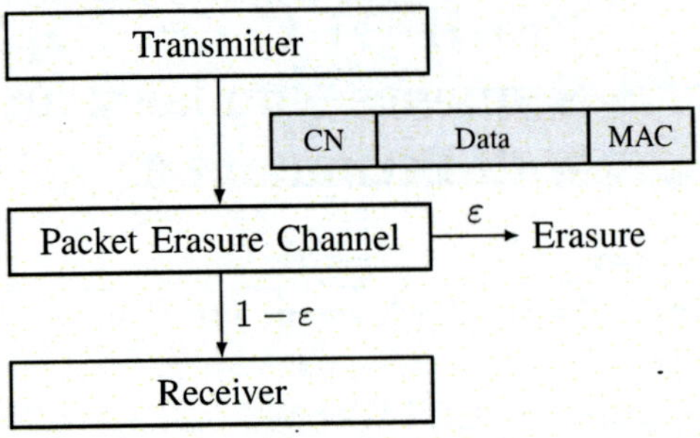

the receiver has to infer the remaining bits from observed uplinks [2]. This scheme is stateful and potentially error-prone.

In our protocol design approach, we replace SNs altogether and merge them with identifiers into a single number, called code number (CN). The same CN is guaranteed to recur as rarely as possible; hence, it can have a dual use as nonce. The gist of the scheme, which we will call Permutation-Based Identification and Order Tracking (PERIDOT), is best explained by two examples based on the following scenario:

We assume that, as shown in Fig. 1, a packet is transmitted over a packet erasure channel (PEC) with erasure probability ε. In our model, the receiver is unaware of the occurrence of an erasure. Let CN be from the alphabet $\mathrm{Z_q} = \{0, \ldots, q-1\}$. Our hard requirement is that, since CNs are used as nonces, once a particular CN has been transmitted, the following $q-1$ packets must not contain the same CN again. Thus, the order in which CNs are transmitted has to be a cyclic permutation of the elements of $\mathrm{Z_q}$. That is, the permutation has only a single cycle whose orbit is $\mathrm{Z_q}$. This permutation $\sigma : \mathrm{Z_q} \to \mathrm{Z_q}$ maps the previously transmitted CN u to the next CN $v = \sigma[u]$. We use $\circ$ to denote composition of permutations and σ^β for β-fold composition. In both examples, it is assumed that there is *only a single transmitter* communicating with a single receiver.

Example 1 For some $q \geq 3$, define an incrementing and a decrementing permutation as

$$\sigma_{\mathrm{inc}} : u \mapsto (u+1) \bmod q \qquad \sigma_{\mathrm{dec}} : u \mapsto (u-1) \bmod q$$

and assign the permutations to one of two transmitters each. The permutations are illustrated in Fig. 2a for $q = 7$. One of the two transmitters is selected to send a packet containing an arbitrary CN $u \in \mathrm{Z_q}$ and then a packet with CN $v = \sigma[u]$, where σ is its assigned permutation. The receiver receives the packets in that order. For erasure probability $\varepsilon = 0$, the receiver is always able to identify the device based on u and v (e.g., $u = 5$ and $v = 6$ can only stem from permutation σ_{inc}). Even for nonzero ε, the receiver can still try to identify the transmitter using two non-erased packets based on which permutation, σ_{inc} or σ_{dec}, and the CNs contained in the packets were more likely drawn from.

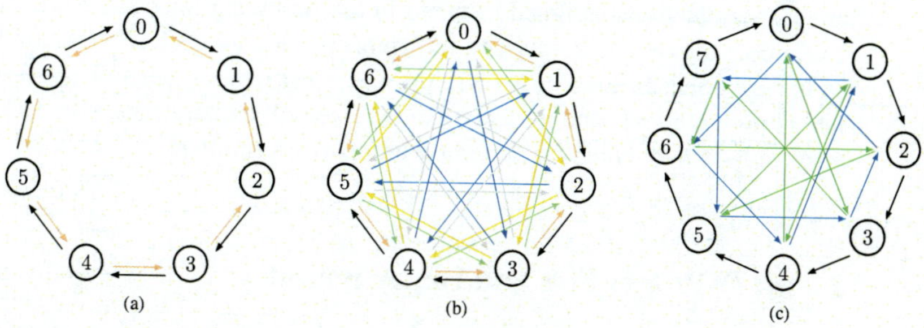

Fig. 2 Illustrations of different permutations (order in which CNs are sent), one color per permutation. (**a**) Permutations in a $(7, 3)$-proper set. (**b**) Permutations in a $(7, 1)$-proper set. (**c**) Permutations in an $(8, 2)$-proper set

Example 2 Suppose $\varepsilon = 0$, i.e., there are no erasures. If q is prime and there are $q - 1$ transmitters or less, we can uniquely assign a number $\Delta \in \mathrm{Z_q} \setminus \{0\}$, called increment, to each of them, so that by knowing Δ, the transmitter is identified. The transmitters are configured to send nonces according to the permutation $\omega_\Delta : u \mapsto (u + \Delta) \bmod q$. Then the receiver can determine a transmitter's identity based on two consecutively received nonces u and v: $\Delta = (v - u) \bmod q$. This can also be seen from the illustration for $q = 7$ in Fig. 2b.

For non-prime q, $\omega_\Delta : u \mapsto (u + \Delta) \bmod q$ is no longer a permutation of $\mathrm{Z_q}$ in case $\gcd(\Delta, q) \neq 1$. For most q, solutions with $q-1$ permutations can nevertheless be found: Formalizing the requirements, we want to find a set of permutations P such that

$$\forall (u, v) \in \mathrm{Z_q^2}\ \exists_{\leq 1} \sigma \in P : \ v = \sigma[u].$$

In other words, the transition from u to v must be unique among all transmitters. Regarding $\mathrm{Z_q}$ as the vertex set of a directed graph, an equivalent problem is to find a set of edge-disjoint directed Hamiltonian cycles in the complete directed graph K_q^*, where the permutations $\sigma \in P$ are given by the order in which cycles visit nodes. From Walecki's construction [3] and from Tillson's paper [4], we find solutions with $q-1$ directed Hamiltonian cycles and thus permutations for all $q \in \mathrm{Z_q} \setminus \{1, 2, 4, 6\}$. The cases $q \in \{1, 2\}$ are trivial, and, by full search, we find that it is not possible to find $q - 1$ directed Hamiltonian cycles for $q \in \{4, 6\}$.

Other than in Example 1, any packet loss between the reception of u and v results in misidentification.

These examples show that CNs may carry both identifying information and information facilitating the detection of lost packets. In Example 1, the receiver can only discern two devices but is guaranteed to tolerate up to $\lfloor (q - 1)/2 \rfloor - 1$ consecutively lost packets, whereas in Example 2, packet loss is not tolerable, but the receiver can identify up to $q - 1$ devices. These examples are two extreme ends in a spectrum of conceivable schemes that permit distinguishing transmitters based on CN permutations assigned to them. In the following Sect. 2, we will formalize this notion, define a necessary and sufficient criterion for such schemes, and provide an upper bound. In Sect. 3, we will provide a particularly suitable construction of sets of permutations, and in Sects. 4 and 5, we will explore how PERIDOT can be practically implemented for systems with many transmitters.

2 (q, l)-Proper Sets and Upper Bound

We assume that the receiver detects two packets from the transmitter with CNs $u, v \in \mathrm{Z_q}$ in that order. We count the number of packets that were transmitted after u, including the one containing v, and call this number β. The number of packets lost between u and v is thus $\beta - 1$, and we always have $\beta \geq 1$. Moreover, we assume that at most $l - 1$ packets are lost over the PEC, such that β can be bounded by $\beta \leq l$.

Definition 1 (**(q, l)-Proper Sets**) A set P of cyclic permutations with orbit $\mathrm{Z_q}$ is called a (q, l)-proper set for $l \in \mathbb{N}_0, l < q$ if, and only if, we have

$$\forall (u, v) \in \mathrm{Z_q^2}\ \forall \beta \in \mathbb{N}, \beta \leq l\ \exists_{\leq 1} \sigma \in P : \ v = \sigma^{\beta}[u].$$

Intuitively, Definition 1 states that a (q, l)-proper set P is a set in which we can *uniquely* identify each permutation $\sigma \in P$ based on an ordered pair (u, v), where this ordered pair is generated by arbitrarily choosing $u \in \mathrm{Z_q}$ and then computing v as $v = \sigma^{\beta}[u]$ with some $\beta \leq l$. This corresponds to a device successfully transmitting the CN u, then losing $\beta - 1 \leq l - 1$ messages (with CNs $\sigma[u]$, $\sigma^2[u], \ldots, \sigma^{\beta-1}[u]$), and then successfully transmitting $v = \sigma^{\beta}[u]$. The receiver can identify the permutation σ and thus the transmitter based on (u, v).

We see that the permutations in a (q, l)-proper set are suitable sequences of CNs. In this notation, Example 1 provides a construction for a $(q, \lfloor (q-1)/2 \rfloor)$-proper set, whereas Example 2 describes $(q, 1)$-proper sets. The following is a nontrivial example for a (q, l)-proper set (for an unrealistically small alphabet size of $q = 8$):

Example 3 A $(8, 2)$-proper set is given by $P = \{\sigma, \rho, \pi\}$ with

$$\sigma = (0, 1, 2, 3, 4, 5, 6, 7) \qquad \rho = (0, 3, 7, 6, 2, 5, 1, 4)$$

$$\pi = (0, 6, 4, 1, 7, 5, 3, 2).$$

σ, ρ, and π are given in cycle notation. There is no $(8, 2)$-proper set containing more than $|P| = 3$ permutations. P was found by computer search and is illustrated in Fig. 2c. For this exemplary set of permutations, if the receiver observes $u = 4$ and $v = 7$ in that order, it would generate the transition paths that σ, ρ, and π

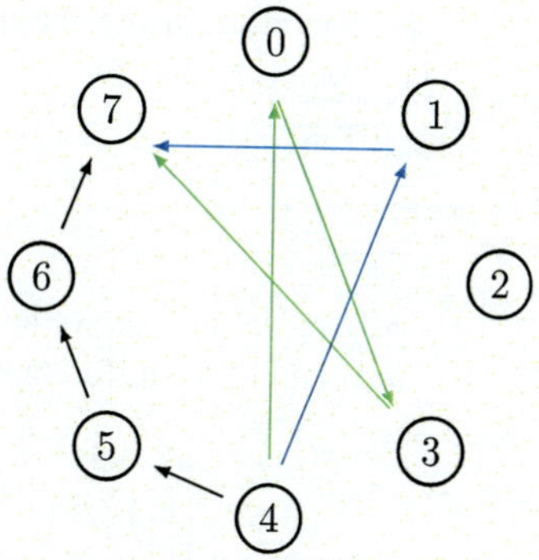

Fig. 3 Transitions from $u = 4$ to $v = 7$ in a (8, 2)-proper set

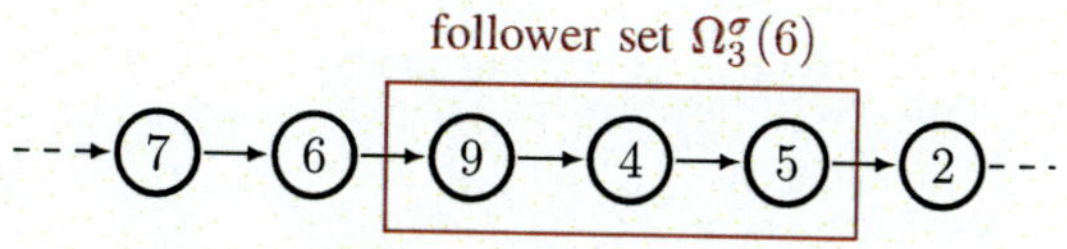

Fig. 4 3-follower set of 6 in some permutation

would have to take to get from u to v as drawn in Fig. 3. From the fact that the path for permutation π is shortest, it concludes that the most likely scenario is that the device to which π was assigned transmitted u and v, and that one packet was lost in between (the one with a CN of 1).

The previous examples suggest that for fixed l and q, there appears to be a maximum number of permutations that fit into a (q, l)-proper set, i.e., a maximum value for $m = |P|$. This has important practical ramifications: For an alphabet size q, which is predetermined by the number of bits allocated to a CN, and $l - 1$ tolerable packet losses, we can only assign m permutations to transmitters and thus only identify up to m devices using PERIDOT. Therefore, we now seek to formulate an upper bound that relates m, l, and q. Such a bound will result as a consequence of an equivalent condition for (q, l)-proper sets based on what we call *l-follower sets*:

Definition 2 (*l*-Follower Sets) The l-follower set of some $u \in Z_\mathrm{q}$ in the permutation σ of Z_q is given by

$$\Omega_l^\sigma(u) = \left\{\sigma^\beta[u] \mid 1 \leq \beta \leq l\right\}.$$

Figuratively, $\Omega_l^\sigma(u)$ contains all the CNs that succeed u in the next l messages, as illustrated in Fig. 4. It is clear that, if we have a set of alphabet permutations P and all l CNs that succeed u are distinct from any permutation to another, then we can identify each permutation $\sigma \in P$ based on (u, v), where v is in the l-follower set of u in the permutation σ.

Therefore, using the definitions of (q, l)-proper sets and follower sets, we can equivalently say that P is a (q, l)-proper set with $l < q$ if, and only if, $\forall u \in \mathbb{Z}_q$ we have

$$\forall \sigma, \pi \in P,\ \sigma \neq \pi :\ \Omega_l^\sigma(u) \cap \Omega_l^\pi(u) = \emptyset. \tag{1}$$

We can now derive an upper bound on $m = |P|$ from Eq. (1). Since the sets Ω_l^σ are disjoint for different $\sigma \in P$ and because $|\Omega_l^\sigma(u)| = l$, we have

$$\left|\bigcup_{\sigma \in P} \Omega_l^\sigma(u)\right| = \sum_{\sigma \in P} |\Omega_l^\sigma(u)| = |P|l \leq |\mathbb{Z}_q \backslash \{u\}| = q - 1. \tag{2}$$

With $m = |P|$, this means

$$lm + 1 \leq q.$$

Adding the restriction that m must be integer and solving for m finally yield the following upper bound:

Theorem 1 *For any (q, l)-proper set P holds:*

$$m = |P| \leq \left\lfloor \frac{q-1}{l} \right\rfloor.$$

A similar upper bound (but for a more restrictive problem and an entirely different application) has previously been derived in [5, eq. 1]. On the basis of Theorem 1, we are now able to assess how "good" a (q, l)-proper set construction is: We will call a (q, l)-proper set *quasiperfect* if it reaches the upper bound, and *perfect* if, additionally, $(q - 1)/l$ is integer. We will see in the following that the bound in Theorem 1 is tight.

3 A Quasiperfect Construction

For some specific values of q, l that are also suitable for practical implementations of PERIDOT, it is possible to construct a quasiperfect (q, l)-proper set. This construction is based on transmitters sending CNs in an order that is determined by an arithmetic progression modulo q, that is, a sequence of the form

$$0,\ \Delta \bmod q,\ 2\Delta \bmod q,\ \ldots . \tag{3}$$

As in Example 2, the mapping from previous to next CN is given by the permutation ω_Δ with *increment* $\Delta \in Z_q \setminus \{0\}$ and

$$\omega_\Delta : u \mapsto (u + \Delta) \bmod q. \tag{4}$$

For ω_Δ to be cyclic with orbit Z_q, we need to have $\gcd(\Delta, q) = 1$. Note that modular arithmetic progressions have the added benefit of being easy to store (only need to store q, Δ at transmitter), which makes the implementation of PERIDOT on low-powered LPWAN end nodes feasible.

Theorem 2 *For prime p, $l \mid (p-1)$ and ω_Δ as in Eq.* (4)*,*

$$P = \{\omega_{1+il} \mid i = 0, \ldots, p-1,\ i \neq (p-1)/l\}$$

is a quasiperfect (pl, l)-proper set with $m = |P| = p - 1$.

Proof We have to prove that:

(i) P is a (q, l)-proper set with $q = pl$ (requiring the elements of P to be cycles with orbit Z_q).
(ii) P is quasiperfect.

Since we have $\gcd(1 + il, q) = 1$ for $i = 0, \ldots, p-1,\ i \neq (p-1)/l$, the elements of P are indeed cyclic permutations with orbit Z_q. The simple, but technical proof for this is omitted, and we will instead focus on proving that P is (l, q)-proper, which we will prove by contradiction and using Definition 1.

Assume to the contrary that P is not a (q, l)-proper set. Then, there are some $i_1, i_2 \in \{0, \ldots, p-1\} \setminus \{(p-1)/l\}$, $i_1 \neq i_2$ such that for some $\beta_1, \beta_2 \in \{1, \ldots, l\}$, we have

$$\omega_{1+i_1 l}^{\beta_1}[u] = \omega_{1+i_2 l}^{\beta_2}[u].$$

By the definition of ω_Δ, we obtain Eq. (5), which can be simplified to Eq. (6):

$$u + (1 + i_1 l)\beta_1 \equiv u + (1 + i_2 l)\beta_2 \pmod q \tag{5}$$

$$\iff \quad (1 + i_1 l)\,\beta_1 \equiv (1 + i_2 l)\,\beta_2 \pmod q. \tag{6}$$

Now since $q = pl$, the equivalence in Eq. (6) must also hold modulo l, from which we can conclude that $\beta_1 \equiv \beta_2 \pmod l$. By the premise, we have $1 \le \beta_1, \beta_2 \le l$, so $\beta_1 = \beta_2$. Simplifying Eq. (6) with $\beta_1 = \beta_2$, we get

$$i_1 \equiv i_2 \pmod q,$$

and because $i_1, i_2 < p \le q$, we conclude $i_1 = i_2$. This is a contradiction.

P is quasiperfect, since $m = |P| = p - 1$ and $q = pl$, so the upper bound from Theorem 1 becomes

$$p - 1 \leq \left\lfloor \frac{pl - 1}{l} \right\rfloor = \left\lfloor p - \frac{1}{l} \right\rfloor,$$

which is fulfilled with equality, implying quasiperfectness (but not perfectness in general). □

A construction similar to, but more restrictive than Theorem 2 has been previously described in [5, Theorem 1], but, again, for a different application. (q, l)-proper sets can also be constructed by adapting techniques applied for the construction of flash memory codes as presented in [5, 6] and [7].

4 Application in Multi-transmitter Systems

Using the previously described PERIDOT code construction, we can now address the practical implementation of the scheme for use with LPWANs. In our LPWAN model, we assume that end devices spontaneously emit uplink-only packets that one or multiple nearby base stations can pick up. As shown in Fig. 5, the channels between any device and the base station are modelled as independent unidirectional packet erasure channels. Furthermore, we assume that all packets originating from all devices globally are being processed by a central service, hereafter called *backend*. The assumption of the backend being centralized is not necessary but will greatly simplify the subsequent explanations. The backend has a database of devices, containing the previously received CNs u for each device as well as each packet's timestamp and a very approximate device location. We assume that if the base station picks up a packet containing the CN v:

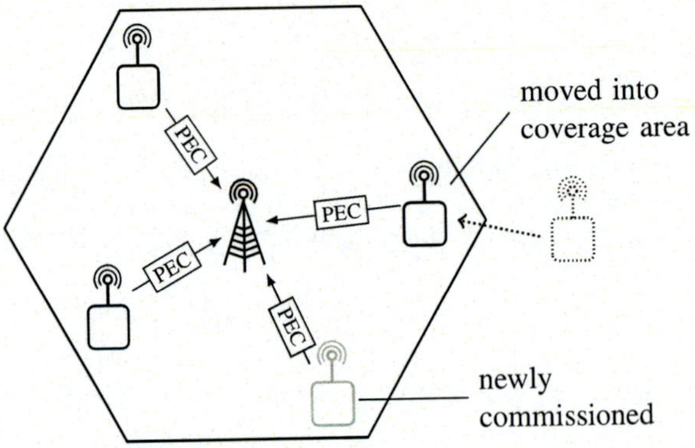

Fig. 5 Model of LPWAN base station's coverage area

Fig. 6 Overview of backend identification algorithm

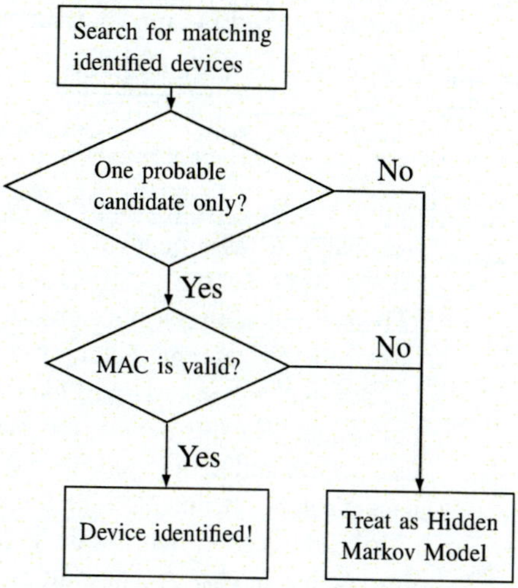

(A) It is very likely that it or a nearby base station has also picked up at least one of the previous l packets from the same device, which has already been identified.
(B) Otherwise, l or more packets were erased by the channel, the device is newly commissioned, or the device is mobile and has moved rapidly.

In all cases, the goal is to identify the transmitter, i.e., find out which permutation σ the CN v was drawn from. The backend can handle all cases using a probability-based algorithm that is illustrated in Fig. 6 and that is only briefly outlined and underpinned by some examples in the scope of this chapter. It treats the vast majority of packets that are easy to identify (case A) in a special way so that the few packets that cannot be identified so easily (case B) can undergo computationally more complex treatment.

In an initial processing step for each received packet, the backend iterates over its database of known, identified devices and for each device (hereafter called *candidate*) computes a probability estimate that it was this particular device that transmitted v. In many cases, candidates can be assigned a zero probability, e.g., in case a device was recently seen in a completely different geographic region. For the candidates that remain, the backend computes the number of packet losses that must have occurred to make the observation of v plausible, given that the last observed packet by the candidate contained the CN u (from backend's database). The number of packets transmitted after u, including v, is denoted by β with $1 \leq \beta$, and we assume $\beta \leq q$. Since the candidate device's assigned permutation σ is known, this number has to fulfill $v = \sigma^{\beta}[u]$. For a modular arithmetic progression-based (q, l)-proper set construction where $\sigma = \omega_{\Delta}$ for some Δ, this simplifies to

$$\beta \equiv \Delta^{-1}(v - u) \pmod{q}. \quad (7)$$

Now, using the PEC model assumption with erasure probability ε, the probability for β packet losses is given by

$$P_\beta = (1 - \varepsilon)\, \varepsilon^{\beta-1}.$$

Intuitively, this means that if v appears soon after u in the permutation σ, it is reasonable to assume that u and v were transmitted by the same device. The accuracy of probability estimates may be further augmented based on methods such as behavioral analyses and signal strength measurements.

For the packets that belong to case A, which we expect to be the vast majority of packets, these criteria will likely let a single device emerge as the only probable transmitter. For this device, the MAC value is checked (i.e., the MAC gets a dual use as both authenticity check and final identity check), and if it also matches, the packet can be attributed to the transmitter.

All other cases, that is case B (no device matches the packet with high enough probability) or case A, but multiple devices match, are treated in the same way: the packet has to be processed by a more sophisticated algorithm that treats the entirety of all yet unidentified devices as a Hidden Markov Model (HMM). It is for this that we constructed (q, l)-proper sets and assigned special permutations to devices—thanks to them, the backend will still be able to identify devices, almost always upon reception of the *second* packet by the same device.

A HMM consists of a Markov chain, i.e., a set of states the system can be in, as well as a set of possible observable values. In our case, the states of the Markov chain are given by the combined state of the *entirety* of all the unidentified LPWAN end devices. This state consists of *all* the previously successfully transmitted CNs of *all* unidentified devices, which is an integer from the alphabet $\mathrm{Z_q}$ for each device. Overall, the state can thus be described by the vector $\underline{u} \in \mathrm{Z}_\mathrm{q}^d$ for d unidentified devices in the whole network, so there are a total of q^d possible states. In our case, an observation in the HMM is made whenever a packet is received, which could not be unambiguously assigned to a device in the previous steps.

In this context, the goal of identifying the transmitter of each packet can be reformulated as guessing the most likely sequence of hidden states from the observations since this sequence also automatically reveals which transitions must have occurred in order to produce it. Trivially, if two state vectors differ in more than one CN, the transition probability between them is zero since packet transmissions can be assumed to occur one at a time. Other transition probabilities between hidden states can be computed as previously explained (compute P_β, possibly augment with metadata).

The *Viterbi algorithm* can be employed to find the most likely sequence of hidden states efficiently. Due to the large number of states in the Markov chain (q^d), a practical implementation has to stop tracking unlikely paths if their probability drops below some threshold. The way the algorithm works is best explained by example. In order to simplify explanations, we will use unrealistically small values for q.

In Examples 4 and 5, we assume a construction as in Theorem 2 with $l = 10$ and $p = 11$, so the set of increments is $\{1, 21, 31, 41, 51, 61, 71, 81, 91, 101\}$ and $q = 110$. Furthermore, we assume that identification is performed only based on CNs and an approximate (within tens to hundreds of kilometers accuracy) device geolocation.

Example 4 The backend has already identified and located the asset tracker device T with increment $\Delta_T = 31$ in Rotterdam. The last CN known to have originated from T is $u = 12$. Several hours later, a base station in Amsterdam, 60 km away, receives a packet with a CN $v = 43$, which is a plausible next CN for T. There are no other identified devices in all of Europe for which $v = 43$ is plausible, so we check the packet's MAC that is also valid. It can be assumed that the device has been successfully identified.

The only reasonable scenario for erroneous identification is if some other previously inactive device that was just commissioned happens to have $v = 43$ as first CN to transmit, happens to be in the same geographic area as T, and the MAC happens to be valid, which is extremely unlikely overall.

With large enough values of p, l, and thereby q, we expect the error scenario in Example 4 to hardly ever occur.

Example 5 The backend has not yet identified the devices with increments 1, 21, and 91. A base station in Singapore receives a packet with CN 77, which is implausible for any of the already identified devices in the area (case B applies). At this point, the packet cannot yet be processed further because the transmitter cannot be identified solely based on this one packet. Thirty minutes later, the same base station receives a packet with CN 9. The only plausible increment for the sequence $77 \to 9$ is 21 with $\beta - 1 = 1$ erasure, since, by Eq. (7), we have $\beta \equiv 21^{-1}\,(9 - 77) \equiv 2 \pmod{q}$. Since the MAC validity checks for both packets succeed, the backend can now process and deliver the current and previous packet.

An increment of 1 would have required 41 erasures ($\beta = 42$), and an increment of 91 would have required 101 erasures to go from 77 to 9. Both of these scenarios are extremely unlikely.

In the following example, we will use the two trivial incrementing / decrementing permutations σ_{inc} and σ_{dec} given in Example 1 for an (unrealistically small) alphabet size $q = 20$.

Example 6 The backend receives the following sequence of CNs: $1 \to 2 \to 10 \to 9 \to 3 \to 7 \to 5 \to 3 \to 4$. It is likely that 1, 2, and 3 originate from σ_{inc} and 10, 9, 7 from σ_{dec}. Upon reception of CN $v_1 = 5$, both σ_{inc} and σ_{dec} are equally likely sources of v_1, so the packet cannot yet be delivered and has to be processed by the HMM for identification. There are now two equally likely current system states: $\underline{u}_{1\text{a}} = (5, 7)$ and $\underline{u}_{1\text{b}} = (3, 5)$, where the CN attributed to σ_{inc} comes first in the tuple.

The value $v_2 = 3$ that the backend receives just afterward provides a hint for identification: the most likely scenario (just a single packet loss) now is that $v_2 = 3$ came from σ_{dec} and that the previous system state was $\underline{u}_{1\text{b}} = (3, 5)$, i.e., the current

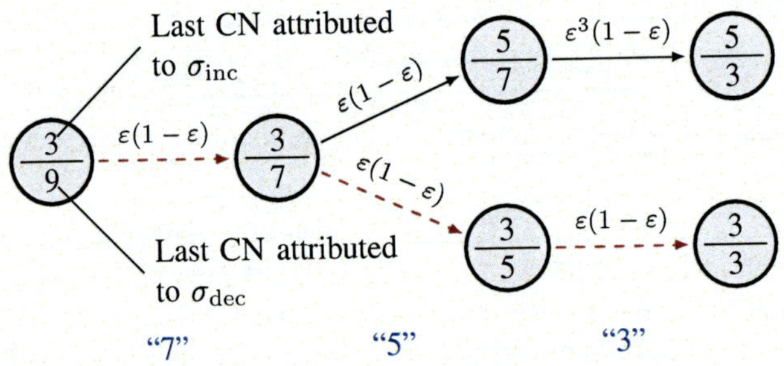

Fig. 7 System states $\underline{u}$ with transition probabilities given observations (blue)

system state is $u_2 = (3, 3)$. So, if the received v_2 is assumed to belong to σ_{dec}, it is significantly more likely that $v_1 = 5$ was also transmitted by this device. Therefore, the value of v_2 that was received at a later point in time influences the backend's interpretation of v_1 and allows it to deliver the packet with v_1, though delayed. This example is also illustrated in Fig. 7, where the most likely path as determined by the Viterbi algorithm is dashed and marked in red. Figure 7 does not show all the extremely unlikely system states and unlikely transitions. The subsequently received CN 4 confirms the identification that was made based on $v_2 = 3$: We conclude that, almost certainly, the device with σ_{inc} has sent the sequence $1 \to 2 \to 3 \to 4$ and the device with σ_{dec} has sent $10 \to 9 \to 7 \to 5 \to 3$.

The kind of resolution presented in Example 6 is not always possible; in case identification is ambiguous, the better course of action can be to drop the packet than to risk erroneous identification. Note that this ambiguity occurs only very rarely if we use a sufficiently large alphabet Z_q. The backend is aware of the ambiguity, and the residual probability of being unable to identify transmitters and having to discard packets becomes negligible compared to the PEC's erasure probability.

5 Exemplary Case: Sigfox

The French company *Sigfox* [8] operates a global LPWAN. Their network protocol uses 32-bit identifiers and 12-bit SNs, thus allowing for up to 2^{32} addressable end devices and a nonce reuse after $2^{12} = 4096$ packets [1]. If Sigfox employed PERIDOT in a new protocol revision, they could increase the theoretical maximum number of devices, prolong nonce reuse cycles, and reduce the number of bits that have to be transmitted all at the same time.

In order to achieve this, we construct a (q, l)-proper set based on Theorem 2. For instance, we may want to encode the CN on a total of 38 bits. We assume that the occurrence of 49 or more consecutive packet losses is extremely unlikely, so we choose $l = 50$. Since we need to have $q = pl \le 2^{38}$ with $l \mid (p-1)$ and p must be prime, we choose the largest such prime, which is $p = 5{,}497{,}554{,}151$. Thus, we get increments for $m = p - 1 = 5{,}497{,}554{,}150$ different devices.

As a result, compared to the current protocol, we:

- Reduced packet size by 6 bit, resulting in a 5% reduction in energy consumption for single-bit payload packets
- Increased the nonce reuse cycle by a factor of $q/2^{12} \approx 67.1 \cdot 10^6$ making replay attacks infeasible
- Increased the theoretical maximum number of identifiable devices by approximately 28%

with negligible implementation costs on the end device side.

6 Conclusion

We introduced the PERIDOT scheme and described how identifiers and SNs can be completely replaced by transmitting CNs based on specially constructed permutations. One particularly suitable construction was presented, though entirely different constructions may also be feasible. We showed that PERIDOT can reduce uplink packet lengths and thus energy consumption, improve security, and increase the theoretical maximum number of addressable devices.

The basic concept of an algorithm for the backend has been outlined, and how exactly to incorporate geolocation information, behavior analysis, and other metadata into state transition probabilities for the HMM remains an open question though. Another research avenue would be to simulate a global LPWAN employing PERIDOT so that the influence of different parameters on the system and its scalability can be tested.

We found and referenced mathematical concepts that are related to PERIDOT in the theory of asymmetric error correction codes. Still, PERIDOT remains to be classified, and its relation to error correction codes needs to be better understood.

References

1. Sigfox connected objects: Radio specifications v1.5, Sigfox (February 2020)
2. LoRaWAN^{TM} 1.1 Specification, LoRa Alliance (October 2017)
3. Alspach, B.: The wonderful Walecki construction. Bull. Instit. Combinat. Appl. **52**, 7–19 (2006)
4. Tillson, T.W.: A Hamiltonian decomposition of K_{2m}^*, $2m \geq 8$. J. Combinat. Theory Ser. B **29**(1), 68–74 (1980)
5. Kløve, T., Bose, B., Elarief, N.: Systematic, single limited magnitude error correcting codes for flash memories. IEEE Trans. Inf. Theory **57**(7), 4477–4487 (2011)
6. Xie, D., Luo, J.: Asymmetric single magnitude four error correcting codes. CoRR **abs/1903.01148** (2019). http://arxiv.org/abs/1903.01148
7. Kløve, T., Luo, J., Naydenova, I., Yari, S.: Some codes correcting asymmetric errors of limited magnitude. IEEE Trans. Inf. Theory **57**, 7459–7472, 12 (2011)
8. Sigfox - The Global Communications Service Provider for the Internet of Things (IoT). https://www.sigfox.com/en. Accessed 20 March 2020

Shannon Capacity of Tensor Channels Under a Family of Power Constraints

Divyanshu Pandey and Harry Leib

1 Introduction

Tensors are multi-way arrays that can be seen as higher-order generalization of vectors and matrices. Tensors have found wide spread applications in various engineering disciplines including signal processing, communications, and multi-linear system theory [1–3]. Each element in a tensor is indexed by multiple indices, known as *modes*. The number of modes is called the *order* of the tensor. A signal spanning multiple domains can be represented as a tensor with each mode corresponding to a specific domain. Thus tensors can be used to model multi-domain communication systems.

Most modern communication systems use multi-domain modulation and coding schemes for effectively harnessing all the available resources in transmission and reception. The domains depend on the specific system and can include space, time, frequency, users, code sequences, transmission devices, and propagation delay. Such a multi-domain nature of the transmit and receive signals can be well represented using tensors. Some examples of using tensor for multi-carrier multi-antenna communication systems can be found in [4, 5]. In this chapter, we present a generic model with input and output as order N and M tensors, respectively, and a multi-linear channel as an order $N + M$ tensor, which can represent several multi-domain communication systems.

The Einstein product of tensors and its associated properties [6, 7] are used in this chapter to develop the system model. The Einstein product has the distinct advantage of allowing to extend many known linear algebra concepts to a multi-linear setting, without having to explicitly transform the tensors to vector or

D. Pandey (✉) · H. Leib
Department of Electrical and Computer Engineering, McGill University, Montreal, QC, Canada
e-mail: divyanshu.pandey@mail.mcgill.ca; harry.leib@mcgill.ca

H. Nguyen et al. (eds.), *30th Biennial Symposium on Communications 2021*,
Signals and Communication Technology, https://doi.org/10.1007/978-3-031-06947-5_2

matrices, thereby retaining the distinction between different modes. The proposed system model in this chapter can be seen as a generalization of multiple-input multiple-output (MIMO) matrix channel model. Capacity achieved by MIMO channel under perfect channel state information and sum-power constraint is well established [8]. In this chapter, we find the capacity of a tensor channel. In particular, we leverage the tensor framework's unique ability to facilitate expressing a family of different power constraints in mathematical terms, of which the sum-power constraint or per-antenna power constraints can be seen as special cases.

In Sect. 2, we present the required tensor basics and establish the system model. Section 3 is dedicated to formulating and finding the Shannon capacity as an optimization problem. Section 4 presents numerical examples to illustrate the concepts. The paper is concluded in Sect. 5.

2 Tensor Basics and System Model

In this chapter, we use calligraphic fonts, e.g., $\mathcal{X} \in \mathbb{C}^{I\times J\times K}$ to denote deterministic tensors, and bold fonts, e.g. $\boldsymbol{\mathcal{X}}$ to denote random tensors. Individual elements of a tensor are denoted by indices in the subscript, e.g., $\mathcal{X}_{i,j,k}$. The operator $\mathbb{E}[\cdot]$ denotes expectation. An all-zero tensor is denoted by $0_{\mathcal{T}}$, and $\log(\cdot)$ represents logarithm to the base 2.

2.1 Basic Tensor Operations and Definitions

Einstein product: For tensors $\mathcal{A}\in\mathbb{C}^{I_1\times\cdots\times I_P\times K_1\times\cdots\times K_N}$ and $\mathcal{B}\in\mathbb{C}^{K_1\times\cdots\times K_N\times J_1\times\cdots\times J_M}$, the Einstein product over N modes, denoted as $\mathcal{C}=\mathcal{A}*_N\mathcal{B}$, is defined as [6]

$$\mathcal{C}_{i_1,\ldots,i_P,j_1,\ldots,j_M}=\sum_{k_1,\ldots,k_N}\mathcal{A}_{i_1,\ldots,i_P,k_1,\ldots,k_N}\mathcal{B}_{k_1,\ldots k_N,j_1,\ldots,j_M}. \tag{1}$$

The outer product between tensors $\mathcal{X}\in\mathbb{C}^{I_1\times\cdots\times I_N}$ and $\mathcal{Y}\in\mathbb{C}^{J_1\times\cdots\times J_M}$, denoted as $\mathcal{X}\circ\mathcal{Y}$ is defined as

$$(\mathcal{X}\circ\mathcal{Y})_{i_1,\ldots,i_N,j_1,\ldots,j_M}=\mathcal{X}_{i_1,\ldots,i_N}\mathcal{Y}_{j_1,\ldots,j_M}=\mathcal{X}*_0\mathcal{Y}. \tag{2}$$

Several linear algebra concepts can be generalized to tensors using the Einstein product and its properties [6]. For instance, a square tensor is defined as an even order tensor $\mathcal{A}\in\mathbb{C}^{I_1\times\cdots\times I_N\times J_1\times\cdots\times J_M}$, where $N=M$ and $I_k=J_k$ for $k=1,\ldots,N$. Furthermore, a square tensor is pseudo-diagonal if its elements $\mathcal{A}_{i_1,\ldots,i_N,j_1,\ldots,j_N}$ are zero except when $i_1=j_1, i_2=j_2,\ldots,i_N=j_N$. The inverse of a square tensor is defined as another square tensor of same size $\mathcal{A}^{-1}\in\mathbb{C}^{I_1\times\cdots\times I_N\times I_1\times\cdots\times I_N}$ such that $\mathcal{A}*_N\mathcal{A}^{-1}=\mathcal{A}^{-1}*_N\mathcal{A}=\mathcal{I}$. The tensor $\mathcal{I}$ is the *identity* tensor that is pseudo-diagonal with $\mathcal{I}_{i_1,\ldots,i_N,i_1\ldots,i_N} = 1$ for all $i_1, \ldots, i_N$. Several algorithms can be used to compute tensor inverse such as higher-order bi-conjugate gradient method [6] or the Newton's method [9]. The *Hermitian* of a tensor $\mathcal{A} \in \mathbb{C}^{I_1\times\cdots\times I_N\times J_1\times\cdots\times J_M}$ is denoted by $\mathcal{A}^H \in \mathbb{C}^{J_1\times\cdots\times J_M\times I_1\times\cdots\times I_N}$, which has entries $(\mathcal{A}^H)^*_{j_1,\ldots,j_M,i_1,\ldots,i_N} = \mathcal{A}_{i_1,\ldots,i_N,j_1,\ldots,j_M}$. A square tensor $\mathcal{A}$ is called *Hermitian* if $\mathcal{A} = \mathcal{A}^H$ and is called unitary if $\mathcal{A}^H *_N \mathcal{A} = \mathcal{A} *_N \mathcal{A}^H = \mathcal{I}$.

The *tensor eigenvalue decomposition* (EVD) of a Hermitian tensor $\mathcal{A} \in \mathbb{C}^{I_1\times\cdots\times I_N\times I_1\times\cdots\times I_N}$ is written as $\mathcal{A} = \mathcal{U} *_N \mathcal{D} *_N \mathcal{U}^H$, where $\mathcal{U} \in \mathbb{C}^{I_1\times\cdots\times I_N\times I_1\times\cdots\times I_N}$ is a unitary tensor. The tensor $\mathcal{D} \in \mathbb{C}^{I_1\times\cdots\times I_N\times I_1\times\cdots\times I_N}$ is pseudo-diagonal where its non-zero values are the eigenvalues of $\mathcal{A}$ as defined in [7, 10]. A square tensor is called *positive semi-definite* ($\mathcal{A} \succeq 0$) if all its eigenvalues are non-negative, and is called *positive definite* ($\mathcal{A} \succ 0$) if all its eigenvalues are strictly positive. *Trace* and *determinant* of a tensor are defined as the sum and product of its eigenvalues, respectively. Using these definitions, several matrix properties can be generalized to tensors. For instance, the Sylvester's matrix property can be extended for tensors $\mathcal{A} \in \mathbb{C}^{I_1\times\cdots\times I_N\times J_1\times\cdots\times J_M}$ and $\mathcal{B} \in \mathbb{C}^{J_1\times\cdots\times J_M\times I_1\times\cdots\times I_N}$ as

$$\det(\mathcal{I}_{2N} + \mathcal{A} *_M \mathcal{B}) = \det(\mathcal{I}_{2M} + \mathcal{B} *_N \mathcal{A}), \tag{3}$$

where $\mathcal{I}_{2N}$ and $\mathcal{I}_{2M}$ are order $2N$ and $2M$ identity tensors, respectively. For more details on tensor algebra properties, refer to [4, 6, 7, 10].

A tensor is called random if its elements are random variables. A random tensor $\mathfrak{X}$ with mean $\mathcal{M} = \mathbb{E}[\mathfrak{X}]$ is called *proper* if its pseudo-covariance defined as $\mathbb{E}[(\mathfrak{X} - \mathcal{M}) \circ (\mathfrak{X} - \mathcal{M})]$ is an all-zero tensor. The pdf of a proper Gaussian complex-valued tensor $\mathfrak{X} \in \mathbb{C}^{I_1\times\cdots\times I_N}$ is given as

$$p_{\mathcal{X}}(\mathcal{X}) = \frac{\exp\big\{-(\mathcal{X} - \mathcal{M})^* *_N \mathcal{Q}^{-1} *_N (\mathcal{X} - \mathcal{M})\big\}}{(\pi)^{I_1 I_2 \ldots I_N} \det(\mathcal{Q})}, \tag{4}$$

where $\mathcal{Q} = \mathbb{E}[(\mathfrak{X} - \mathcal{M}) \circ (\mathfrak{X} - \mathcal{M})^*]$ is the order $2N$ covariance tensor. A complex random tensor $\mathfrak{X}$ is called *circular* if for any real α, the tensors $\mathfrak{X}$ and $\mathcal{Y} = e^{j\alpha}\mathfrak{X}$ have the same probability distribution. Furthermore, using the definition of circularity, it can be shown that a complex Gaussian tensor is circular if and only if it is zero mean and proper. The differential entropy, defined as $\mathcal{H}(\mathfrak{X}) = \mathbb{E}\big[-\log p_{\mathcal{X}}(\mathfrak{X})\big]$, for a circular symmetric complex Gaussian tensor is given as

$$\mathcal{H}(\mathfrak{X}) = \mathbb{E}\Bigg[-\log\Bigg\{\frac{\exp\big(-\mathfrak{X}^* *_N \mathcal{Q}^{-1} *_N \mathfrak{X}\big)}{(\pi)^{I_1 I_2 \ldots I_N} \det(\mathcal{Q})\,\cdot}\Bigg\}\Bigg]$$

$$= \log\Big((\pi)^{I_1 I_2 \ldots I_N} \det(\mathcal{Q})\Big) + (\log e)\,\mathrm{tr}\,\big(\mathbb{E}[\mathfrak{X} \circ \mathfrak{X}^*] *_N \mathcal{Q}^{-1}\big)$$

$$= \log\Big((e\pi)^{I_1 I_2 \ldots I_N} \det(\mathcal{Q})\Big). \tag{5}$$

The following lemma regarding circular symmetric complex Gaussian tensor is proven in [11].

Lemma 1 *If* $\mathcal{X} \in \mathbb{C}^{I_1 \times \cdots \times I_N}$ *is a circular symmetric complex Gaussian tensor with covariance* $\mathcal{Q}$ *and* $\mathcal{Y} \in \mathbb{C}^{I_1 \times \cdots \times I_N}$ *is another zero-mean random tensor with same covariance tensor, then* $\mathcal{H}(\mathcal{X}) \geq \mathcal{H}(\mathcal{Y})$.

2.2 System Model

Input and output in a multi-domain communication system can be defined as tensors of orders N and M, respectively. Assume that $\mathcal{X} \in \mathbb{C}^{I_1 \times \cdots \times I_N}$ denotes the input (transmitted) tensor symbol where $I_1, I_2, \ldots, I_N$ are the dimensions of the N transmit domains. Each element $\mathcal{X}_{i_1,\ldots,i_N}$ is a discrete complex symbol. Similarly, $\mathcal{Y} \in \mathbb{C}^{J_1 \times \cdots \times J_M}$ represents the output tensor where $J_1, J_2, \ldots, J_M$ are the dimensions of the M receive domains. A multi-linear channel between input and output can be defined as a tensor of order $M + N$ represented by $\mathcal{H} \in \mathbb{C}^{J_1 \times \cdots \times J_M \times I_1 \times \cdots \times I_N}$. The system model can be specified by using the Einstein product as

$$\mathcal{Y} = \mathcal{H} *_N \mathcal{X} + \mathcal{N}, \tag{6}$$

where $\mathcal{N} \in \mathbb{C}^{J_1 \times \cdots \times J_M}$ is the received noise tensor. The discrete-time MIMO matrix channel model is a specific case of the proposed tensor model where input and output are order 1 tensors (vectors), and thus the channel is an order 2 tensor (matrix). In such a case, the Einstein product reduces to regular matrix multiplication. The proposed system model is generic, and the number of modes along with the physical interpretation of the individual modes is system-specific. The modes can represent different domains of communications such as space, time, frequency, propagation delay, users, and spreading sequence. Thus this system model can be used for representing multi-domain communication systems such as MIMO Orthogonal Frequency-Division Multiplexing (OFDM) [9] and Generalized Frequency-Division Multiplexing (GFDM) [11]. For the system model defined in (6), assuming $\mathcal{X}$ and $\mathcal{N}$ to be zero mean and independent, the output covariance tensor and the output entropy conditioned on input can be written as

$$\mathcal{Q}_{\mathcal{Y}} = \mathrm{Cov}(\mathcal{Y}) = \mathbb{E}[\mathcal{Y} \circ \mathcal{Y}^*] = \mathcal{H} *_N \mathcal{Q}_{\mathcal{X}} *_N \mathcal{H}^H + \mathcal{Q}_{\mathcal{N}}, \tag{7}$$

$$\mathcal{H}(\mathcal{Y}|\mathcal{X}) = \mathcal{H}(\mathcal{H} *_N \mathcal{X} + \mathcal{N}|\mathcal{X}) = \mathcal{H}(\mathcal{N}|\mathcal{X}) = \mathcal{H}(\mathcal{N}), \tag{8}$$

respectively, where $\mathcal{Q}_{\mathcal{X}}$ and $\mathcal{Q}_{\mathcal{N}}$ are the input and noise covariance tensors, respectively. Subsequently, the mutual information between input and output tensors is given as

$$\mathcal{I}(\mathcal{X};\mathcal{Y}) = \mathcal{H}(\mathcal{Y}) - \mathcal{H}(\mathcal{Y}|\mathcal{X}) = \mathcal{H}(\mathcal{Y}) - \mathcal{H}(\mathcal{N})$$
$$= \mathcal{H}(\mathcal{Y}) - \log\left((e\pi)^{J_1 J_2 \cdots J_M} \det(\mathcal{Q}_{\mathcal{N}})\right). \tag{9}$$

Using Lemma 1 and (7), we can write

$$\mathcal{H}(\mathcal{Y}) \leq \log\left((e\pi)^{J_1 J_2 \cdots J_M} \det(\mathcal{H} *_N \mathcal{Q}_{\mathcal{X}} *_N \mathcal{H}^H + \mathcal{Q}_{\mathcal{N}})\right) \tag{10}$$

$$\Rightarrow \mathcal{I}(\mathcal{X};\mathcal{Y}) \leq \log\left[\frac{\det\left(\mathcal{H} *_N \mathcal{Q}_{\mathcal{X}} *_N \mathcal{H}^H + \mathcal{Q}_{\mathcal{N}}\right)}{\det(\mathcal{Q}_{\mathcal{N}})}\right]. \tag{11}$$

Note that in (11), equality is achieved only if $\mathcal{Y}$ is circularly symmetric Gaussian distributed.

3 Capacity Under Different Power Constraints

To find the tensor channel capacity, we need to maximize the mutual information between the input and output tensors under given input constraints. Here we assume that the channel is known, and we will consider different power constraints. In the system model, if we assume that noise is zero-mean circular symmetric with covariance as identity tensor, i.e., $\mathcal{Q}_{\mathcal{N}} = \mathcal{I}$, it can be shown that $\mathcal{Y}$ will also be zero-mean circular symmetric complex Gaussian if $\mathcal{X}$ is so. Hence, for maximizing the mutual information, we take $\mathcal{X}$ as circular complex Gaussian with covariance $\mathcal{Q}$. Thus based on (11), we can write capacity as

$$C = \max_{\mathcal{Q}: f(\mathcal{Q}) \leq 0,\ \mathcal{Q} \succeq 0} \log\left[\det\left(\mathcal{H} *_N \mathcal{Q} *_N \mathcal{H}^H + \mathcal{I}\right)\right], \tag{12}$$

where the inequality constraint $f(\mathcal{Q}) \leq 0$ can represent a family of power constraints. The positive semi-definiteness of the covariance tensor ensures that $(\mathcal{H} *_N \mathcal{Q} *_N \mathcal{H}^H + \mathcal{I})$ is a positive definite tensor, which in turn implies that the objective function is a concave function.

3.1 *Different Power Constraints*

In a multi-domain communication system, power constraints can span multiple domains. For example, if the input signal is transmitted using multiple space, time, and frequency domains, then the input may have power constraint for each antenna, or for each antenna and time slot, or each antenna, time slot, and frequency bin, thereby creating different constraints for each element. The tensor framework provides a structured way of mathematically representing any such power constraint. To define such constraints, we use a simplified notation where we denote the sequence of indices $(i_1, i_2, \ldots, i_N)$ as $\underline{i}$. Let $\underline{i}_c$ denote the sequence of indices corresponding to power constraint and $\underline{i}_r$ represent the rest of the indices in $\underline{i}$ that are not in $\underline{i}_c$. For instance, in an order 5 tensor of size $I_1 \times I_2 \times I_3 \times I_4 \times I_5$, $\underline{i} = (i_1, i_2, i_3, i_4, i_5)$. Assume that the domains under power constraints are 2 and 4; then $\underline{i}_c = (i_2, i_4)$ and $\underline{i}_r = (i_1, i_3, i_5)$. In this notation, we can write $\sum_{i_2=1}^{I_2} \sum_{i_4=1}^{I_4}$ as simply $\sum_{\underline{i}_c}$. We will now describe a family of optimization problems to find capacity under different types of power constraints, as follows:

$$
\begin{aligned}
&\max_{\mathcal{Q}} \ \log\det\left(\mathcal{H} *_N \mathcal{Q} *_N \mathcal{H}^H + \mathcal{I}\right) \\
&s.t. \ \sum_{\underline{i}_r} \mathcal{Q}_{\underline{i},\underline{i}} \leq P_{\underline{i}_c} \quad \forall \underline{i}_c, \quad \mathcal{Q} \succeq 0.
\end{aligned}
\tag{13}
$$

Suppose $\underline{i}_c = \underline{i}$, and thus $\underline{i}_r$ is empty. This represents the case of power constraint per element where $P_{\underline{i}_c} = P_{i_1,\ldots,i_N}$, and the inequality constraint in (13) becomes $\mathcal{Q}_{i_1,\ldots,i_N,i_1,\ldots,i_N} \leq P_{i_1,\ldots,i_N}, \forall i_1, \ldots, i_N$. Now suppose $\underline{i}_c = i_K$, where $K \leq N$. This represents the case with per-domain element constraint for the Kth domain where each element i_K has a different budget of P_{i_K} (for instance, per-antenna power constraint), i.e., $P_{\underline{i}_c} = P_{i_K}$. The inequality constraint in (13) becomes

$$
\sum_{i_1=1}^{I_1} .. \sum_{i_{K-1}=1}^{I_{K-1}} \sum_{i_{K+1}=1}^{I_{K+1}} .. \sum_{i_N=1}^{I_N} \mathcal{Q}_{i_1,..,i_K,..,i_N,i_1,..,i_K,...,i_N} \leq P_{i_K},
\tag{14}
$$

for all i_K. Similarly, the tensor framework can represent power constraints on any combination of different domains. Last, consider the case when $\underline{i}_c$ is empty. Hence, $\underline{i}_r = \underline{i}$, and the power constraint translates to the sum-power constraint, i.e., $P_{\underline{i}_c} = P$. Then inequality constraint in (13) becomes

$$
\sum_{i_1=1}^{I_1} \cdots \sum_{i_N=1}^{I_N} \mathcal{Q}_{i_1,\ldots,i_N,i_1,\ldots,i_N} = \operatorname{tr}(\mathcal{Q}) \leq P.
\tag{15}
$$

Any such power constraint is a linear constraint, and the objective function in (13) is a concave function. Hence, the optimization problem defined in (13) is a convex optimization problem that can be solved using the Karush–Kuhn–Tucker (KKT) conditions.

3.2 Solution Using the KKT Conditions

Let $\mathcal{M} \succeq 0$ represent the Lagrangian multiplier for the positive semi-definite constraint of size $I_1 \times \ldots I_N \times I_1 \times \ldots I_N$. Let $\mu_{\underline{i}_c} \geq 0$ for all $\underline{i}_c$ be the Lagrange multipliers corresponding to all the linear constraints. Then the Lagrangian functional is defined as

$$\mathcal{L}(\mathcal{Q}, \{\mu_{\underline{i}_c}\}, \mathcal{M}) = -\log\det(\mathcal{H} *_N \mathcal{Q} *_N \mathcal{H}^H + \mathcal{I}) + \sum_{\underline{i}_c} \mu_{\underline{i}_c} (\sum_{\underline{i}_r} \mathcal{Q}_{\underline{i},\underline{i}} - P_{\underline{i}_c}) - \operatorname{tr}(\mathcal{M} *_N \mathcal{Q}). \tag{16}$$

We consider a pseudo-diagonal tensor $\mathcal{B}$ of same size as the input covariance such that its non-zero elements are $\mathcal{B}_{\underline{i},\underline{i}} = \mu_{\underline{i}_c}, \forall \underline{i}_r$. For instance, if $\underline{i}_c = (i_1, i_2)$, then $\mathcal{B}_{i_1,\ldots,i_N,i_1,\ldots,i_N} = \mu_{i_1,i_2}$ for all $(i_3, \ldots, i_N)$. Then we have

$$\sum_{\underline{i}_c} \sum_{\underline{i}_r} \mu_{\underline{i}_c} \cdot \mathcal{Q}_{\underline{i},\underline{i}} = \sum_{\underline{i}} \mathcal{B}_{\underline{i},\underline{i}} \cdot \mathcal{Q}_{\underline{i},\underline{i}} = \operatorname{tr}(\mathcal{B} *_N \mathcal{Q}). \tag{17}$$

Using (17), we can re-write the Lagrangian from (16) as

$$\mathcal{L}(\mathcal{Q}, \{\mu_{\underline{i}_c}\}, \mathcal{M}) = -\log\det(\mathcal{H} *_N \mathcal{Q} *_N \mathcal{H}^H + \mathcal{I}) - \sum_{\underline{i}_c} \mu_{\underline{i}_c} P_{\underline{i}_c} + \operatorname{tr}(\mathcal{B} *_N \mathcal{Q}) - \operatorname{tr}(\mathcal{M} *_N \mathcal{Q}). \tag{18}$$

The KKT conditions corresponding to the optimization problem in (13) are given as

$$\mathcal{H}^H *_M (\mathcal{H} *_N \mathcal{Q} *_N \mathcal{H}^H + \mathcal{I})^{-1} *_M \mathcal{H} = \mathcal{B} - \mathcal{M}, \tag{19}$$

$$\operatorname{tr}(\mathcal{M} *_N \mathcal{Q}) = 0 \text{ or } \mathcal{Q}^{1/2} *_N \mathcal{M} *_N \mathcal{Q}^{1/2} = 0_{\mathcal{T}}, \tag{20}$$

$$\mu_{\underline{i}_c} (\sum_{\underline{i}_r} \mathcal{Q}_{\underline{i},\underline{i}} - P_{\underline{i}_c}) = 0, \quad \forall \underline{i}_c. \tag{21}$$

Note that all the elements of $\mathcal{B}$, i.e., $\mu_{\underline{i}_c}$, will be strictly greater than 0 at optimum because the inequality constraint must be met with equality at optimum. So $\mathcal{B}$ is positive definite and invertible. Also since $\mu_{\underline{i}_c} > 0$, (21) can be written as

$$\sum_{\underline{i}_r} \mathcal{Q}_{\underline{i},\underline{i}} - P_{\underline{i}_c} = 0, \quad \forall \underline{i}_c. \tag{22}$$

Let us define $\bar{\mathcal{H}} \in \mathbb{C}^{I_1 \times \cdots \times I_N \times I_1 \times \cdots \times I_N}$ and its EVD as

$$\bar{\mathcal{H}} \triangleq \mathcal{B}^{-1/2} *_N (\mathcal{H}^H *_M \mathcal{H}) *_N \mathcal{B}^{-1/2} = \mathcal{V} *_N \bar{\mathcal{D}} *_N \mathcal{V}^H. \tag{23}$$

Using a generalization of Theorem 1 from [12] to tensors, the KKT conditions (19) and (20) can be solved subject to $\mathcal{Q} \succeq 0$, $\mathcal{M} \succeq 0$ and $\mathcal{B} \succ 0$ to obtain the optimal $\mathcal{Q}$ as

$$\mathcal{Q}^{(opt)} = \mathcal{B}^{-1/2} *_N \mathcal{V} *_N \left(\mathcal{I} - \bar{\mathcal{D}}^{-1}\right)^+ *_N \mathcal{V}^H *_N \mathcal{B}^{-1/2}, \tag{24}$$

where $\bar{\mathcal{D}}$ and $\mathcal{V}$ are obtained through the tensor EVD of $\bar{\mathcal{H}}$ and $()^+$ denotes a pseudo-diagonal tensor whose all the entries are non-negative, i.e., $(a)^+ = \max\{0, a\}$. Based on (3), we can write the capacity as $C = \log[\det(\mathcal{H} *_N \mathcal{Q}^{(opt)} *_N \mathcal{H}^H + \mathcal{I})] = \log[\det(\mathcal{Q}^{(opt)} *_N \mathcal{H}^H *_M \mathcal{H} + \mathcal{I})]$. On substituting $\mathcal{Q}^{(opt)}$ in C and using (3), we get

$$\begin{aligned} C &= \log\det(\mathcal{V} *_N (\mathcal{I} - \bar{\mathcal{D}}^{-1})^+ *_N \mathcal{V}^H *_N \underbrace{\mathcal{B}^{-1/2} *_N \mathcal{H}^H *_M \mathcal{H} *_N \mathcal{B}^{-1/2}}_{\bar{\mathcal{H}}} + \mathcal{I}) \\ &= \log\det(\mathcal{V} *_N ((\mathcal{I} - \bar{\mathcal{D}}^{-1})^+ *_N \bar{\mathcal{D}} + \mathcal{I}) *_N \mathcal{V}^H). \end{aligned} \tag{25}$$

Since the determinant is the product of eigenvalues, we get

$$C = \sum_{i_1,\ldots,i_N} (\log(\bar{d}_{i_1,\ldots,i_N}))^+, \tag{26}$$

where $\bar{d}_{i_1,\ldots,i_N}$ are the non-zero eigenvalues of $\bar{\mathcal{H}}$. Hence, (22), (24), and (26) can be used to find the capacity. We can substitute (24) in (22) to solve for $\mathcal{B}$, which in turn is used to find $\bar{\mathcal{H}}$ and subsequently calculate capacity.

Under sum-power constraint of (15), $\underline{i}_c$ is empty, and we only have a single Lagrange multiplier μ associated with the constraint (15). Hence, all the pseudo-diagonal elements of $\mathcal{B}$ are μ, making $\mathcal{B}$ a multiple of identity tensor. Thus substituting $\mathcal{B} = \mu\mathcal{I}$ in (23) gives us $\bar{\mathcal{H}} = \mu^{-1}.(\mathcal{H}^H *_M \mathcal{H}) = \mathcal{V} *_N (\mu^{-1}.\mathcal{D}) *_N \mathcal{V}^H$. Thus, (24) and (26) become

$$\mathcal{Q} = \mathcal{V} *_N \left(\mu^{-1}\mathcal{I} - \mathcal{D}^{-1}\right)^+ *_N \mathcal{V}^H \tag{27}$$

$$C = \sum_{i_1,\ldots,i_N} \left(\log\left(\frac{d_{i_1,\ldots,i_N}}{\mu}\right)\right)^+, \tag{28}$$

respectively, which give us the capacity and the optimal covariance to achieve the capacity under sum-power constraint. The eigenvalues of $\mathcal{H}^H *_M \mathcal{H}$ are denoted by $d_{i_1,\ldots,i_N}$. Notice that (27) can be seen as a higher-order generalization of the classical water-filling solution. Also, the capacity and optimum covariance from (24) and (26) can be seen as a generalization of matrix-field water-filling from [13] or capacity with per-antenna power constraint for MIMO channels in [14].

Assuming the inverse of $(\mathcal{H}^H *_M \mathcal{H})$ exists and high SNR, such that we can ignore $()^+$, (24) can be simplified as

$$\mathcal{Q} = \mathcal{B}^{-1/2} *_N \Big(\mathcal{I} - \underbrace{\mathcal{V} *_N \bar{\mathcal{D}}^{-1} *_N \mathcal{V}^H}_{\bar{\mathcal{H}}^{-1}}\Big) *_N \mathcal{B}^{-1/2} \quad (29)$$

$$= \mathcal{B}^{-1} - (\mathcal{H}^H *_M \mathcal{H})^{-1} \quad (30)$$

and (25) as

$$C = \log\det(\mathcal{V} *_N \bar{\mathcal{D}} *_N \mathcal{V}^H) = \log\det(\bar{\mathcal{H}}) \quad (31)$$

$$= \log(\det(\mathcal{B}^{-1}) \cdot \det(\mathcal{H}^H *_M \mathcal{H})). \quad (32)$$

Subsequently, $\mu_{\underline{i}_c}$, which are entries of $\mathcal{B}$, can be found by substituting $\mathcal{Q}$ from (30) into (22) to get

$$\sum_{\underline{i}_r} \mathcal{B}^{-1}_{\underline{i},\underline{i}} - \sum_{\underline{i}_r} ((\mathcal{H}^H *_M \mathcal{H})^{-1})_{\underline{i},\underline{i}} = P_{\underline{i}_c} \quad \forall \underline{i}_c. \quad (33)$$

We use $N_{\underline{i}_r}$ to denote the number of values $\underline{i}_r$ can take. For example, if $\underline{i}_r = (i_1, \ldots, i_N)$, then $N_{\underline{i}_r} = I_1 \cdots I_N$; if $\underline{i}_r = (i_K, i_L)$, then $N_{\underline{i}_r} = I_K \cdot I_L$. Since the pseudo-diagonal elements of $\mathcal{B}$ are $\mu_{\underline{i}_c}$ and each $\mu_{\underline{i}_c}$ appears exactly $N_{\underline{i}_r}$ times on the pseudo-diagonal, from (33), we get

$$N_{\underline{i}_r} \cdot \mu^{-1}_{\underline{i}_c} - \sum_{\underline{i}_r} ((\mathcal{H}^H *_M \mathcal{H})^{-1})_{\underline{i},\underline{i}} = P_{\underline{i}_c} \quad \forall \underline{i}_c \quad (34)$$

$$\mu_{\underline{i}_c} = \frac{N_{\underline{i}_r}}{P_{\underline{i}_c} + \sum_{\underline{i}_r} ((\mathcal{H}^H *_M \mathcal{H})^{-1})_{\underline{i},\underline{i}}} \quad \forall \underline{i}_c, \quad (35)$$

which gives us all the elements of $\mathcal{B}$. The proposed solution in (35) and (30) assumes high SNR as we have ignored the $()^+$ operation. To make the solution applicable to any setting, we suggest similar scaling technique as used in [13] for matrix-field water-filling. If the tensor obtained from (30) has negative eigenvalues, then we force the negative eigenvalues of $\mathcal{Q}$ to be zero. This brute force approach will result into slightly larger values at the pseudo-diagonal of $\mathcal{Q}$, which can make the solution infeasible, i.e., the power constraint $P_{\underline{i}_c}$ might not be met, and we may get a different power allocated say $P'_{\underline{i}_c}$, where $P'_{\underline{i}_c} \geq P_{\underline{i}_c}$. Note that $P'_{\underline{i}_c} = \sum_{\underline{i}_r} \mathcal{Q}_{\underline{i},\underline{i}}, \forall \underline{i}_c$. So we scale the resulting $\mathcal{Q}$ by using another pseudo-diagonal tensor $\mathcal{S}$ such that the power constraint remains satisfied, i.e., $\mathcal{Q}^{(\text{scaled})} = \mathcal{S} *_N \mathcal{Q} *_N \mathcal{S}^H$, where the pseudo-diagonal entries of the scaling tensor $\mathcal{S}$ are given as $\mathcal{S}_{\underline{i},\underline{i}} = \sqrt{P_{\underline{i}_c}/P'_{\underline{i}_c}}, \forall \underline{i}_c$. This method ensures the positive semi-definiteness of the covariance, and the scaling ensures that the power constraints are satisfied. Note that this scaling approach simplifies the computation of the input covariance but also makes it sub-optimal and hence gives an approximation to the capacity, where the approximation gets better as SNR grows and is exact at sufficiently high SNR.

3.3 Comparing Capacity Under Various Constraints

Let the set of all possible positive semi-definite tensors of size $I_1 \times \cdots \times I_N \times I_1 \times \cdots \times I_N$ be represented by $\mathbb{Q}$. Let the feasible set for the optimization problem in (13) for two different settings $\underline{i}_{c1}$ and $\underline{i}_{c2}$ be $\mathbb{Q}_1$ and $\mathbb{Q}_2$, respectively. Assume $\underline{i}_{c1}$ is a sub-sequence of $\underline{i}_{c2}$, and the power budget for each $\underline{i}_{c2}$ when summed over indices in $\underline{i}_{c2} - \underline{i}_{c1}$ (indices included in $\underline{i}_{c2}$ but not in $\underline{i}_{c1}$) gives the power budget for the domains in $\underline{i}_{c1}$. Mathematically, we can represent it as $\sum_{\underline{i}_{c2}-\underline{i}_{c1}} P_{\underline{i}_{c2}} = P_{\underline{i}_{c1}}$, which implies that for the same power budget we have more domains under individual constraint when moving from $\underline{i}_{c1}$ to $\underline{i}_{c2}$. In such case, we have $\mathbb{Q}_2 \subset \mathbb{Q}_1$. From the principles of convex optimization, it can be shown that for a concave objective function $g(\mathcal{Q})$ where $\mathcal{Q} \in \mathbb{Q}$, if $\mathbb{Q}_2 \subset \mathbb{Q}_1 \subset \mathbb{Q}$, we have

$$\max_{\mathcal{Q}\in\mathbb{Q}_2} g(\mathcal{Q}) \leq \max_{\mathcal{Q}\in\mathbb{Q}_1} g(\mathcal{Q}). \tag{36}$$

Hence, capacity achieved when $\underline{i}_{c2}$ domains are under individual constraints cannot be larger than the capacity when $\underline{i}_{c1}$ domains are under individual constraints. For instance, consider a 2×2 input where the two domains are antenna and time slots. Let the capacity under the total power constraint P be C_1. Let the capacity under per-antenna power constraints of P_1 for antenna 1 and P_2 for antenna 2 such that $P_1 + P_2 = P$ be C_2. Then $C_2 \leq C_1$. Similarly, capacity achieved under power constraints per element, i.e., $P_{11}, P_{12}, P_{21}, P_{22}$, where P_{ij} represents power budget for the ith antenna and the jth time slot, such that $P_{11} + P_{12} = P_1$, $P_{21} + P_{22} = P_2$, be C_3, then $C_3 \leq C_2$.

We can also characterize the capacity contribution by each constrained domain separately. For a fixed tensor channel, the tensor $\mathcal{B}^{-1}$ is pseudo-diagonal with entries $\mu_{\underline{i}_c}^{-1}$; hence, $\det(\mathcal{B}^{-1}) = \prod_{\underline{i}} \mu_{\underline{i}_c}^{-1}$. Also, assuming all the eigenvalues of $(\mathcal{H}^H *_M \mathcal{H})$ denoted by $d_{\underline{i}}$ are non-zero, $\det(\mathcal{H}^H *_M \mathcal{H}) = \prod_{\underline{i}} d_{\underline{i}}$. Hence, (32) can be written as

$$C = \sum_{\underline{i}} \log \frac{d_{\underline{i}}}{\mu_{\underline{i}_c}} = \sum_{\underline{i}_c} \Big(\sum_{\underline{i}_r} \log \frac{d_{\underline{i}}}{\mu_{\underline{i}_c}} \Big) = \sum_{\underline{i}_c} C_{\underline{i}_c}, \tag{37}$$

where $C_{\underline{i}_c}$ can be seen as the contribution of the $\underline{i}_c^{th}$ element of the constrained domains to the overall capacity. Based on $\mu_{\underline{i}_c}$ from (35), we have

$$C_{\underline{i}_c}(P_{\underline{i}_c}) = \sum_{\underline{i}_r} \log \Big(\frac{P_{\underline{i}_c} + \sum_{\underline{i}_r} ((\mathcal{H}^H *_M \mathcal{H})^{-1})_{\underline{i},\underline{i}}}{N_{\underline{i}_r}} d_{\underline{i}} \Big)$$

$$\Rightarrow \lim_{P_{\underline{i}_c} \to \infty} C_{\underline{i}_c}(P_{\underline{i}_c}) = \sum_{\underline{i}_r} \Big[\log \big(P_{\underline{i}_c} \big) + \log \Big(\frac{d_{\underline{i}}}{N_{\underline{i}_r}} \Big) \Big]. \tag{38}$$

The second term in this expression is independent of $P_{\underline{i}_c}$. Hence, the multiplexing gain achieved by putting constraints on domains in $\underline{i}_c$ is given by

$$\chi_{\underline{i}_c} = \lim_{P_{\underline{i}_c}\to\infty} \frac{C_{\underline{i}_c}(P_{\underline{i}_c})}{\log(P_{\underline{i}_c})} = N_{\underline{i}_r}. \tag{39}$$

Since $N_{\underline{i}_r}$ increases exponentially with the number of unconstrained domains, we can see that for tensor channels, the multiplexing gain increases exponentially with increase in the number of unconstrained domains.

4 Numerical Examples

We compute the capacity for a system with order two input of size 2×2, order two output of size 2×2, and channel as an order 4 tensor of size $2 \times 2 \times 2 \times 2$ in the presence of circular symmetric Gaussian noise with variance σ^2. The individual elements of the channel are i.i.d. circular complex Gaussian with zero mean and unit variance as in [8, 14]. The SNR is defined as P/σ^2, where P is the total power constraint and $\sigma^2 = 1$. The results are averaged over 100 channel realizations, which are known at the transmitter and receiver. Figure 1 presents the capacity under per-element power constraints and compares it with sum-power constraint. If the total available power is P, then $P_1 = x \cdot P$ and $P_2 = (1 - x) \cdot P$. Further, $P_{11} = y \cdot P_1$, $P_{12} = (1 - y) \cdot P_1$, $P_{21} = y \cdot P_2$, and $P_{22} = (1 - y) \cdot P_2$. Thus, with $0 \leq x, y \leq 1$, $P_{11}, P_{12}, P_{21}, P_{22}$ denote the individual power constraints on all the four elements of input tensor such that $P_{11} + P_{12} + P_{21} + P_{22} = P$. To put it in context, let the two domains represent time and antenna, and then P_{ij} represents the power constraint on the ith antenna and the jth time slot. With different choices of x and y, we achieve different per-element power constraints such that the total power remains P. The capacity with per-element power constraints against x and y at SNR of 10 dB is presented in Fig. 1. It can be seen that for different values of x and y, the capacity achieved under per-element constraints can be significantly lower than the capacity achieved under sum-power constraint, and as $x \approx y \approx 0.5$, which would mean uniform power constraint on all the elements, the capacity under per-element power constraints is almost same as that of the capacity under sum-power constraint.

Figure 2 presents the case for per-domain element power constraints, where only one of the two available domains is under power constraints. To put it in context, assume that the two domains are space and time, and we have power constraints P_1 and P_2 on each antenna, such that $P_1 + P_2 = P$. Figure 2 plots the capacity against $x = P_1/P$ at SNR of 15 dB. In this case also, a significant difference in capacity can be observed as compared to the sum-power constraint, especially when difference between P_1 and P_2 is large.

Figure 3 presents the capacity for sum-power constraint, per-domain element power constraints where one of the two domains are constrained with power budgets P_1, P_2, and per-element power constraints where all the four elements have different

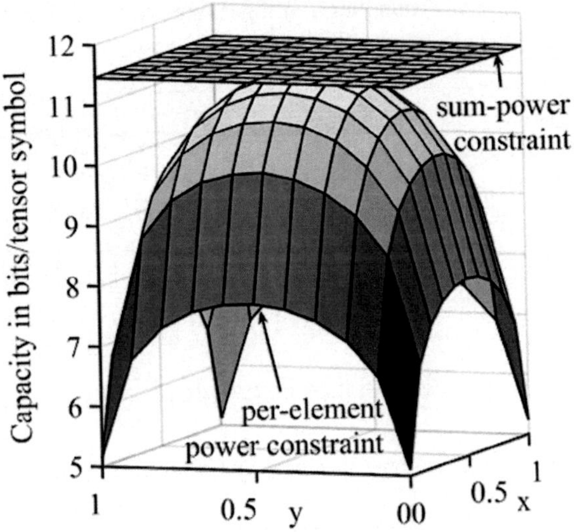

Fig. 1 Capacity for per-element power constraint vs. x vs. y

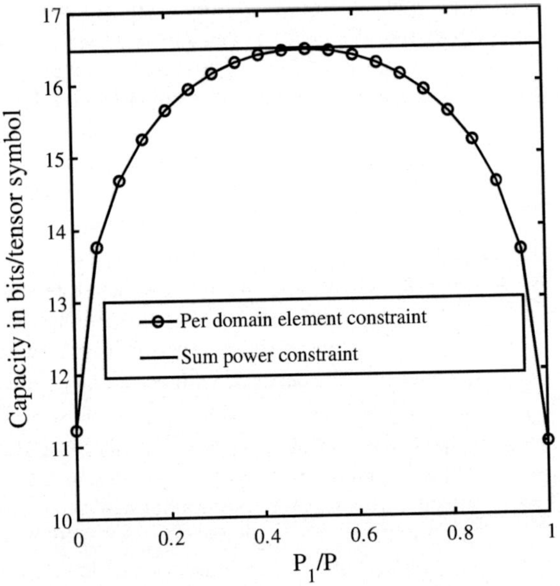

Fig. 2 Capacity for per-domain element constraint vs. P_1/P

power budgets P_{11}, P_{12}, P_{21}, P_{22} against SNR for $x = y = 0.1$. It can be seen that the capacity under per-domain element and per-element constraints is always upper bounded by the capacity under sum-power constraint. The capacity increases with SNR for all three cases, but the performance difference also gradually increases between the three cases, with sum-power constraint performing the best, followed by the per-domain element and lastly the per-element constraints.

Fig. 3 Capacity for different constraints vs. SNR

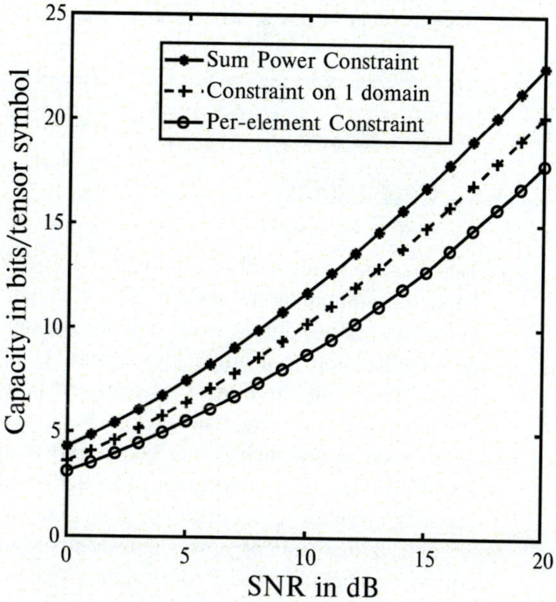

Fig. 4 Capacity for different SNRs vs. input order

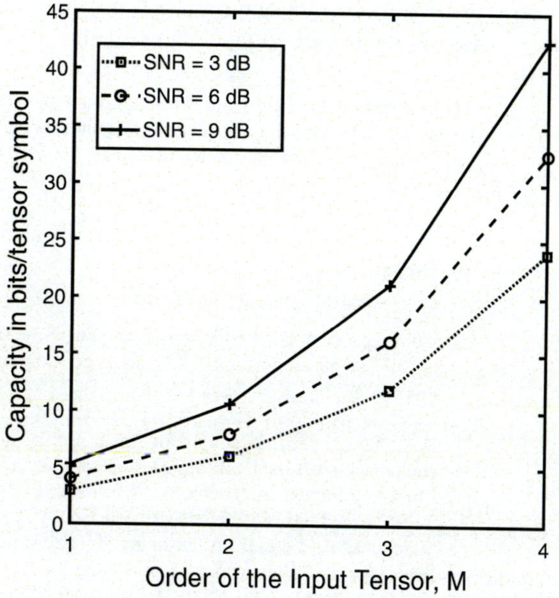

Figure 4 shows the capacity of order $2M$ tensor channels under sum-power constraint with input and output of order M. The dimensions of all the individual domains are 2, and the plot is presented against M for various SNR values. It can be observed that as SNR increases, the capacity increase is almost exponential in M.

Since under sum-power constraint the number of unconstrained input domains is M, hence the increase in capacity can be attributed to the exponential multiplexing gain achieved by increasing the number of domains M.

5 Conclusions

The Einstein product of tensors was used to propose a system model for a multi-domain communication system. The channel was represented as a higher-order tensor, and information-theoretic analysis extended the classical water-filling input power allocation to a multi-domain setting. In particular, the tensor framework's unique suitability to express a family of power constraints across domains was exploited. We derived a general expression for capacity and optimum input covariance tensor under such power constraints assuming the tensor channel is known at the transmitter and receiver. Through numerical examples, we analyzed the capacity behavior under different power constraints such as sum-power constraint, a single domain under power constraints, and per-element power constraints. A comparison between these constraints revealed that for the same total power, as the number of domains under individual constraints increases, the capacity decreases. It was also shown that the multiplexing gain achieved by a tensor channel increases exponentially with increase in the number of unconstrained domains.

Acknowledgments This work was supported by grant RGPIN-2016-03647 entitled "Tensor modulation for space-time-frequency communication systems," awarded by the Natural Sciences and Engineering Research Council of Canada (NSERC).

References

1. Sidiropoulos, N.D., Lathauwer, L.D., Fu, X., Huang, K., Papalexakis, E.E., Faloutsos, C.: Tensor decomposition for signal processing and machine learning. IEEE Trans. Signal Process. **65**(13), 3551–3582 (2017)
2. Cichocki, A., Mandic, D., De Lathauwer, L., Zhou, G., Zhao, Q., Caiafa, C., Phan, H.A.: Tensor decompositions for signal processing applications: from two-way to multiway component analysis. IEEE Signal Process. Mag. **32**(2), 145–163 (2015)
3. Chen, C., Surana, A., Bloch, A., Rajapakse, I.: Multilinear time invariant system theory. In: 2019 Proceedings of the Conference on Control and its Applications, pp. 118–125 (2019)
4. Venugopal, A., Leib, H.: A tensor based framework for multi-domain communication systems. IEEE Open J. Commun. Soc. **1**, 606–633 (2020)
5. Favier, G., de Almeida, A.L.F.: Tensor space-time-frequency coding with semi-blind receivers for MIMO wireless communication systems. IEEE Trans. Signal Process. **62**(22), 5987–6002 (2014)
6. Brazell, M., Li, N., Navasca, C., Tamon, C.: Solving multilinear systems via tensor inversion. SIAM J. Matrix Analy. Appl. **34**(2), 542–570 (2013)
7. Liang, M.L., Zheng, B., Zhao, R.J.: Tensor inversion and its application to the tensor equations with Einstein product. Linear Multilinear Algebra **67**(4), 843–870 (2019)

8. Telatar, I.E.: Capacity of multi-antenna Gaussian channels. Eur. Trans. Telecom. **10**, 585–595 (1999)
9. Pandey, D., Leib, H.: Tensor multi-linear MMSE estimation using the Einstein product. In: Advances in Information and Communication (FICC 2021), pp. 47–64. Springer International Publishing, Berlin (2021)
10. Cui, L.-B., Chen, C., Li, W., Ng, M.K.: An eigenvalue problem for even order tensors with its applications. Linear Multilinear Algebra **64**(4), 602–621 (2016)
11. Pandey, D., Leib, H.: A tensor based precoder and receiver for MIMO GFDM systems. In: 2021 IEEE International Conference on Communications. IEEE (2021)
12. Xing, C., Jing, Y., Wang, S., Wang, J., An, J.: A general framework for covariance matrix optimization in MIMO systems (2018). Preprint arXiv:1711.04449v8
13. Xing, C., Fei, Z., Zhou, Y., Pan, Z.: Matrix-field water-filling architecture for MIMO transceiver designs with mixed power constraints. In: 2015 IEEE 26th Annual International Symposium on Personal, Indoor, and Mobile Radio Communications (PIMRC), pp. 392–396 (2015)
14. Vu, M.: MIMO capacity with per-antenna power constraint. In: 2011 IEEE Global Telecommunications Conference - GLOBECOM 2011, pp. 1–5 (2011)

On Downlink Interference Decoding in Multi-cell Massive MIMO Systems

Meysam Shahrbaf Motlagh, Subhajit Majhi, Patrick Mitran, and Hideki Ochiai

1 Introduction

Massive MIMO systems with base stations (BSs) simultaneously serving several users via a large number of antennas are considered to be an integral part of 5G standards [1]. Moreover, the benefits of massive MIMO systems reported in [1], e.g., high spectral efficiency, increased capacity due to aggressive spatial multiplexing, and high energy efficiency, can all be achieved with simple linear precoding techniques.

The use of time-division duplex (TDD) mode has been suggested as the "canonical form" of massive MIMO [1], where downlink and uplink channels are assumed to be reciprocal. When using TDD mode, channel estimates are obtained at BSs using uplink training pilots, which are then employed for precoding/decoding. However, as the length of the channel coherence time is finite, the sequences of orthogonal pilots are limited in number. Hence, in a multi-cell system, one way to address this limitation is to re-use a set of orthogonal pilots across all cells. Thus, the channel estimate of a user in one cell is contaminated by the channels of users

Subhajit Majhi contributed to this work while he was affiliated with the University of Waterloo, ON, Canada.

M. Shahrbaf Motlagh (✉) · P. Mitran
University of Waterloo, Waterloo, ON, Canada
e-mail: mshahrba@uwaterloo.ca

S. Majhi
Qualcomm India Pvt. Ltd, Hyderabad, India

H. Ochiai
Yokohama National University, Yokohama, Japan

H. Nguyen et al. (eds.), *30th Biennial Symposium on Communications 2021*,
Signals and Communication Technology, https://doi.org/10.1007/978-3-031-06947-5_3

in other cells that use the same pilot sequence as the former [2], which is referred to as pilot contamination (PC).

Interestingly, when the channels of users to BSs in a massive MIMO system are assumed to be independent Rayleigh fading, as the number of BS antennas M grows, the effects of multi-user interference, additive noise, and small-scale fading all vanish owing to channel hardening and asymptotic favorable propagation [3]. However, inter-cell interference due to PC does not vanish [3], which constitutes a bottleneck in multi-cell massive MIMO systems. Extensive works have been carried out in the literature, such as those in [4–6], which have proposed solutions to tackle the PC problem, but under some constraints that may be difficult to realize in practice. Specifically, these include among others, the assumptions of: (1) channel coherence time approaching infinity [4], (2) the presence of a central unit and cumbersome backhaul overhead to share all data symbols for centralized processing [5], and (3) channels of pilot-sharing users in different cells with asymptotically linearly independent covariance matrices [6].

Most studies in the literature treat the PC interference as noise (TIN) and the resulting rates, which are suboptimal in many cases [7], saturate, and do not grow with M. In this chapter, we address the PC interference problem in the downlink of a multi-cell massive MIMO system arising from re-using the same set of orthogonal pilots across cells. In particular, we depart from the viewpoint of treating the PC interference as noise and propose full/partial interference decoding schemes that attain unbounded rates for large M.

The major contributions of this chapter are summarized below. First, for two well-known linear precoding techniques at the BS, maximum ratio transmission (MRT) and zero forcing (ZF), we derive achievable rates for schemes that decode the PC interference either fully such as the simultaneous unique/non-unique decoding (SD/SND) or partially (PD). Specifically, the PC interference is fully decoded uniquely in SD and non-uniquely in SND. We similarly study achievable rates for a PD scheme, where messages at each BS are encoded into two independent layers; then only one layer of PC interference is decoded non-uniquely, while the other layer is treated as noise. The per-user achievable rates for SD/SND/PD, derived using a worst-case uncorrelated noise technique, are shown to scale as $\mathcal{O}(\log M)$, thus eliminating the rate saturation problem. Focusing on the 2-cell case, we study the problem of maximum symmetric rate, i.e., maximizing the worst user's rate, for TIN/SD/SND/PD schemes, and observe that the full interference decoding scheme of SND outperforms TIN, while the partial decoding scheme of PD outperforms even SND.

Note that in [2], full PC interference decoding schemes were proposed for the uplink of a multi-cell massive MIMO system using maximum ratio combining (MRC). In contrast, this chapter focuses on the downlink setting with both MRT and ZF and proposes not only full (SD/ SND) but also partial interference decoding (PD) schemes based on rate splitting. In particular, we find that the use of ZF that better mitigates multi-user interference in conjunction with either SND or PD results in notably larger rates such that the number of BS antennas required for interference decoding schemes to outperform TIN now reduces by more than a factor of ≈100 (i.e., within the practical range), which is a significant improvement over the results of [2].

2 Preliminaries

2.1 System Model

A TDD multi-cell system with L cells is considered, where each cell comprises a BS that serves K single antenna users with M transmitting antennas ($M \gg K$). The channel between the BS in cell j (BS_j) and users in cell l is denoted as $\boldsymbol{G}_{jl} = \boldsymbol{H}_{jl}\boldsymbol{D}_{jl}^{1/2}$, where $\boldsymbol{H}_{jl} = [\boldsymbol{h}_{j1l}, \boldsymbol{h}_{j2l}, \ldots, \boldsymbol{h}_{jKl}] \in \mathbb{C}^{M\times K}$ is the channel matrix corresponding to the channel vectors $\boldsymbol{h}_{jkl} \in \mathbb{C}^{M\times 1}$ of small-scale fading coefficients between BS_j and the kth user in cell l (U_{kl}), and $\boldsymbol{D}_{jl} = \text{diag}(\beta_{j1l}, \beta_{j2l}, \ldots, \beta_{jKl})$ is the matrix containing the large-scale fading coefficients. Assuming a block fading model, the large-scale fading coefficient β_{jkl}, which models path loss and shadowing between BS_j and U_{kl}, is taken to be a constant that is known *a priori* at BS_j, remains unchanged over multiple coherence time intervals, and is independent of the antenna index of BS_j. The small-scale fading coefficients $\boldsymbol{h}_{jkl} \sim \mathcal{CN}(\boldsymbol{0}, \boldsymbol{I}_M)$ are also independent across coherence intervals but remain constant in a given coherence interval, with $\boldsymbol{I}_M$ denoting the identity matrix of order M. The channel between BS_j and U_{kl} can also be expressed as $\boldsymbol{g}_{jkl} = \sqrt{\beta_{jkl}}\boldsymbol{h}_{jkl}$. Finally, the downlink and uplink channels are assumed reciprocal.

Downlink Data Transmission The baseband signal received at U_{il} is given by

$$y_{il} = \sum\nolimits_{j=1}^{L} \sqrt{\rho_{\text{d}}}\boldsymbol{g}_{jil}^{\dagger}\boldsymbol{x}_j + w_{il}, \tag{1}$$

where $\boldsymbol{x}_j = \left[x_j[1], x_j[2], \ldots, x_j[M]\right]^T$ is the signal transmitted by BS_j, $w_{il} \sim \mathcal{CN}(0, 1)$ is the AWGN at U_{il}, $(.)^{\dagger}$ denotes conjugate transpose, and ρ_{d} can be interpreted as the *downlink* transmission SNR per user of the BS. Defining $\boldsymbol{u}_{jk} \in \mathbb{C}^{M\times 1}$ to be the precoding vector of BS_j, we have $\boldsymbol{x}_j = 1/\sqrt{\lambda_j}\sum_{k=1}^{K}\boldsymbol{u}_{jk}s_j[k] = \boldsymbol{U}_j\boldsymbol{s}_j/\sqrt{\lambda_j}$, where $\boldsymbol{s}_j = \left[s_j[1], s_j[2], \ldots, s_j[K]\right]^T$ is the vector of data symbols intended for the users in cell j, $\boldsymbol{U}_j = \left[\boldsymbol{u}_{j1}, \boldsymbol{u}_{j2}, \ldots, \boldsymbol{u}_{jK}\right] \in \mathbb{C}^{M\times K}$, and λ_j is a normalization constant such that the downlink SNR per user for BS_j is ρ_{d}, i.e., $\mathbb{E}[\rho_{\text{d}}\boldsymbol{x}_j^{\dagger}\boldsymbol{x}_j]/K = \rho_{\text{d}}$.

CSI Estimation at BS Similar to [5], a set of orthogonal pilot sequences is re-used in all cells, and thus the channel estimate will be contaminated with the PC interference from users in other cells that are using the same pilot. During the uplink training phase, U_{kj} transmits a pilot to BS_j, $k = 1, \ldots, K$, $j = 1, \ldots, L$, and then BS_j estimates the local channels $\boldsymbol{G}_{jj}$, given by $\hat{\boldsymbol{G}}_{jj}$.

The minimum mean squared error (MMSE) estimate $\hat{\boldsymbol{g}}_{jkj}$ of $\boldsymbol{g}_{jkj}$ can be shown to be [5]

$$\hat{\boldsymbol{g}}_{jkj} = \alpha_{jkj}\left(\sum\nolimits_{l=1}^{L}\sqrt{\rho_{\mathrm{p}}}\boldsymbol{g}_{jkl} + \tilde{\boldsymbol{z}}_{jk}\right), \tag{2}$$

where $\alpha_{jkj} := \sqrt{\rho_{\mathrm{p}}}\beta_{jkj}/(1+\rho_{\mathrm{p}}\sum_{l=1}^{L}\beta_{jkl})$, $\tilde{\boldsymbol{z}}_{jk} \sim \mathcal{CN}(\boldsymbol{0},\ \boldsymbol{I}_M)$, and ρ_{p} is the pilot SNR. Also note that the channel vector $\boldsymbol{g}_{jkj}$ can be decomposed as $\boldsymbol{g}_{jkj} = \hat{\boldsymbol{g}}_{jkj} + \boldsymbol{\epsilon}_{jkj}$, where the estimation error $\boldsymbol{\epsilon}_{jkj} \sim \mathcal{CN}\left(\boldsymbol{0},\ \beta_{jkj}\left(1-\sqrt{\rho_{\mathrm{p}}}\alpha_{jkj}\right)\boldsymbol{I}_M\right)$ is independent of the MMSE estimate $\hat{\boldsymbol{g}}_{jkj} \sim \mathcal{CN}\left(\boldsymbol{0},\ \sqrt{\rho_{\mathrm{p}}}\beta_{jkj}\alpha_{jkj}\boldsymbol{I}_M\right)$ [5].

Treating Interference as Noise (TIN) For MRT at BS_j, the precoding vector is $\boldsymbol{u}_{jk} = \hat{\boldsymbol{g}}_{jkj}$, and the normalization factor λ_j^{mrt}, as found from the condition $\mathbb{E}[\rho_{\mathrm{d}}\boldsymbol{x}_j^{\dagger}\boldsymbol{x}_j]/K = \rho_{\mathrm{d}}$, is $\lambda_j^{\mathrm{mrt}} = \frac{M}{K}\sum_{k=1}^{K}\sqrt{\rho_{\mathrm{p}}}\beta_{jkj}\alpha_{jkj}$. Thus, rewriting (1), we have

$$y_{il}^{\mathrm{mrt}} = \sum\nolimits_{j=1}^{L}\sqrt{\rho_{\mathrm{d}}/\lambda_j^{\mathrm{mrt}}}\sum\nolimits_{k=1}^{K}\boldsymbol{g}_{jil}^{\dagger}\hat{\boldsymbol{g}}_{jkj}s_j[k] + w_{il}. \tag{3}$$

Using (2) and applying the strong law of large numbers to (3), we obtain

$$\frac{y_{il}^{\mathrm{mrt}}}{\sqrt{M}} \overset{\text{a.s.}}{\to} \left(\frac{\sqrt{K\rho_{\mathrm{d}}\rho_{\mathrm{p}}}\beta_{lil}\alpha_{lil}s_l[i]}{\sqrt{\sum_{k=1}^{K}\sqrt{\rho_{\mathrm{p}}}\beta_{lkl}\alpha_{lkl}}} + \sum_{j=1, j\neq l}^{L}\frac{\sqrt{K\rho_{\mathrm{d}}\rho_{\mathrm{p}}}\beta_{jil}\alpha_{jij}s_j[i]}{\sqrt{\sum_{k=1}^{K}\sqrt{\rho_{\mathrm{p}}}\beta_{jkj}\alpha_{jkj}}}\right), \tag{4}$$

as $M \to \infty$, i.e., a channel hardening effect is observed.

Assuming TIN in the downlink, U_{il} decodes only the desired signal $s_l[i]$ while treating the interfering signals $s_j[i],\ \ j \neq l$ as noise. Hence, defining R_{il} as the downlink rate associated with decoding $s_l[i]$, any rate tuple $(R_{i1}, \ldots, R_{iL})$ is achievable if the following conditions hold:

$$R_{il} \leq I\left(y_{il}^{\mathrm{mrt}};\, s_l[i]\right), \tag{5}$$

for $l = 1, \ldots, L,\ \ i = 1, \ldots, K$, where $I(X; Y)$ is the mutual information between X and Y [8]. Using (4), it is verified that the right-hand side (r.h.s.) in (5) converges to a constant independent of M. A similar phenomenon is observed when ZF precoding is used at the BS. In particular, considering ZF precoding matrix at BS_j given by $\boldsymbol{V}_{jj} = \left[\boldsymbol{v}_{j1j}, \ldots, \boldsymbol{v}_{jKj}\right] = \hat{\boldsymbol{G}}_{jj}\left(\hat{\boldsymbol{G}}_{jj}^{\dagger}\hat{\boldsymbol{G}}_{jj}\right)^{-1}$, it can be verified that $\boldsymbol{V}_{jj}^{\dagger}\hat{\boldsymbol{G}}_{jj} = \boldsymbol{I}_K$, and thereby $\boldsymbol{v}_{jkj}^{\dagger}\hat{\boldsymbol{g}}_{jmj} = \delta_{mk}$, where δ_{mk} is the Kronecker delta function. The normalization factor λ_j^{zf} is also derived to be $\lambda_j^{\mathrm{zf}} = \frac{1}{K(M-K)}\sum_{k=1}^{K}1/(\sqrt{\rho_{\mathrm{p}}}\beta_{jkj}\alpha_{jkj})$, from which an analysis similar to that of TIN in MRT results in the saturation of the rate obtained by ZF, i.e., $I\left(y_{il}^{\mathrm{zf}};\, s_l[i]\right)$ converges to a constant almost surely, as $M \to \infty$. Hence, the benefits of increasing

M saturate since the TIN rates obtained by MRT/ZF converge to a constant as M becomes large.

3 Decoding the PC Interference in Downlink

We propose schemes that treat the PC interference terms as individual users and thus decode them fully or partially, as opposed to treating them as noise. This results in new achievable rates that grow unbounded with M.

3.1 Simultaneous Unique Decoding (SD)

In SD, the PC interference terms are decoded uniquely, details of which for MRT and ZF are given below.

MRT Note that in (4), which expresses the received signal after performing MRT and letting $M \to \infty$, the first term is the desired signal and the other non-vanishing terms are the inter-cell PC interference. Treating PC interference terms as users to be uniquely decoded, (4) becomes identical to the output of a noise-free multiple-access channel (MAC). More specifically, after performing MRT at the BSs and letting $M \to \infty$, a noise-free L-user MAC is obtained at the user's side in downlink: for this MAC, unbounded per-user rates can be obtained by uniquely jointly decoding all signals $\{s_l[i]\}_{l=1}^{L}$.

We provide an analysis that establishes the achievable rates for finite M. We add and subtract a term associated with the mean of the effective channel $\boldsymbol{g}_{jil}^{\dagger}\hat{\boldsymbol{g}}_{jij}$ in (3), which results in the following:

$$
\begin{aligned}
y_{il}^{\text{mrt}} &= \underbrace{\sum_{j=1}^{L} \sqrt{\frac{\rho_{\text{d}}}{\lambda_j^{\text{mrt}}}} \mathbb{E}\left[\boldsymbol{g}_{jil}^{\dagger}\hat{\boldsymbol{g}}_{jij}\right] s_j[i]}_{\text{Desired signals}} \\
&\quad + \underbrace{\sum_{j=1}^{L} \sqrt{\frac{\rho_{\text{d}}}{\lambda_j^{\text{mrt}}}} \left(\boldsymbol{g}_{jil}^{\dagger}\hat{\boldsymbol{g}}_{jij} - \mathbb{E}\left[\boldsymbol{g}_{jil}^{\dagger}\hat{\boldsymbol{g}}_{jij}\right]\right) s_j[i]}_{\text{Interference due to beamforming gain uncertainty}} \\
&\quad + \underbrace{\sum_{j=1}^{L} \sqrt{\frac{\rho_{\text{d}}}{\lambda_j^{\text{mrt}}}} \sum_{k=1,k\neq i}^{K} \boldsymbol{g}_{jil}^{\dagger}\hat{\boldsymbol{g}}_{jkj} s_j[k]}_{\text{Inter-cell interference}} + \underbrace{w_{il}}_{\text{Noise}} \qquad (6) \\
&= \sum_{j=1}^{L} \theta_{jij}^{\text{mrt}} s_j[i] + w_{il}', \qquad (7)
\end{aligned}
$$

where $\theta_{jij}^{\mathrm{mrt}} = \sqrt{\rho_{\mathrm{d}}/\lambda_j^{\mathrm{mrt}}}\mathbb{E}\left[\mathbf{g}_{jil}^{\dagger}\hat{\mathbf{g}}_{jij}\right]$ and w'_{il} is the effective noise, incorporating the last three terms in (6).

Consider the L-user MAC in (7) associated with the estimate of $s_l[i]$ at U_{il}. Using the usual definitions of [8], messages $m_{il} \in \left[1 : 2^{nR_{il}}\right], l = 1, \ldots, L$ (distributed uniformly) are communicated over this MAC by encoding them into codewords $s_l[i](m_{il})$ of length n, where $s_l[i] \sim \mathcal{CN}(0, 1)$, i.i.d. Using the standard random coding analysis [8] and decoding all messages uniquely as in SD, decoding error probability can be shown to approach zero as $n \to \infty$, if

$$\sum\nolimits_{l\in\Omega} R_{il} \leq I\left(y_{il}^{\mathrm{mrt}};\ s_{\Omega}\ \middle|\ s_{\Omega^c}\right), \tag{8}$$

for all $\Omega \subseteq S = \{1, 2, \ldots, L\}$, and s_{Ω} is the vector with entries $s_l[i],\ l \in \Omega$. Thus, the rate tuple $(R_{i1}, \ldots, R_{iL})$ in (8) is achievable. Finally, the network-wide achievable region is obtained by taking the intersection of the achievable regions over all receivers.

As opposed to an ordinary MAC, where noise is typically Gaussian and independent of users' signals, for the MAC in (7), it is neither Gaussian nor independent of the users' signals, which makes it hard to characterize exact expressions for $I(.;.|.)$ terms in (8). However, the effective noise term in (7) is uncorrelated from the desired signals. Thus, we are able to establish lower bounds for these mutual information terms based on a worst-case uncorrelated noise technique as follows.

Theorem 1 *For $\mathbf{s}_l = [s_l[1], s_l[2], \ldots, s_l[K]]^T \sim \mathcal{CN}(\mathbf{0},\ \mathbf{I}_K), l \in \{1, 2, \ldots, L\}$, a set of lower bounds to (8) that can be achieved with MRT for the MAC given in (7), associated with U_{il}, is given as follows:*

$$I\left(y_{il}^{mrt};\ s_{\Omega}\middle|s_{\Omega^c}\right) \geq C\left(P_1^{mrt}/\left(P_2^{mrt} + P_3^{mrt} + P_4^{mrt}\right)\right), \tag{9}$$

where

$$P_1^{mrt} = M^2 \sum\nolimits_{j\in\Omega} \rho_d \rho_p \beta_{jil}^2 \alpha_{jij}^2 / \lambda_j^{mrt} \tag{10}$$

$$P_2^{mrt} = M \sum\nolimits_{j=1}^{L} \rho_d \sqrt{\rho_p} \beta_{jij} \alpha_{jij} \beta_{jil} / \lambda_j^{mrt} \tag{11}$$

$$P_3^{mrt} = M \sum_{j=1}^{L} \sum\nolimits_{k=1, k\neq i}^{K} \rho_d \sqrt{\rho_p} \beta_{jkj} \alpha_{jkj} \beta_{jil} / \lambda_j^{mrt}, \tag{12}$$

$$P_4^{mrt} = 1. \tag{13}$$

Proof The proof is relegated to Appendix. □

The terms $P_{(.)}^{\mathrm{mrt}}$ in Theorem 1 can be interpreted as follows. P_1^{mrt} is the power of the desired signals associated with s_{Ω}, whereas P_2^{mrt} is the power of the interference due to beamforming gain uncertainty, P_3^{mrt} is the power of the interference of other users, and P_4^{mrt} is the power of the noise. Moreover, using (10)–(13), the r.h.s. of (9) is expressed as

$$C_{\text{LB}}^{\text{mrt}} = C\left(\frac{M^2 \sum_{j\in\Omega} \rho_{\text{d}}\rho_{\text{p}}\beta_{jil}^2\alpha_{jij}^2/\lambda_j^{\text{mrt}}}{M\sum_{j=1}^{L}\sum_{k=1}^{K}\left(\rho_{\text{d}}\sqrt{\rho_{\text{p}}}\beta_{jkj}\alpha_{jkj}\beta_{jil}\right)/\lambda_j^{\text{mrt}}+1}\right). \tag{14}$$

Since for fixed K and large M, $\lambda_j^{\text{mrt}} \propto M$, the rates in (14) grow as $\mathcal{O}(\log M)$. Hence, by uniquely jointly decoding signals $\{s_l[i]\}_{l=1}^{L}$, unbounded rates are obtained as $M \to \infty$.

ZF Similar to MRT, by performing ZF at the BS and letting $M \to \infty$, a noise-free L-user MAC is obtained, which is similar to that in (4) for MRT. We now provide an analysis for ZF with finite M, similar to (6)–(7) for MRT:

$$y_{il}^{\text{zf}} = \sum_{j=1}^{L}\sqrt{\frac{\rho_{\text{d}}}{\lambda_j^{\text{zf}}}}\sum_{k=1}^{K}\left(\hat{\boldsymbol{g}}_{jil}^{\dagger}+\boldsymbol{\epsilon}_{jil}^{\dagger}\right)\boldsymbol{v}_{jkj}s_j[k]+w_{il} \tag{15}$$

$$\begin{aligned}
&= \sum_{j=1}^{L}\sqrt{\frac{\rho_{\text{d}}}{\lambda_j^{\text{zf}}}}\left(\frac{\beta_{jil}}{\beta_{jij}}\right)\hat{\boldsymbol{g}}_{jij}^{\dagger}\boldsymbol{v}_{jij}s_j[i] \\
&\quad +\sum_{j=1}^{L}\sqrt{\frac{\rho_{\text{d}}}{\lambda_j^{\text{zf}}}}\sum_{k=1,k\neq i}^{K}\left(\frac{\beta_{jil}}{\beta_{jij}}\right)\hat{\boldsymbol{g}}_{jij}^{\dagger}\boldsymbol{v}_{jkj}s_j[k] \\
&\quad +\sum_{j=1}^{L}\sqrt{\frac{\rho_{\text{d}}}{\lambda_j^{\text{zf}}}}\sum_{k=1}^{K}\boldsymbol{\epsilon}_{jil}^{\dagger}\boldsymbol{v}_{jkj}s_j[k]+w_{il}
\end{aligned} \tag{16}$$

$$\begin{aligned}
&\overset{(a)}{=} \underbrace{\sum_{j=1}^{L}\sqrt{\frac{\rho_{\text{d}}}{\lambda_j^{\text{zf}}}}\left(\frac{\beta_{jil}}{\beta_{jij}}\right)s_j[i]}_{\text{Desired signals}} + \underbrace{\sum_{j=1}^{L}\sqrt{\frac{\rho_{\text{d}}}{\lambda_j^{\text{zf}}}}\sum_{k=1}^{K}\boldsymbol{\epsilon}_{jil}^{\dagger}\boldsymbol{v}_{jkj}s_j[k]}_{\text{Interference due to channel estimation error}} + \underbrace{w_{il}}_{\text{Noise}} \\
&= \sum_{j=1}^{L}\theta_{jij}^{\text{zf}}s_j[i]+w_{il}'',
\end{aligned} \tag{17}$$

where (a) follows from the fact that $\boldsymbol{v}_{jkj}^{\dagger}\hat{\boldsymbol{g}}_{jmj} = \delta_{mk}$, w_{il}'' incorporates the last two terms after the third equality, and $\theta_{jij}^{\text{zf}} = \sqrt{\rho_{\text{d}}/\lambda_j^{\text{zf}}}(\beta_{jil}/\beta_{jij})$.

We now have an L-user MAC in (17) associated with the estimate of $s_l[i]$. Following the treatment of the MRT case, and using standard random coding arguments as in [8], it can be shown that the SD scheme achieves the rate tuple $(R_{i1},\ldots,R_{iL})$ that satisfies the set of conditions, obtained by replacing y_{il}^{mrt} in (8) with the equivalent received signal y_{il}^{zf}. The network-wide achievable rates are then obtained by taking the intersection of the achievable regions over all receivers.

Similar to the MRT case, the noise term in (17) is neither Gaussian nor independent of the users' signals, nevertheless, it is uncorrelated from the desired signals. Thus, similar to those for MRT in Theorem 1, we establish the following lower bounds for ZF using a worst-case uncorrelated noise technique.

Theorem 2 *For* $s_l = [s_l[1], s_l[2], \ldots, s_l[K]]^T \sim \mathcal{CN}(\mathbf{0}, \boldsymbol{I}_K)$, $l \in \{1, 2, \ldots, L\}$, *a set of lower bounds that can be achieved with ZF for the MAC given in* (17), *associated with* U_{il}, *is given as follows:*

$$I\left(y_{il}^{zf};\ s_\Omega \,\middle|\, s_{\Omega^c}\right) \geq C\left(P_1^{zf} / \left(P_2^{zf} + P_3^{zf}\right)\right), \tag{18}$$

where

$$P_1^{zf} = \sum\nolimits_{j\in\Omega} \left(\rho_d \beta_{jil}^2\right) / \left(\lambda_j^{zf} \beta_{jij}^2\right) \tag{19}$$

$$P_2^{zf} = \sum\nolimits_{j=1}^{L} \frac{\rho_d}{\lambda_j^{zf}} \sum\nolimits_{k=1}^{K} \frac{\beta_{jil} - \sqrt{\rho_p}\beta_{jil}\alpha_{jil}}{(M-K)\sqrt{\rho_p}\beta_{jkj}\alpha_{jkj}}, \tag{20}$$

$$P_3^{zf} = 1. \tag{21}$$

The proof is similar to that of Theorem 1 and hence is omitted.

Similar to those for MRT, P_1^{zf} is the power of the desired signals associated with s_Ω, while P_2^{zf} is the power of the interference due to the channel estimation error, and P_3^{zf} is the noise power. Using (19)–(21), the r.h.s. of (18) is given by

$$C_{\text{LB}}^{\text{zf}} = C\left(\frac{\sum_{j\in\Omega} \left(\rho_\text{d} \beta_{jil}^2\right) / \left(\lambda_j^{\text{zf}} \beta_{jij}^2\right)}{\sum_{j=1}^{L} \frac{\rho_\text{d}}{\lambda_j^{\text{zf}}} \sum_{k=1}^{K} \frac{\beta_{jil} - \sqrt{\rho_\text{p}}\beta_{jil}\alpha_{jil}}{(M-K)\sqrt{\rho_\text{p}}\beta_{jkj}\alpha_{jkj}} + 1}\right). \tag{22}$$

Since $\lambda_j^{\text{zf}} = \frac{1}{K(M-K)} \sum_{k=1}^{K} 1/(\sqrt{\rho_\text{p}}\beta_{jkj}\alpha_{jkj})$, for fixed K and large M, $\frac{1}{\lambda_j^{\text{zf}}} \propto M$. Thus, $C_{\text{LB}}^{\text{zf}}$ grows as $\mathcal{O}(\log M)$.

3.2 *Simultaneous Non-unique Decoding (SND)*

It is important to note that U_{il} is only interested in correctly decoding $s_l[i]$. Thus, incorrectly decoding $s_j[i]$, $j \neq l$, should not penalize the rates achievable at U_{il}. Furthermore, the power of the received signal for the users located in distant cells is very small, and thus trying to decode signals of such users can reduce achievable rates considerably. This motivates the need for *non-unique* interference decoding (SND), which achieves higher rates than SD since the stringent condition of *uniquely* decoding interference is now relaxed.

In contrast to the SD scheme for which we focused on the MAC formed at U_{il}, we now consider an L-user interference channel (IC) that consists of L senders (i.e., $\{\mathsf{BS}_l\}_{l=1}^L$) and L receivers (i.e., $\{\mathsf{U}_{il}\}_{l=1}^L$). In particular, U_{il} decodes only the intended message $s_l[i]$ uniquely, while the interference signals $s_j[i],\ j \neq l$ are decoded non-uniquely in the sense that incorrect decoding of the latter does not incur any penalty [8, Chap. 6]. When point-to-point random codes are used, as in our setting, SND has been shown to achieve the capacity region of such L-user IC [9]. In what follows, we only focus on the case of a 2-cell system for the sake of simplicity. From [9], the rate region for the 2-cell system achieved with SND at U_{i1} is given by

$$R_{i1} \leq I\left(y_{i1}^{\text{mrt/zf}}; s_1[i] \middle| s_2[i]\right) \tag{23}$$

$$R_{i1} + \min\left\{R_{i2}, I\left(y_{i1}^{\text{mrt/zf}}; s_2[i] \middle| s_1[i]\right)\right\} \leq I\left(y_{i1}^{\text{mrt/zf}}; s_1[i], s_2[i]\right).$$

The $I(.; .|.)$ terms above can be readily lower bounded using Theorem 1 and Theorem 2, which scale as $\mathcal{O}(\log M)$ as in the SD case. Similarly, the rate region achieved at the other user, U_{i2}, is obtained by replacing $y_{i1}^{\text{mrt/zf}}$ with $y_{i2}^{\text{mrt/zf}}$ and swapping appropriate indices in the above rate expressions. Hence, the network-wide rate region is given by the intersection of regions achieved at both users. Note that the SND region in (23) can be viewed as the union of TIN and SD, i.e., TIN and SD regions are both strictly contained in SND, showing the advantages of non-uniquely decoding the interference. It should be pointed out that practical implementations of interference decoding in cellular networks using SND have been thoroughly studied in [10, 11].

3.3 *Partial Decoding (PD)*

In SD and SND, interfering messages are decoded either uniquely or non-uniquely but in their *entirety*. Nevertheless, it may be useful to decode only a *part* of the message. Such flexible encoding/decoding is obtained by the celebrated Han–Kobayashi (HK) scheme [12], which provides the best rates for the IC. While the HK scheme is complex in general, we characterize a PD scheme for our 2-cell setting based on the simplified rate splitting scheme of [13].

In SND, a message from BS_l, $m_{il} \in \left[1 : 2^{nR_{il}}\right], l = 1, 2$, is encoded into a single codeword. In contrast, encoding in the PD scheme proceeds as follows: message m_{il} is first partitioned into two parts $m_{il}^{[1]}$ and $m_{il}^{[2]}$; part $m_{il}^{[1]}$ is then encoded into codeword $s_l^{[1]}[i](m_{il}^{[1]})$ of length n, while part $m_{il}^{[2]}$ is encoded into another codeword $s_l^{[2]}[i](m_{il}^{[1]}, m_{il}^{[2]})$ of length n; finally, the latter codeword is superimposed (or layered) on the former to produce a single codeword for transmission $s_l[i](m_{il}^{[1]}, m_{il}^{[2]}) = s_l^{[1]}[i](m_{il}^{[1]}) + s_l^{[2]}[i](m_{il}^{[1]}, m_{il}^{[2]})$, $l = 1, 2$. The total transmit power budget at BS_l is split into two fixed parts according to the power splitting variable $\mu_{il} \in [0, 1]$: the fraction μ_{il} of the budget is allocated to the "outer" layer $s_l^{[2]}[i]$, while the fraction $(1 - \mu_{il})$ of the budget is allotted to the "inner" layer $s_l^{[1]}[i]$. Finally, $s_l^{[1]}[i]$ and $s_l^{[2]}[i]$ are chosen to be i.i.d., zero-mean circularly symmetric complex Gaussian, with powers determined by μ_{il}.

Note that while full information $(m_{il}^{[1]}, m_{il}^{[2]})$ is carried in $s_l[i]$, the inner layer $s_l^{[1]}[i]$ carries only coarse information $m_{il}^{[1]}$ that can be decoded without decoding $m_{il}^{[2]}$ in the outer layer. Hence, when the channel condition is poor, a receiver may only decode $m_{il}^{[1]}$, while for strong channels, it may decode $(m_{il}^{[1]}, m_{il}^{[2]})$ from both layers. Such adaptability helps PD to outperform SND, where messages are not layered, and thus full interference is either non-uniquely decoded or treated as noise.

The decoding at U_{i1} proceeds as follows: it decodes both the inner and outer layers of the intended message $(m_{i1}^{[1]}, m_{i1}^{[2]})$ uniquely and only tries to non-uniquely decode the inner layer of the interfering message $m_{i2}^{[1]}$, while treating the outer layer $m_{i2}^{[2]}$ as noise. The other user, U_{i2}, decodes similarly with the roles of the interfering and intended messages exchanged. The resulting rate region for the user pair $\mathsf{U}_{il}, l = 1, 2$, is given by 7 inequalities found by changing the variables in [8, Theorem 6.4] to those in our setting as follows: $X_1 := s_1[i], X_2 := s_2[i], U_1 := s_1^{[1]}[i], U_2 := s_2^{[1]}[i], Y_1 := y_{i1}^{\text{mrt/zf}}, Y_2 := y_{i2}^{\text{mrt/zf}}, Q := \emptyset$.

Note that similar to those in the MRT/ZF cases for the SD scheme, the rate expressions involve $I(, ; , |.)$ terms that cannot be computed exactly; hence, we characterize lower bounds of these $I(, ; , |.)$ terms following similar steps to those in Theorems 1 and 2. Due to space limitations, we skip the expressions and detailed derivations and instead show that the achievable rates of PD outperform TIN/SD/SND, through simulation results presented in Sect. 5.

4 Maximum Symmetric Rate Allocation

From (7) and (17), it is evident that users with relatively weak effective channel gain obtain smaller rates than those with a stronger channel. In order to ensure fairness among users, we study the maximum symmetric rate allocation problem for schemes TIN/SD/SND/PD, where the minimum achievable user rate is maximized. The following analysis is for $\{\mathsf{U}_{il}\}_{l=1}^{L}$ across multiple cells that employ the same pilots. Since i is arbitrary, the same analysis holds for other pilot-sharing user sets.

When rate allocation is symmetric, all users are assigned the same rate. The maximum symmetric rate for a decoding scheme is given by $R_{\text{Sym}} = \max R$ such that the rate vector $(R, R, \ldots, R)$ belongs to the rate region achieved by that scheme at U_{il}, for $l = 1, \ldots, L$. Hence, the maximum symmetric rate vector $(R_{\text{Sym}}, R_{\text{Sym}}, \ldots, R_{\text{Sym}})$ of a given scheme is found at the intersection of the diagonal $(R_{i1} = \ldots = R_{iL})$ with the boundary of the achievable region obtained by that scheme.

5 Simulation Results and Discussion

To illustrate the performance of the different interference management schemes, TIN/SD/SND/PD, with symmetric spectral efficiency (SE in units of bits/sec/Hz) allocation, we simulate the downlink of a 2-cell massive MIMO system suffering from PC. In particular, we consider two hexagonal cells, as in Fig. 1, with a radius of $r = 400$ m, and $K = 15$ users are uniformly distributed at random within the area of each cell, and at least 35 m away from their BSs. To evaluate the performance, the average of maximum symmetric SEs is calculated over 150 random realizations of user locations. The downlink transmit power of each BS is taken to be 40 W (i.e., 46 dBm), and to model large-scale fading coefficients, β_{jkl}, a path-loss model adopted from [14] is considered:

$$\begin{aligned}[\beta_{jkl}]_{\text{dB}} = &- 13.54 - 39.08 \log_{10}\left(d_{jkl}^{3D}\right) \\ &- 20 \log_{10}(f_c) + 0.6\,(h_{UT} - 1.5)\,,\end{aligned} \tag{24}$$

where d_{jkl}^{3D} is the 3D distance (in meters) from U_{kl} to BS_j, the carrier frequency is $f_c = 3.5$ GHz, $h_{UT} = 1.5$ m is the user height, and the BS height is 25 m. Also, with a system bandwidth of 20 MHz, the noise power is assumed to be -101 dBm. The SEs are calculated using the closed-form expressions of (14) and (22) for MRT and ZF, respectively. In Fig. 2a, the achieved symmetric SEs of the different schemes TIN/SD/SND/PD are shown with MRT precoding, as a function of M. Note that the SEs for the PD scheme depend on the power splitting coefficients $\mu_{i1}, \mu_{i2} \in [0, 1]$, and hence, the best SE for each value of M is obtained by numerically optimizing $\mu_{i1}, \mu_{i2} \in [0, 1]$. First, notice that the SD scheme performs poorly compared to the other schemes, and unless M becomes truly large (not shown in this figure), it cannot outperform TIN. Moreover, for the range of M shown, as M increases, the SEs achieved by TIN, SND, and PD improve, while the interference decoding schemes, SND and PD, outperform TIN for moderately large values of M (i.e., $M \geq 128$). Moreover, since the PD scheme enables partial decoding of the PC interference with the remaining part treated as noise, this flexibility produces better rates than SND, such that PD outperforms both TIN and SND even for a small number of antennas M (i.e., $M \geq 32$). More specifically, when using MRT, SND provides performance gains of about 8 and 10% over TIN for $M = 128$ and $M = 256$, respectively,

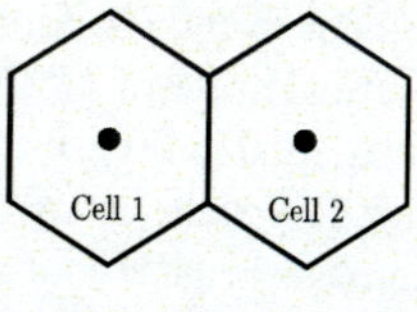

Fig. 1 A cellular system with two hexagonal cells, where BSs are located at the center of the cell and K users are randomly positioned within the area of each cell

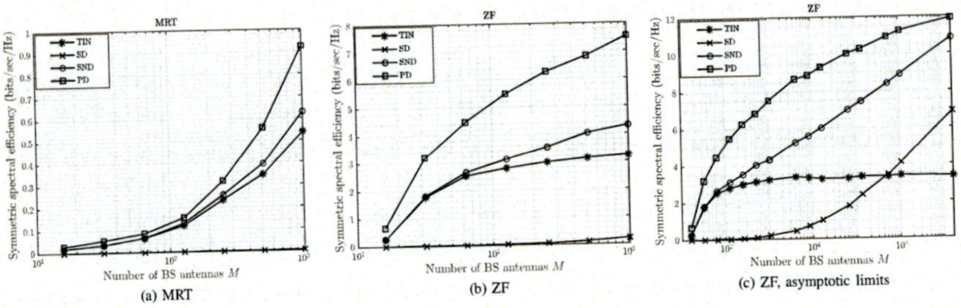

Fig. 2 (**a**) Symmetric SEs achieved with MRT under TIN/SD/SND/PD, (**b**) Symmetric SEs achieved with ZF under TIN/SD/SND/PD, for a range of moderately large M, (**c**) Asymptotic limits of symmetric SEs with ZF under TIN/SD/SND/PD for truly large values of M

while these gains improve to more than 17% when $M = 1024$. Furthermore, the gains offered by PD over TIN are about 29 and 36% for $M = 128$ and $M = 256$, respectively, and reach approximately 72% when $M = 1024$. While not shown in the figure, as M continues to grow and becomes truly large, the performances of SND and PD eventually converge, whereas the performance of TIN saturates to a constant value independent of M.

The performance of TIN/SD/SND/PD with ZF is demonstrated in Fig. 2b and c. Figure 2b shows the achieved symmetric SEs with ZF for a range of moderately large values of M, while Fig. 2c shows the same for a range of truly large M. While the latter covers a range of M that is beyond practical, the results of Fig. 2c can be used to confirm asymptotic performance limits as $M \to \infty$. One can notice in Fig. 2b that all schemes achieve better SEs compared to MRT, which is due to the fact that ZF better mitigates multi-user interference. As such, compared to MRT, for the same number of antennas M, both SND and PD provide larger gains over TIN. More precisely, SND provides performance gains of 11 and 17% over TIN for $M = 128$ and $M = 256$, respectively, and these gains grow to more than 34% when $M = 1024$. On the other hand, due to the additional flexibility of partial interference decoding provided by PD, the gains offered by this scheme over TIN are about 96 and 108% for $M = 128$ and $M = 256$, respectively, while these gains improve to approximately 133% when $M = 1024$. Hence, these results confirm that the use of ZF precoding in conjunction with either SND or PD can potentially be considered as an attractive solution to combat the PC interference in multi-cell massive MIMO systems. Particularly in the case of PD, the achieved SEs are increased by more than a factor of 2 for a practical range of antennas M, which is a significant improvement over TIN. Similar to the MRT case, from Fig. 2b, it is observed that for the practical range of M, SD is outperformed by the other schemes. The asymptotic limits presented in Fig. 2c show that as M continues to grow without limit, the SE obtained by TIN saturates, whereas the achieved SEs of both PD and SND monotonically increase with M and will eventually converge. This is due to the fact that, as M grows, and thus the strength of the PC interference increases, the power of the inner layer can be increased. This improvement is continued until the entire part of an interference signal can be decoded non-uniquely under PD (i.e., $\mu_{i1}, \mu_{i2} \rightarrow 0$), thus achieving a performance identical to that of SND. Last, Fig. 2c shows that SD starts to outperform TIN at approximately $M \approx 4 \times 10^5$.

Note that the SND scheme can be readily extended to more than two cells in downlink in a similar manner discussed in [2] for uplink. With small modifications, the message splitting technique for PD can also be extended to more than two cells. Simulation results (not shown here due to the space limitations) have revealed that doing so again results in significant gains over TIN and SND for the case of $L > 2$.

6 Conclusion

In this chapter, for the downlink of a multi-cell massive MIMO system, schemes were proposed that decode the interference caused by PC, as opposed to treating it as noise. It was shown that for two linear precoding techniques, MRT and ZF, by decoding the PC interference, the achievable rates grow unbounded as $M \rightarrow \infty$. Furthermore, using a worst-case uncorrelated noise technique, new closed-form achievable rates for the interference decoding schemes with MRT and ZF were obtained, which are shown to scale as $\mathcal{O}(\log M)$. Simulation results for the maximum symmetric SEs in a 2-cell system revealed that non-uniquely decoding the PC interference provides a better performance compared to TIN. Moreover, a partial interference decoding scheme based on rate splitting was proposed, outperforming *all other schemes* for a practical range of M.

Acknowledgments This work was supported in part by the Natural Sciences and Engineering Research Council of Canada and Cisco.

Appendix

We employ the worst-case uncorrelated noise technique of [2, Lemma 1], i.e., an achievable lower bound on the capacity of a MAC with uncorrelated additive non-Gaussian noise, w'_{il}, is obtained by replacing the noise term with an independent zero-mean Gaussian noise having the same variance.

We start by computing P_1^{mrt}. Note that

$$
\begin{aligned}
P_1^{\mathrm{mrt}} &= \sum\nolimits_{j\in\Omega}^{L} \rho_{\mathrm{d}}/\lambda_j^{\mathrm{mrt}} \left| \mathbb{E}\left[\hat{\boldsymbol{g}}_{jil}^{\dagger}\hat{\boldsymbol{g}}_{jij}\right] + \mathbb{E}\left[\boldsymbol{\epsilon}_{jil}^{\dagger}\hat{\boldsymbol{g}}_{jij}\right]\right|^2 \\
&\overset{(a)}{=} \sum\nolimits_{j\in\Omega}^{L} \rho_{\mathrm{d}}/\lambda_j^{\mathrm{mrt}} \left| \mathbb{E}\left[\hat{\boldsymbol{g}}_{jil}^{\dagger}\hat{\boldsymbol{g}}_{jij}\right]\right|^2 \\
&\overset{(b)}{=} \sum\nolimits_{j\in\Omega}^{L} \left(M\sqrt{\rho_{\mathrm{p}}}\beta_{jij}\alpha_{jij}\right)^2 \rho_{\mathrm{d}}\beta_{jil}^2/\lambda_j^{\mathrm{mrt}}\beta_{jij}^2
\end{aligned}
$$

$$= M^2 \sum\nolimits_{j\in\Omega} \rho_{\mathrm{d}}\rho_{\mathrm{p}}\beta_{jil}^2\alpha_{jij}^2/\lambda_j^{\mathrm{mrt}}, \tag{25}$$

where (a) is due to the fact that $\boldsymbol{\epsilon}_{jil}$ and $\hat{\boldsymbol{g}}_{jij}$ are independent and (b) follows from $\hat{\boldsymbol{g}}_{jil} = (\beta_{jil}/\beta_{jij})\hat{\boldsymbol{g}}_{jij}$ and the distribution of $\hat{\boldsymbol{g}}_{jkj}$ explained below (2). Since all three terms in the effective noise in (7) are uncorrelated, we have $\mathrm{var}\left[w'_{il}\right] = P_2^{\mathrm{mrt}} + P_3^{\mathrm{mrt}} + P_4^{\mathrm{mrt}}$. For the power of the interference due to beamforming gain uncertainty, P_2^{mrt}, we write

$$\begin{aligned} P_2^{\mathrm{mrt}} = &\sum\nolimits_{j=1}^{L} \rho_{\mathrm{d}}/\lambda_j^{\mathrm{mrt}}\ \mathbb{E}\left[\left|\hat{\boldsymbol{g}}_{jil}^{\dagger}\hat{\boldsymbol{g}}_{jij} - \mathbb{E}\left[\hat{\boldsymbol{g}}_{jil}^{\dagger}\hat{\boldsymbol{g}}_{jij}\right]\right|^2\right] \\ &+ \sum\nolimits_{j=1}^{L} \rho_{\mathrm{d}}/\lambda_j^{\mathrm{mrt}}\ \mathbb{E}\left[\left|\boldsymbol{\epsilon}_{jil}^{\dagger}\hat{\boldsymbol{g}}_{jij}\right|^2\right]. \end{aligned} \tag{26}$$

For the first term in (26), it is obtained

$$\begin{aligned} &\sum\nolimits_{j=1}^{L} \frac{\rho_{\mathrm{d}}}{\lambda_j^{\mathrm{mrt}}}\mathbb{E}\left[\left|\hat{\boldsymbol{g}}_{jil}^{\dagger}\hat{\boldsymbol{g}}_{jij} - \mathbb{E}\left[\hat{\boldsymbol{g}}_{jil}^{\dagger}\hat{\boldsymbol{g}}_{jij}\right]\right|^2\right] \\ &= \sum\nolimits_{j=1}^{L} \frac{\rho_{\mathrm{d}}}{\lambda_j^{\mathrm{mrt}}}\mathrm{var}\left[\hat{\boldsymbol{g}}_{jil}^{\dagger}\hat{\boldsymbol{g}}_{jij}\right] = M\sum\nolimits_{j=1}^{L} \frac{\rho_{\mathrm{d}}}{\lambda_j^{\mathrm{mrt}}}\rho_{\mathrm{p}}\beta_{jil}^2\alpha_{jij}^2. \end{aligned} \tag{27}$$

Similarly, for the second term in (26), we obtain

$$\begin{aligned} &\sum_{j=1}^{L} \frac{\rho_{\mathrm{d}}}{\lambda_j^{\mathrm{mrt}}}\mathbb{E}\left[\left|\boldsymbol{\epsilon}_{jil}^{\dagger}\hat{\boldsymbol{g}}_{jij}\right|^2\right] = \sum_{j=1}^{L} \frac{\rho_{\mathrm{d}}}{\lambda_j^{\mathrm{mrt}}}\mathbb{E}\left[\hat{\boldsymbol{g}}_{jij}^{\dagger}\boldsymbol{\epsilon}_{jil}\boldsymbol{\epsilon}_{jil}^{\dagger}\hat{\boldsymbol{g}}_{jij}\right] \\ &= M\sum\nolimits_{j=1}^{L} \rho_{\mathrm{d}}\beta_{jil}\left(1-\sqrt{\rho_{\mathrm{p}}}\alpha_{jil}\right)\left(\sqrt{\rho_{\mathrm{p}}}\beta_{jij}\alpha_{jij}\right)/\lambda_j^{\mathrm{mrt}}. \end{aligned} \tag{28}$$

Using (27) and (28), P_2^{mrt} is obtained as in (11). Next,

$$\begin{aligned} P_3^{\mathrm{mrt}} &= \sum\nolimits_{j=1}^{L} \frac{\rho_{\mathrm{d}}}{\lambda_j^{\mathrm{mrt}}}\sum\nolimits_{k=1,k\neq i}^{K}\left(\mathbb{E}\left[\left|\hat{\boldsymbol{g}}_{jil}^{\dagger}\hat{\boldsymbol{g}}_{jkj}\right|^2\right] + \mathbb{E}\left[\left|\boldsymbol{\epsilon}_{jil}^{\dagger}\hat{\boldsymbol{g}}_{jkj}\right|^2\right]\right) \\ &= \sum\nolimits_{j=1}^{L} \rho_{\mathrm{d}}/\lambda_j^{\mathrm{mrt}}\sum\nolimits_{k=1,k\neq i}^{K}\Big(M\left(\sqrt{\rho_{\mathrm{p}}}\beta_{jil}\alpha_{jil}\right)\left(\sqrt{\rho_{\mathrm{p}}}\beta_{jkj}\alpha_{jkj}\right) \\ &\quad + M\left(\sqrt{\rho_{\mathrm{p}}}\beta_{jkj}\alpha_{jkj}\right)\left(\beta_{jil} - \sqrt{\rho_{\mathrm{p}}}\beta_{jil}\alpha_{jil}\right)\Big) \\ &= M\sum\nolimits_{j=1}^{L} \rho_{\mathrm{d}}/\lambda_j^{\mathrm{mrt}}\sum\nolimits_{k=1,k\neq i}^{K}\sqrt{\rho_{\mathrm{p}}}\beta_{jkj}\alpha_{jkj}\beta_{jil}. \end{aligned}$$

Finally, the power of the noise w_{il} in (6) is $P_4^{\mathrm{mrt}} = 1$.

References

1. Björnson, E., Sanguinetti, L., et al.: Massive MIMO is a reality-what is next? five promising research directions for antenna arrays. Digital Signal Process. **94**, 3–20 (2019)
2. Motlagh, M.S., Majhi, S., Mitran, P.: Performance of multi-cell massive MIMO systems with interference decoding: IEEE Trans. Commun. **68**(2), 930–946 (2019)
3. Marzetta, T.L.: Noncooperative cellular wireless with unlimited numbers of base station antennas. IEEE Trans. Wirel. Commun. **9**(11), 3590–3600 (2010)
4. Yin, H., Cottatellucci, L., et al.: Robust pilot decontamination based on joint angle and power domain discrimination. IEEE Trans. Signal Process. **64**(11), 2990–3003 (2016)
5. Adhikary, A., Ashikhmin, A., Marzetta, T.L.: Uplink interference reduction in large-scale antenna systems. IEEE Trans. Commun. **65**(5), 2194–2206 (2017)
6. Björnson, E., Hoydis, J., Sanguinetti, L.: Massive MIMO has unlimited capacity. IEEE Trans. Wirel. Commun. **17**(1), 574–590 (2018)
7. Bandemer, B., El Gamal, A.: Interference decoding for deterministic channels. IEEE Trans. Inf. Theory **57**(5), 2966–2975 (2011)
8. El Gamal, A., Kim, Y.-H.: Network Information Theory. Cambridge University Press, Cambridge (2011)
9. Bandemer, B., El Gamal, A., Kim, Y.-H.: Optimal achievable rates for interference networks with random codes. IEEE Trans. Inf. Theory **61**(12), 6536–6549 (2015)
10. Kim, K.T., Ahn, S.-K., et al.: Adaptive sliding-window coded modulation in cellular networks. In: IEEE Conference and Exhibition on Global Telecommunications (GLOBECOM) (2015)
11. Kim, K.T., Ahn, S.-K., et al.: Interference management via sliding-window coded modulation for 5G cellular networks. IEEE Commun. Mag. **54**(11), 82–89 (2016)
12. Han, T., Kobayashi, K.: A new achievable rate region for the interference channel. IEEE Trans. Inf. Theory **27**(1), 49–60 (1981)
13. Etkin, R., Tse, D., Wang, H.: Gaussian interference channel capacity to within one bit. IEEE Trans. Inf. Theory **54**(12), 5534–5562 (2008)
14. 3GPP: Technical specification group radio access network; study on channel model for frequencies from 0.5 to 100 GHz (release 16) (2020). [Online]. Available: https://portal.3gpp.org/

Compute-and-Forward for Uplink Massive MIMO-NOMA

A. Malik Nasser Aljalai, Chen Feng, Victor C. M. Leung, and Rabab K. Ward

1 Introduction

Compute-and-forward (CaF) started as a relaying strategy where each relay computes a linear combination of the transmitted signals and then forwards it to the ultimate receiver [1]. Then, the applications of CaF were extended into several systems including multiple-input multiple-output (MIMO) wireless systems, with promising performance gains [2]. Besides, CaF applications in MIMO systems generalize the conventional linear receivers used in MIMO systems into a class that is well known as Integer-Forcing Liner Receivers (IFLRs).

In MIMO wireless communications, the base station (BS) usually utilizes a linear receiver architecture to avoid the implementation complexity of a maximum likelihood receiver. IFLR recovers linearly independent combinations of the transmitted codewords by multiple users, with each integer-forcing decoder recovering a different linear combination. Therefore, a full rank matrix of the integer coefficients of these linear combinations is required. However, in conventional linear receivers such as Minimum Mean Squared Error (MMSE), the receiver recovers

A. M. N. Aljalai (✉) · R. K. Ward
Department of Electrical and Computer Engineering, The University of British Columbia, Vancouver, BC, Canada
e-mail: aaljalai@ece.ubc.ca; rababw@ece.ubc.ca

C. Feng
School of Engineering, The University of British Columbia, Kelowna, BC, Canada
e-mail: chen.feng@ubc.ca

V. C. M. Leung
College of Computer Science and Software Engineering, Shenzhen University, Shenzhen, China

Department of Electrical and Computer Engineering, The University of British Columbia, Vancouver, BC, Canada
e-mail: vleung@ieee.org

H. Nguyen et al. (eds.), *30th Biennial Symposium on Communications 2021*,
Signals and Communication Technology, https://doi.org/10.1007/978-3-031-06947-5_4

the codewords transmitted by multiple users directly. Thus linear MMSE receiver, which is widely used in massive MIMO, is a special case of IFLR where the integer coefficients matrix of IFLR is the identity matrix. Interestingly, IFLR offers an optimal diversity multiplexing trade-off over the entire range of the signal-to-noise ratios (SNR) compared to conventional linear receivers such as zero forcing (ZF) and MMSE [3].

Massive MIMO is an essential technology for the realization of fifth-generation (5G) wireless networks, in which the BS is equipped with a very large number of antennas. Massive MIMO offers massive access connectivity to serve a very large number of users and devices. The classical multiple access technique used with massive MIMO is Orthogonal Multiple Access (OMA). Compared with OMA, Non-orthogonal Multiple Access (NOMA) can significantly improve spectrum efficiency, reduce latency, and scale up the number of users [4]. As such, NOMA is considered as an important 5G technology that compliments massive MIMO [5]. Several configurations of NOMA have been proposed. Power-domain NOMA (PD-NOMA) is the most promising due to its simple superposition of multiple signals over the same bandwidth, which preserves and enhances spectral efficiency.

In a typical uplink massive MIMO-NOMA scenario, the separation of user signals at the BS receiver is achieved by using successive interference cancelation (SIC). While SIC improves the performance of MMSE linear receivers, it cannot guarantee the optimal performance offered by IFLR. Also, SIC does not work well when users have comparable received power levels at the BS [6]. As a starting point to address this challenge, we have proposed a new decoding method based on the application of CaF for uplink NOMA in our previous work [7], which has the potential to achieve better performance at essentially the same computational cost as SIC. However, this previous work has several limitations. First, it only considered the case where the BS was equipped with a *single* antenna, while the usage of a massive number of antennas at the BS is widely anticipated for 5G deployment. Second, it assumed availability of perfect channel state information (CSI) at the BS.

In this chapter, we aim to overcome these limitations by taking our proposed decoding method one step closer to the reality. First, we move from a BS with a single-antenna setup to a massive multi-antenna setup. This allows us to derive new results in terms of achievable rates for the uplink massive MIMO-NOMA scenario. We evaluate the fairness and outage probability offered by CaF for the new configuration. Then, we explore other limitations that may arise in practical deployment of massive MIMO-NOMA with CaF decoding. Specifically, we consider the effects of imperfect CSI on the CaF performance for uplink MIMO-NOMA.

We make *two contributions* in this work. First, we characterize the achievable rates of CaF for uplink massive MIMO-NOMA. Based on this characterization, we are able to illustrate the advantage of CaF over the standard SIC decoding. Second, we investigate the practical limitations of CaF decoding for massive MIMO-NOMA. In this regard, we examine through simulations the effects of imperfect CSI on the achievable rates of CaF decoding for uplink MIMO-NOMA and illustrate their impacts.

The rest of this chapter is organized as follows. Section 2 presents the system model and characterizes the best achievable rates of CaF. Section 3 extends CaF performance analyses into the massive MIMO-NOMA uplink scenario. Section 4 characterizes the practical limitations and impacts of imperfect CSI on CaF decoding in massive MIMO-NOMA. Section 5 presents our numeral results. Section 6 concludes the chapter.

Notation We use $(\cdot)^\dagger$ to denote the complex conjugate transpose (Hermitian), $(\cdot)^T$ to denote the transpose, $\|\cdot\|_2$ to denote the second (Euclidean) norm, $|\cdot|$ to denote the determinant of a matrix, $gcd(\cdot)$ to denote the greatest common divisor operator, bold upper-case letters to denote matrices (e.g., **H** and **P**), bold lower-case letters to denote vectors (e.g., **h** and **x**), $\mathbb{C}^{M\times n}$ to denote the field of complex numbers with $M\times n$ dimensions, $\mathbb{R}^{M\times n}$ to denote the field of real numbers with $M\times n$ dimensions, and $\mathbb{Z}^{M\times n}$ to denote the field of integer numbers with $M \times n$ dimensions.

2 System Model

Consider an uplink scenario with two single-antenna users and one BS with M antennas. The BS receives

$$\mathbf{Y} = \mathbf{h}_1\mathbf{x}_1 + \mathbf{h}_2\mathbf{x}_2 + \mathbf{Z}, \tag{1}$$

where $\mathbf{x}_\ell \in \mathbb{R}^{1\times n}$ is the transmitted codeword for user ℓ, $\mathbf{h}_\ell \in \mathbb{R}^{M\times 1}$ is the channel-coefficients vector for user ℓ, and $\mathbf{Z} \in \mathbb{R}^{M\times n}$ is the channel noise with zero-mean and unit variance entries. Note that $\{\mathbf{h}_\ell\}$ are column vectors and $\{\mathbf{x}_\ell\}$ are row vectors corresponding to the transmitted codewords. We assume that the channel vectors $\{\mathbf{h}_\ell\}$ are known to the BS but unknown to the users.

Decoding From a high-level view, the BS first computes one integer-linear combination $a_1\mathbf{x}_1 + a_2\mathbf{x}_2$ and then subtracts it from the received signal $\mathbf{Y}$ in order to compute another integer-linear combination $b_1\mathbf{x}_1 + b_2\mathbf{x}_2$.[1] The BS can recover the transmitted messages as long as these two combinations are linearly independent. Clearly, this decoding strategy generalizes the standard successive interference cancelation (SIC) decoding, where $\mathbf{a} \triangleq (a_1, a_2)^T = (1, 0)^T$ and $\mathbf{b} \triangleq (b_1, b_2)^T = (0, 1)^T$, or $\mathbf{a} = (0, 1)^T$ and $\mathbf{b} = (1, 0)^T$.

The detailed decoding steps can be found in [8]. Here, we only present the achievable rates of this decoding strategy. We need to introduce some additional notation. Let **P** be the diagonal matrix of the power constraints

$$\mathbf{P} \triangleq \text{diag}(P_1, P_2),$$

[1] a_1, a_2, b_1, and b_2 are integer coefficients. Their optimal values are selected by the CaF decoding algorithm as explained in Sect. 3.

where P_ℓ is the average power constraint for user ℓ. Let **H** be the channel-coefficient matrix

$$\mathbf{H} \triangleq \left[\mathbf{h}_1, \mathbf{h}_2\right].$$

Let

$$\sigma^2(\mathbf{H}, \mathbf{a}) \triangleq \mathbf{a}^T \left(\mathbf{P}^{-1} + \mathbf{H}^T\mathbf{H}\right)^{-1} \mathbf{a}$$

and

$$\sigma^2(\mathbf{H}, \mathbf{b}|\mathbf{a}) \triangleq \frac{P_1 P_2}{\sigma^2(\mathbf{H}, \mathbf{a})} \cdot \frac{1}{\left|\mathbf{I}_M + \mathbf{HPH}^T\right|}.$$

Operationally, $\sigma^2(\mathbf{H}, \mathbf{a})$ is the effective noise variance associated with the first integer-linear combination $a_1\mathbf{x}_1 + a_2\mathbf{x}_2$, and $\sigma^2(\mathbf{H}, \mathbf{b}|\mathbf{a})$ is the effective noise variance associated with the second integer-linear combination $b_1\mathbf{x}_1 + b_2\mathbf{x}_2$ (by using the first combination as side information), as explained in [8].

Proposition 1 *The following rate pairs are achievable:*

- *When* $a_1 \bmod p \neq 0$ *and* $a_2 \bmod p = 0$,

$$R_1 = \frac{1}{2}\log_2\left(1 + P_1\mathbf{h}_1^T(\mathbf{I}_M + P_2\mathbf{h}_2\mathbf{h}_2^T)^{-1}\mathbf{h}_1\right)$$
$$R_2 = \frac{1}{2}\log_2\left(1 + P_2\|\mathbf{h}_2\|_2^2\right).$$

- *When* $a_1 \bmod p = 0$ *and* $a_2 \bmod p \neq 0$,

$$R_1 = \frac{1}{2}\log_2\left(1 + P_1\|\mathbf{h}_1\|_2^2\right)$$
$$R_2 = \frac{1}{2}\log_2\left(1 + P_2\mathbf{h}_2^T(\mathbf{I}_M + P_1\mathbf{h}_1\mathbf{h}_1^T)^{-1}\mathbf{h}_2\right).$$

- *When* $a_1 \bmod p \neq 0$ *and* $a_2 \bmod p \neq 0$,

$$R_1 = \min\left\{\frac{1}{2}\log_2^+\left(\frac{P_1}{\sigma^2(\mathbf{H}, \mathbf{a})}\right), \frac{1}{2}\log_2^+\left(\frac{P_1}{\sigma^2(\mathbf{H}, \mathbf{b}|\mathbf{a})}\right)\right\}$$
$$R_2 = \frac{1}{2}\log_2^+\left(\frac{P_2}{\sigma^2(\mathbf{H}, \mathbf{a})}\right)$$

or

$$R_1 = \frac{1}{2}\log_2^+\left(\frac{P_1}{\sigma^2(\mathbf{H},\mathbf{a})}\right),$$

$$R_2 = \min\left\{\frac{1}{2}\log_2^+\left(\frac{P_2}{\sigma^2(\mathbf{H},\mathbf{a})}\right), \frac{1}{2}\log_2^+\left(\frac{P_2}{\sigma^2(\mathbf{H},\mathbf{b}|\mathbf{a})}\right)\right\},$$

where $\log_2^+(.) = \max\{0, \log_2(.)\}$. *In particular, when* $\sigma^2(\mathbf{H},\mathbf{a}) \le \sigma^2(\mathbf{H},\mathbf{b}|\mathbf{a})$,

$$R_1 + R_2 = \frac{1}{2}\log_2\left|\mathbf{I}_M + \mathbf{HPH}^T\right|. \tag{2}$$

Here, p is a large prime number used for code construction.

The proof of Proposition 1 is essentially the same as the proof of Theorem 5 in [8]. Compared to the statement of Theorem 5, we have managed to get rid of the complicated admissible mappings by *explicitly* evaluating all the possibilities of successive cancelation for the two-user case. These explicit expressions in Proposition 1 may be of independent interest.

2.1 Extension to the Complex Field

As stated in [8], all of its results can be extended from real-valued channel models to complex-valued channel models either via a real-valued decomposition of the channel or by building nested lattices directly over the complex field using the algebraic framework in [9]. This extension leads to the following achievable rates for the channel-coefficient matrix $\mathbf{H}$ over the complex field.

Proposition 2 *The following rate pairs are achievable:*

- *When* $a_1 \bmod p \neq 0$ *and* $a_2 \bmod p = 0$,

$$R_1 = \log_2\left(1 + P_1\mathbf{h}_1^\dagger(\mathbf{I}_M + P_2\mathbf{h}_2\mathbf{h}_2^\dagger)^{-1}\mathbf{h}_1\right)$$

$$R_2 = \log_2\left(1 + P_2\|\mathbf{h}_2\|_2^2\right).$$

- *When* $a_1 \bmod p = 0$ *and* $a_2 \bmod p \neq 0$,

$$R_1 = \log_2\left(1 + P_1\|\mathbf{h}_1\|_2^2\right)$$

$$R_2 = \log_2\left(1 + P_2\mathbf{h}_2^\dagger(\mathbf{I}_M + P_1\mathbf{h}_1\mathbf{h}_1^\dagger)^{-1}\mathbf{h}_2\right)$$

- *When* $a_1 \bmod p \neq 0$ *and* $a_2 \bmod p \neq 0$,

$$R_1 = \min\left\{\log_2^+\left(\frac{P_1}{\sigma^2(\mathbf{H},\mathbf{a})}\right), \log_2^+\left(\frac{P_1}{\sigma^2(\mathbf{H},\mathbf{b}|\mathbf{a})}\right)\right\}$$

$$R_2 = \log_2^+\left(\frac{P_2}{\sigma^2(\mathbf{H},\mathbf{a})}\right)$$

or

$$R_1 = \log_2^+\left(\frac{P_1}{\sigma^2(\mathbf{H},\mathbf{a})}\right),$$

$$R_2 = \min\left\{\log_2^+\left(\frac{P_2}{\sigma^2(\mathbf{H},\mathbf{a})}\right), \log_2^+\left(\frac{P_2}{\sigma^2(\mathbf{H},\mathbf{b}|\mathbf{a})}\right)\right\}.$$

In particular, when $\sigma^2(\mathbf{H},\mathbf{a}) \leq \sigma^2(\mathbf{H},\mathbf{b}|\mathbf{a})$, *we have*

$$R_1 + R_2 = \log_2\left|\mathbf{I}_M + \mathbf{HPH}^\dagger\right|. \quad (3)$$

Here, **a** and **b** are over Gaussian integers[2] $\mathbb{Z}[i] \triangleq \{a + bi : a, b \in \mathbb{Z}\}$, and p should be a Gaussian prime (i.e., $p \in \mathbb{Z}$ is a prime number of the form $4m+3$), as explained in [9]. The new expressions of $\sigma^2(\mathbf{H},\mathbf{a})$ and $\sigma^2(\mathbf{H},\mathbf{b}|\mathbf{a})$ are given below:

$$\sigma^2(\mathbf{H},\mathbf{a}) \triangleq \mathbf{a}^\dagger\left(\mathbf{P}^{-1} + \mathbf{H}^\dagger\mathbf{H}\right)^{-1}\mathbf{a}$$

and

$$\sigma^2(\mathbf{H},\mathbf{b}|\mathbf{a}) \triangleq \frac{P_1P_2}{\sigma^2(\mathbf{H},\mathbf{a})} \cdot \frac{1}{\left|\mathbf{I}_M + \mathbf{HPH}^\dagger\right|}.$$

Again, the explicit expressions in Proposition 2 may be of independent interest.

3 CaF for Massive MIMO-NOMA

In this section, we characterize the best possible achievable rates of CaF for uplink massive MIMO-NOMA. We consider an uplink transmission scenario, as shown in Fig. 1, for massive MIMO-NOMA in which up to $\frac{3K}{2}$ NOMA pairs[3] of users, each

[2] When real and imaginary parts of a complex number are integers, the number is called a Gaussian integer.

[3] A NOMA pair can involve one user from the near region and up to two users from the far region.

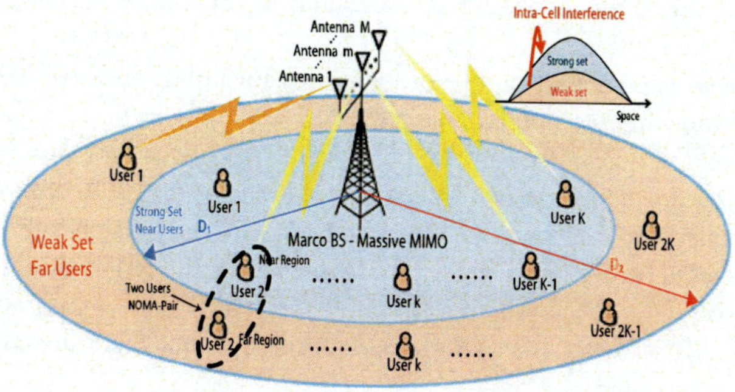

Fig. 1 The uplink of massive MIMO-NOMA with more far users than near users

with user equipment (UE) that is equipped with a single antenna, are transmitting signals to a BS equipped with M antennas, whereas in massive MIMO $M \gg K$. We assume, without loss of generality, that the UEs are ordered by their distances to the BS so that $d_1 \le d_2 \le \cdots \le d_{3K}$. We allow *any* pairing method in our system model as long as different pairs work in OMA mode[4] so that a pair does not interfere another pair. Hence, it suffices for us to focus on a particular pair with a near user, say user k, and a far user, say user l (where $d_k < d_l$). Under this model, the received signal $\mathbf{Y} \in \mathbb{C}^{M\times n}$ observed by the BS across the M antennas over n channel uses can be written as

$$\mathbf{Y} = \mathbf{h}_k\mathbf{x}_k + \mathbf{h}_l\mathbf{x}_l + \mathbf{Z}, \tag{4}$$

where $\mathbf{x}_k \in \mathbb{C}^{1\times n}$ and $\mathbf{x}_l \in \mathbb{C}^{1\times n}$ are the transmitted signals of length n at kth UE and lth UE, respectively, $\mathbf{h}_k \in \mathbb{C}^{M\times 1}$ and $\mathbf{h}_l \in \mathbb{C}^{M\times 1}$ are the channel-coefficients vectors for user k and user l respectively, and $\mathbf{Z} \in \mathbb{C}^{M\times n}$ is the channel noise with zero-mean and unit variance entries. As before, $\mathbf{P} \triangleq \text{diag}(P_k, P_l)$, where P_k and P_l are the average power constraints for user k and user l, respectively, and $\mathbf{H} \triangleq \left[\mathbf{h}_k, \mathbf{h}_l\right]$.

As stated before, the BS first computes one Gaussian-integer combination (say, $a_k\mathbf{x}_k + a_l\mathbf{x}_l$, where $a_k, a_l \in \mathbb{Z}[i]$), scales it with a multiplying coefficient $C \in \mathbb{C}$, and then subtracts it from the projection of received signal $\boldsymbol{\beta}\mathbf{Y}$, $\boldsymbol{\beta} \in \mathbb{C}^{M\times 1}$, in order to compute another Gaussian-integer combination (say, $b_k\mathbf{x}_k + b_l\mathbf{x}_l$), which is linearly independent of the first combination.[5] With two linearly independent combinations, the BS can recover the transmitted signals $\mathbf{x}_k$ and $\mathbf{x}_l$. The achievable rates of this decoding strategy are given in Proposition 2, which clearly includes

[4] It is important to mention that user pairing methods are a key research topic in energy-efficient NOMA cellular networks in 5G and beyond, and our concluding outcomes of this section can apply to any of them.

[5] The multiplying quantization vector $\boldsymbol{\beta}$ and scaling factor C are determined by the CaF algorithm.

the standard SIC decoding as a special case. In other words, CaF decoding enlarges the achievable rates of user k and user l with SIC decoding.[6] This leads to better fairness and smaller outage probabilities for uplink massive MIMO-NOMA as we will show in our simulation results.

Next, a question naturally arises: *How to choose the "best possible" Gaussian-integer coefficients* $\mathbf{a} = (a_k, a_l)^T$ *and* $\mathbf{b} = (b_k, b_l)^T$? To answer this question, we consider two rate pairs (R_k, R_l) and (R'_k, R'_l) associated with $(\mathbf{a}, \mathbf{b})$, and $(\mathbf{a}', \mathbf{b}')$, respectively. We say (R_k, R_l) is *better than* (R'_k, R'_l) if $R_k > R'_k$ and $R_l > R'_l$. We say (R_k, R_l) is a *Pareto-optimal solution* if there exists no (R'_k, R'_l) such that (R'_k, R'_l) is better than (R_k, R_l). It turns out that there are at most three Pareto-optimal solutions.

Theorem 1 *Let* $\boldsymbol{a}^*$ *be an optimal solution to*

$$\min_{\boldsymbol{a}\neq \boldsymbol{0}} \boldsymbol{a}^\dagger \left(\mathbf{P}^{-1} + \mathbf{H}^\dagger \mathbf{H}\right)^{-1} \boldsymbol{a}. \tag{5}$$

Let $\boldsymbol{b}^*$ *be an optimal solution to* (5) *under the constraint that* $\boldsymbol{a}^*$ *and* $\boldsymbol{b}^*$ *are linearly independent. Then, any Pareto-optimal solution is associated with* $[(1,0)^T, (0,1)^T]$, $[(0,1)^T, (1,0)^T]$, *or* $[\boldsymbol{a}^*, \boldsymbol{b}^*]$.

Proof First, we show that the rate pairs associated with $[(1,0)^T, (0,1)^T]$, $[(0,1)^T, (1,0)^T]$, and $[\mathbf{a}^*, \mathbf{b}^*]$ are indeed Pareto-optimal solutions. When $\mathbf{a} = (1,0)^T$ and $\mathbf{b} = (0,1)^T$, by Proposition 2, the associated rate pair (R_k, R_l) satisfies

$$R_k + R_l = \log_2 \left|\mathbf{I}_M + \mathbf{HPH}^\dagger\right|, \tag{6}$$

which is the sum capacity. Hence, (R_k, R_l) is a Pareto-optimal solution. Similarly, when $\mathbf{a} = (0,1)^T$ and $\mathbf{b} = (1,0)^T$, the associated rate pair (R_k, R_l) is a Pareto-optimal solution. We proceed to the rate pair (R_k, R_l) associated with $[\mathbf{a}^*, \mathbf{b}^*]$. Without loss of generality, we assume that $a_k \bmod p \neq 0$ and $a_l \bmod p \neq 0$ because otherwise it reduces to the previous two cases. We consider two sub-cases as follows:

(1) $\sigma^2(\mathbf{H}, \mathbf{a}^*) \leq \sigma^2(\mathbf{H}, \mathbf{b}^*|\mathbf{a}^*)$. In this case, by Proposition 2, we have $R_k + R_l = \log_2 \left|\mathbf{I}_M + \mathbf{HPH}^\dagger\right|$. Hence, (R_k, R_l) is a Pareto-optimal solution.
(2) $\sigma^2(\mathbf{H}, \mathbf{a}^*) > \sigma^2(\mathbf{H}, \mathbf{b}^*|\mathbf{a}^*)$. In this case, the rate pair (R_k, R_l) is given by

$$R_k = \log_2^+ \left(\frac{P_k}{\sigma^2(\mathbf{H}, \mathbf{a}^*)}\right)$$
$$R_l = \log_2^+ \left(\frac{P_l}{\sigma^2(\mathbf{H}, \mathbf{a}^*)}\right).$$

[6] This decoding strategy can be extended to the case of more than two users, where the BS decodes linearly independent combinations by applying a lattice reduction algorithm [8, 9].

We can show that (R_k, R_l) is a Pareto-optimal solution by using the "proof by contradiction" technique developed in our previous work [7].

Next, we consider a Pareto-optimal rate pair $(\tilde{R}_k, \tilde{R}_l)$ associated with $[\tilde{\mathbf{a}}, \tilde{\mathbf{b}}]$, and we show that $(\tilde{R}_k, \tilde{R}_l)$ is also associated with $[(1,0)^T, (0,1)^T]$, $[(0,1)^T, (1,0)^T]$, or $[\mathbf{a}^*, \mathbf{b}^*]$. To this end, we consider the following cases:

(1) The rate pair $(\tilde{R}_k, \tilde{R}_l)$ is a "corner point" of the capacity region. In this case, $(\tilde{R}_k, \tilde{R}_l)$ is also associated with $[(1,0)^T, (0,1)^T]$ or $[(0,1)^T, (1,0)^T]$.
(2) The rate pair $(\tilde{R}_k, \tilde{R}_l)$ is associated with $[\tilde{\mathbf{a}}, \tilde{\mathbf{b}}]$ with $\tilde{a}_k \bmod p \neq 0$ and $\tilde{a}_l \bmod p \neq 0$. In this case, we will show that $(\tilde{R}_k, \tilde{R}_l)$ is also associated with $[\mathbf{a}^*, \mathbf{b}^*]$. We assume, without loss of generality, that the greatest common divisor $\gcd(\tilde{a}_k, \tilde{a}_l) = 1$ because otherwise we can achieve a strictly better rate pair than $(\tilde{R}_k, \tilde{R}_l)$ by first decoding the linear combination associated with $\frac{1}{d}\mathbf{a}^*$, where $d = \gcd(\tilde{a}_k, \tilde{a}_l)$ and $|d| > 1$. We now consider two sub-cases as follows.

Sub-case (1) $\tilde{\mathbf{a}}$ and $\mathbf{a}^*$ are linearly dependent. In this case, we have $\tilde{\mathbf{a}} = u\mathbf{a}^*$ for some $|u| = 1$ because $\gcd(a_k^*, a_l^*) = 1$ and $\gcd(\tilde{a}_k, \tilde{a}_l) = 1$. Hence, we have $\sigma^2(\mathbf{H}, \tilde{\mathbf{a}}) = \sigma^2(\mathbf{H}, \mathbf{a}^*)$, and $\sigma^2(\mathbf{H}, \tilde{\mathbf{b}}|\tilde{\mathbf{a}}) = \sigma^2(\mathbf{H}, \mathbf{b}^*|\mathbf{a}^*)$. That is, $(\tilde{R}_k, \tilde{R}_l)$ is associated with $[\mathbf{a}^*, \mathbf{b}^*]$ as well.

Sub-case (2) $\tilde{\mathbf{a}}$ and $\mathbf{a}^*$ are linearly independent. Recall that $\mathbf{b}^*$ is an optimal solution to (5) under the constraint that $\mathbf{a}^*$ and $\mathbf{b}^*$ are linearly independent. Hence, we have $\sigma^2(\mathbf{H}, \tilde{\mathbf{a}}) \geq \sigma^2(\mathbf{H}, \mathbf{b}^*)$, and

$$\begin{cases} \tilde{R}_k \leq \log_2\left(\frac{P_k}{\sigma^2(\mathbf{H}, \tilde{\mathbf{a}})}\right) \leq \log_2\left(\frac{P_k}{\sigma^2(\mathbf{H}, \mathbf{b}^*)}\right) \\ \tilde{R}_l \leq \log_2\left(\frac{P_l}{\sigma^2(\mathbf{H}, \tilde{\mathbf{a}})}\right) \leq \log_2\left(\frac{P_l}{\sigma^2(\mathbf{H}, \mathbf{b}^*)}\right). \end{cases} \tag{7}$$

On the other hand, we have $\sigma^2(\mathbf{H}, \mathbf{b}^*) \geq \sigma^2(\mathbf{H}, \mathbf{a}^*)$ and $\sigma^2(\mathbf{H}, \mathbf{b}^*) \geq \sigma^2(\mathbf{H}, \mathbf{b}^*|\mathbf{a}^*)$ (as shown in [8]). Hence, if we compare (7) with the rate expressions in Proposition 2, we conclude that the rate pair $(\tilde{R}_k, \tilde{R}_l)$ is either the same as the pair associated with $[\mathbf{a}^*, \mathbf{b}^*]$ or strictly worse than it. This completes the proof of Theorem 1. ∎

Theorem 1 states that the BS only needs to evaluate three rate pairs (associated with $[(1,0)^T, (0,1)^T]$, $[(0,1)^T, (1,0)^T]$, and $[\mathbf{a}^*, \mathbf{b}^*]$) and then pick the best one based on a particular criterion such as Jain's fairness index and targeted outage probability. If the rate pair associated with $[\mathbf{a}^*, \mathbf{b}^*]$ is better than others, then the new decoding strategy outperforms the standard SIC decoding.

CaF decoding has essentially the same complexity as SIC decoding. The only additional computational cost is computing $(\mathbf{a}^*, \mathbf{b}^*)$, which can be done very efficiently via an algorithm that involves a fast computational lattice reduction algorithm, as explained in [10], that is preceded by a Singular Value Decomposition (SVD) algorithm. SVD algorithm only executed once (i.e., not every iteration) over a low-dimensional square matrix $\boldsymbol{M} = (\mathbf{P}^{-1} + \mathbf{H}^{\dagger}\mathbf{H})^{-1}$. The dimension of $\boldsymbol{M}$ is determined by the number of UEs in the NOMA pair.

4 Practical Considerations

In the previous sections, we have assumed perfect CSI (i.e., the BS knows the channel-coefficients matrix $\mathbf{H}$). In reality, the BS can only estimate $\mathbf{H}$, which leads to some estimation errors. Besides, in practical deployment, the cell is surrounded by adjacent cells, which induces inter-cell interference. In practical deployment, acquiring perfect CSI is very challenging. On top of these challenges, one of the inherent limitations of massive MIMO is pilot contamination that worsens the channel estimation errors [11]. Therefore, the channel estimation error matrix $\mathbf{E} = \mathbf{H} - \hat{\mathbf{H}} \neq 0$, which impacts the overall system performance.

Thus, in NOMA with SIC decoding, one of the biggest challenges from a practical perspective is the lack of accurate CSI. Furthermore, imperfect CSI leads to error propagation in SIC and degradation of the achieved performance. However, in massive MIMO-NOMA that employs CaF decoding, CSI accuracy has multi-fold effects. On the one hand, accurate CSI is essential to achieve a reliable integration of NOMA into massive MIMO. On the other hand, CaF is sensitive to channel estimation errors, as explained in [12]. Therefore, in particular, imperfect CSI may lead to a significant rate loss for CaF decoding [12, 13]. As mentioned in Sect. 3, computing $[\mathbf{a}^*, \mathbf{b}^*]$ depends on the CSI. Thus, without accurate CSI, finding the optimal values of $[\mathbf{a}^*, \mathbf{b}^*]$ cannot be achieved. Moreover, accurate CSI prevents any error prorogation to the computation of second integer combination. Therefore, while CaF decoding is theoretically promising, its application in practical massive MIMO-NOMA systems needs further assessments. Our simulation results in the next section highlight the effects of imperfect CSI on the performance of CaF decoding in massive MIMO-NOMA.

5 Numerical Results

In this section, we present results of MATLAB simulations to demonstrate the effectiveness of employing CaF decoding at the BS located at the center of a cell employing massive MIMO-NOMA, as shown in Fig. 1. We follow the communication channels modeling used in [14]. First, we demonstrate how CaF decoding outperforms SIC by providing additional rate pairs, increasing flexibility at the BS, improving fairness index, and lowering outage probability margins. Then, we show the impact of imperfect CSI on the reduction of the achievable rates and shrinking the rate region by comparing the ideal (perfect) channel estimation with the channel estimation using the conventional Single Pilot Scheme (SPS) that suffers from pilot contamination [15].

5.1 Fairness Index and Outage Probability

The capacity region of a multiple access channel (MAC) represents the closure of the set of all possible achievable rate pairs of multiple users and gives the best possible achievable rates under all decoding methods and perfect CSI. Figure 2 shows the MAC capacity region as well as four dominant rate pairs using CaF decoding at a BS with massive MIMO-PD-NOMA. Figure 2 shows that CaF offers two additional achievable rate pairs at the boundary of MAC capacity region. The positioning of these additional achievable rate pairs at the boundary of the capacity region implies that CaF achieves the maximum possible rate pair. Besides, these additional rate pairs are closer to the fairness index line, which is the trajectory that includes the positions of achievable rates that ensure all UEs in the NOMA pair achieve their maximum targeted rates according to their channel conditions. Actually, the fairness index line passes through one of them. The fairness index line in Fig. 2 illustrates that CaF decoding offers a better fairness and outperforms SIC decoding. Additionally, as shown in Fig. 2, CaF decoding lowers the outage probability margins. The conventional *Jain's fairness index*, J, is defined by

$$J = \frac{\left(\sum_{k=1}^{K} UE_k^{Achievable-Rate}\right)^2}{K\sum_{k=1}^{K}\left(UE_k^{Achievable-Rate}\right)^2}. \tag{8}$$

The outage probability of UE_k is defined as the probability that the achievable rate by the UE is below certain threshold.

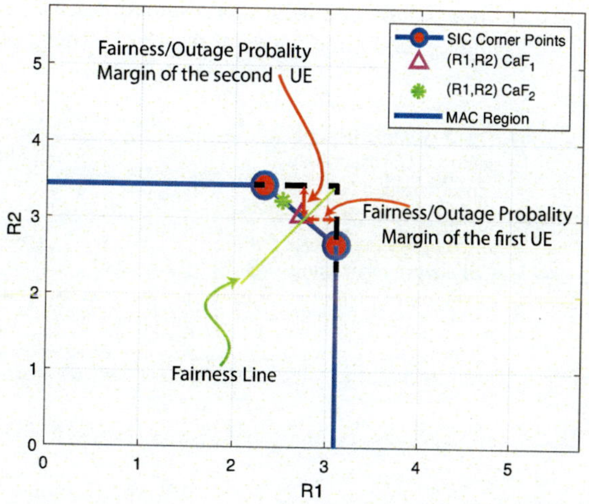

Fig. 2 Gaussian MAC capacity region for massive MIMO-NOMA that shows SIC corner points, and CaF dominant rate pairs when $M = 16$, and powers $P1 = 0.6866$ and $P2 = 0.9832$. We assume the BS employs an antenna selection mechanism, and each NOMA pair is being served by $M = 16$ antennas

5.2 *Effects of Imperfect CSI*

Using the same configurations, we assess the performance of CaF decoding in two scenarios: (1) the existence of perfect CSI at the BS and (2) the BS estimates the CSI using SPS. We computed the achievable rate region at each case. Then, we compare the obtained rate region with the capacity region of the ideal channel estimation. Figure 3 shows the obtained rate regions using perfect CSI and imperfect CSI that have been obtained by using SPS with the existence of pilot contamination. Figure 3 also illustrates the implications of channel estimation errors on CaF decoding, where estimation errors resulted in shrinking the rate region in a proportional relation where higher estimation error leads to a higher shrinking in size. Figure 3 also shows how the sensitivity of CaF decoding to estimation errors results in shifting the achievable rates pairs from the boundary line of MAC capacity region into the interior area with a higher diffraction from the fairness index line. Therefore, imperfect CSI results in an average sum-rate loss measured in bits per channel use (BPCU), as shown in Fig. 3, which consequently minimizes the cell's sum-rate capacity.

Furthermore, Fig. 4 accentuates the effects of imperfect CSI on computing $\mathbf{a}^*$. As mentioned in Sect. 2, the BS computes the first integer-linear combination $a_1\mathbf{x}_1 + a_2\mathbf{x}_2$. The optimal value of $\mathbf{a}^*$ determines the optimal values of a_1 and a_2, which essentially depends on the CSI accuracy. Finding an optimal value of $\mathbf{a}^*$ affects not only the first integer-linear combination (i.e., a_1 and a_2), but actually these effects propagate into decoding the second integer-linear combination (i.e., after subtracting, the first integer-linear combination from the received signal is not guaranteed to obtain optimal integer coefficients for decoding the second integer-linear combination). In our simulations, we model the effects of imperfect CSI on finding an optimal value of $\mathbf{a}^*$ by $\boldsymbol{\epsilon}_\alpha$. Therefore, when $\boldsymbol{\epsilon}_\alpha = 0$ that means optimal

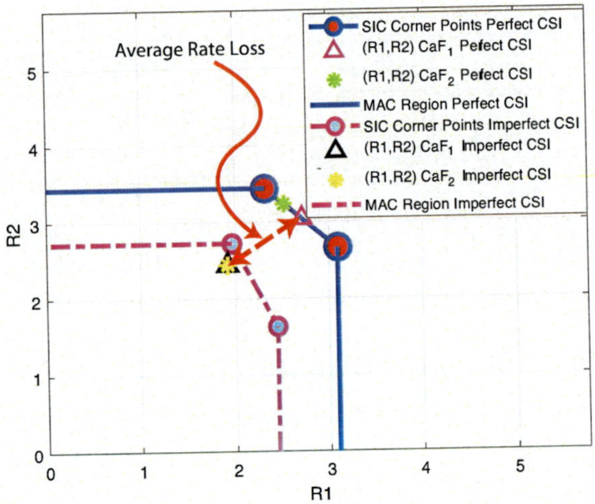

Fig. 3 Comparison of achieved Gaussian MAC capacity region for massive MIMO-NOMA in perfect CSI environment and non-perfect CSI environment using SPS

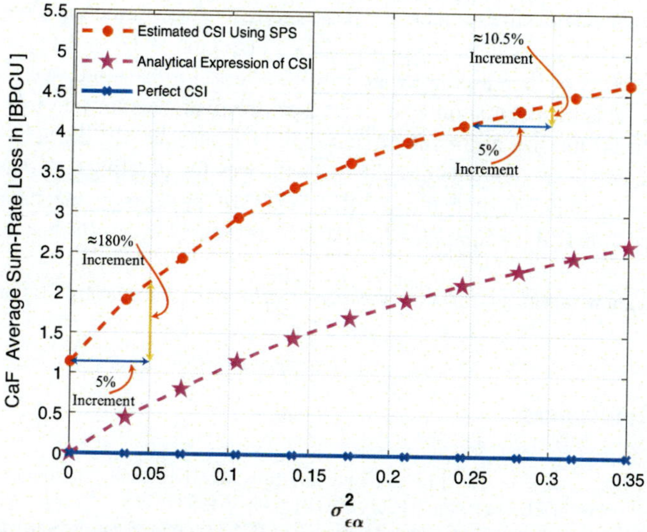

Fig. 4 Performance comparisons of the average sum-rate loss between perfect CSI and estimating CSI using SPS over a range of $\sigma^2_{\epsilon_\alpha}$

value of $\mathbf{a}^*$ is computed. Higher values of the entries of channel estimation error matrix $\mathbf{E}$ lead to a higher value of σ_{ϵ_α}.

Figure 4 illustrates the effects of varying the variance of ϵ_α on the average sum-rate loss, which consequently reveals the effects of imperfect CSI. Figure 4 shows that as the variance of ϵ_α increases (indicating an increase in the CSI estimation errors) the average sum-rate loss increases. It also shows how CaF is very sensitive to the estimation errors, where a very slight change by 5% increment in σ_{ϵ_α} leads to an increment in the average sum-rate loss by 180% (i.e., average sum-rate loss increased from 1.2 BPCU into 2.15 BPCU), when SPS is used in estimating the CSI. However, the CaF sensitivity reaches a saturation point for large values of σ_{ϵ_α}, where the incremental percentage of the average sum-rate loss drops into around 10.5% for each 5% increment of σ_{ϵ_α}.

6 Conclusion

We have shown that CaF decoding outperforms SIC decoding and provides promising advantages in terms of offering flexibility, enhancing achievable rates, improving fairness, and lowering outage potability when it is deployed in massive MIMO-NOMA systems. However, to achieve all of these targets, accurate CSI is needed at the BS. We have further evaluated the impact of imperfect CSI on the achieved performance and demonstrated how CaF decoding is very sensitive to the channel estimation errors. Through extensive simulations, we have shown that imperfect CSI causes shrinking of the rate region and degrading of the performance in terms of increased average sum-rate loss. Furthermore, we have found that as the channel estimation error increases, the average sum-rate loss increases, which shrinks the size of the rate region rapidly. Consequently, the achievable sum rate of the cell decreases. Therefore, to pave the way for adopting CaF decoding in practical implementation of massive MIMO-NOMA systems, a mechanism to ensure an accurate CSI is required, which will be presented in our further work.

Acknowledgement This work was supported by grants and scholarships from NSERC Canada.

References

1. Nazer, B., Gastpar, M.: Compute-and-forward: Harnessing interference through structured codes. IEEE Trans. Inf. Theory **57**(10), 6463–6486 (2011)
2. Zhan, J., Nazer, B., Gastpar, M., Erez, U.: MIMO compute-and-forward. In: IEEE International Symposium on Information Theory, Seoul, 28 June–3 July 2009, pp. 2848–2852
3. Zhan, J., Nazer, B., Erez, U., Gastpar, M.: Integer-forcing linear receiver. IEEE Trans. Inf. Theory **60**(12), 7661–7685 (2014)
4. Islam, S.M.R., Avazov, N., Dobre, O.A., Kwak, K.: Power-domain non-orthogonal multiple access (NOMA) in 5G systems: Potentials and challenges. IEEE Commun. Surveys Tutorials **19**(2), 721–742 (2017)
5. Ding, Z., Poor, H.V.: Design of Massive-MIMO-NOMA with limited feedback. IEEE Signal Process. Lett. **23**(5), 629–633 (2016)
6. Pan, H., Lu, L., Liew, S.C.: Practical power-balanced non-orthogonal multiple access. IEEE J. Selected Areas Commun. **35**(10), 2312–2327 (2017)
7. Guo, C., Aljalai, A.N., Feng, C., Zhao, L., Leung, V.C.M., Ward, R.K.: Compute-and-forward for uplink non-orthogonal multiple access. IEEE Wirel. Commun. Lett. **7**(6), 986–989 (2018)
8. Nazer, B., Cadambe, V.R., Ntranos, V., Caire, G.: Expanding the compute-and-forward framework: unequal powers, signal levels, and multiple linear combinations. IEEE Trans. Inf. Theory **62**(9), 4879–4909 (2016)
9. Feng, C., Silva, D., Kschischang, F.R.: An algebraic approach to physical-layer network coding. IEEE Trans. Inf. Theory **59**(11), 7576–7596 (2013)
10. Yao, H., Wornell, G.W.: Lattice-reduction-aided detectors for MIMO communication systems. In: Global Telecommunications Conference (GLOBECOM), Taipei, 17–21 Nov. 2002, pp. 424–428
11. Jose, J., Ashikhmin, A., Marzetta, T.L., Vishwanath, S.: Pilot contamination and precoding in multi-cell TDD systems. IEEE Trans. Wirel. Commun. **10**(8), 2640–2651 (2011)
12. Pappi, K.N., Karagiannidis, G.K., Schober, R.: How sensitive is compute-and-forward to channel estimation errors? In: IEEE International Symposium on Information Theory, Istanbul, 7–12 July 2013, pp. 3110–3114
13. Feng, C., Silva, D., Kschischang, F.R.: Blind compute-and-forward. IEEE Trans. Commun. **64**(4), 1451–1463 (2016)
14. Jeruchim, M.C., Balaban, P., Shanmugan, K.S.: Simulation of Communication Systems: Modeling, Methodology and Techniques, 2nd edn. Springer Science & Business Media, Berlin (2000)
15. Aljalai, A.N., Feng, C., Leung, V.C.M., Ward, R.K.: Dual pilot scheme (DPS) and its application in massive MIMO. IEEE Trans. Commun. **69**(3), 1431–1444 (2020)

Regularized Classification-Aware Quantization

Daniel Severo, Elad Domanovitz, and Ashish Khisti

1 Introduction

As the worldwide estimate of active edge devices surpasses tens of billions, reducing data storage and transmission costs becomes vital. When dealing with continuous data, most pipelines include quantization, where the data is discretized in some way. Traditionally, quantization is designed to minimize the reconstruction error with respect to some notion of distortion, such as the mean squared error (MSE).

Despite small reconstruction error being an intuitive goal, it might not be the optimal measure when considering downstream tasks such as binary classification [1]. For example, there is no need to store data with high precision if it will be thresholded afterward for decision-making. Quantizing near the threshold value will suffice.

Following [1], we assume the fusion center has access to a pre-trained classifier, and it wishes to apply it over data features collected at d distributed sensor nodes. The focus of our work is on designing distributed quantizers to be used for classification. Thus, our goal is not to improve the performance of the classifier, but rather, to find a good quantization scheme that would result in minimal performance degradation when used in conjunction with the pre-trained classifier.

The assumption of a pre-trained classifier can be motivated by considering a two-phase approach in systems design. When training the classifier, data is sent at high resolution over rate-constrained channels, i.e., several channel uses can be used to send a single sample. After, real-time decisions are required, and the data has to be quantized to meet these constraints.

D. Severo (✉) · E. Domanovitz · A. Khisti
Electrical and Computer Engineering, University of Toronto, Toronto, ON, Canada
e-mail: d.severo@mail.utoronto.ca; elad.domanovitz@utoronto.ca; akhisti@ece.utoronto.ca

H. Nguyen et al. (eds.), *30th Biennial Symposium on Communications 2021*,
Signals and Communication Technology, https://doi.org/10.1007/978-3-031-06947-5_5

In practical settings, quantization schemes must be learned from a finite set of data points, as the true data generating distribution is not known. In production systems, re-learning with incoming data can be cumbersome, or even impossible, as this would require bidirectional communication between the decoder and encoders to agree on a new codebook. It is therefore desirable for the performance of a learned quantization scheme to not deteriorate once applied to new, but similar, data. This begets the question: *how well do learned quantization schemes generalize to out-of-sample data?*

In this chapter, we propose a quantization scheme called *Regularized Classification-Aware Quantization* (RCAQ) that is learned from data. Experiments show that RCAQ generalizes with few training examples on synthetic Gaussian data. We evaluate RCAQ and previous work by Hanna et al. [1] called on-the-line, on both in-sample (training) and out-of-sample (testing) datasets to estimate generalization. Results indicate that both methods have comparable performance. However, RCAQ has significantly lower computational complexity, as it does not require iterating twice over the dataset as in on-the-line. All code is available at https://github.com/dsevero/rcaq.

2 Problem Formulation

This chapter focuses on a distributed setting, where d encoders must quantize and transmit data to a single decoder tasked with performing linear binary classification. Learning is done with complete information, i.e., the decoder and encoders can communicate to decide on a codebook. Once deployed, encoders do not coordinate with each other, and communication with the decoder is unidirectional. Data points are in $\mathbb{R}^d$, but each encoder has access only to one dimension in $\mathbb{R}$. Quantization indices are transmitted losslessly over a noiseless channel. The classifier is considered to be given, i.e., pre-trained.

Our goal is *not* to improve the classifier, but to learn a quantization scheme at the encoders that will minimize the additional classification error introduced by quantization. The quantization error is measured by comparing the output of the classifier with and without quantization. Note that, in some sense, the classifier is our ground truth.

We extend the formulation of [2], by emphasizing generalization performance. Let $\mathbf{x} = (x_1, \ldots, x_d) \in \mathbb{R}^d$ represent a d-dimensional data point and $y(\mathbf{x}) \in \{-1, +1\}$ the label assigned by the pre-trained classifier *without* quantization. The k-th encoder must learn a quantization function $\mathcal{E}_k : \mathbb{R} \to [2^{R_k}]$, where R_k is the rate and $[2^{R_k}] = \{1, 2, \ldots, 2^{R_k}\}$ are the set of integer quantization indices. The decoder receives an integer d-tuple $\mathcal{E}(\mathbf{x}) = (\mathcal{E}_1(x_1), \ldots, \mathcal{E}_d(x_d))$ and must output a classification $\hat{y}(\mathcal{E}(\mathbf{x})) \in \{-1, +1\}$.

Remark 1 Note that, unlike traditional quantization, there is no need for reconstruction points as the decoder can directly map the quantization integers to a class label.

If $\mathbf{x} \sim P$ is the true data generating distribution, the objective is to solve the following optimization problem:

$$\begin{aligned} \underset{\mathcal{E},\hat{y}}{\text{minimize}} \quad & \mathbb{E}[\mathbb{1}[\hat{y}(\mathcal{E}(\mathbf{x})) \neq y(\mathbf{x})]] \\ \text{subject to} \quad & R_1, \ldots, R_d \text{ and } P, \end{aligned} \tag{1}$$

where the expectation is taken over $\mathbf{x} \sim P$ and $\mathbb{1}$ the indicator function.

The objective function in Eq. (1) is known as the *0-1 loss* and differs from [1] only in the choice of P. Here, P is the (unknown) data generating distribution, while [1] uses the empirical distribution of a set of points $\mathcal{T} = \{\mathbf{x}^{(i)}\}_{i=1}^{N}$ sampled i.i.d. from P (i.e., they optimize the training loss).

In practice, only $\mathcal{T}$ is available, not P. Therefore, we minimize a regularized estimator of Eq. (1) by partitioning $\mathcal{T}$ into training, validation, and test sets. Quantization is learned on the training set, while performance is evaluated on the test set. The validation set is used to tune the number of bins, as discussed in Sect. 5. Note that it is always possible to achieve zero training error by setting $\mathcal{E}_1(\mathbf{x}) = y(\mathbf{x})$. This is shown in Fig. 1. Clearly, this scheme overfits.

Throughout this chapter, we assume that the classifier available at the decoder performs binary linear classification. Formally, a weight vector $\mathbf{w} = (w_1, \ldots, w_d) \in \mathbb{R}^d$ is available such that the class label is assigned as $\hat{y}(\mathbf{x}) = \text{sign}(\langle \mathbf{w}, \mathbf{x} \rangle)$. We assume $\mathbf{w}$ is a 45° hyperplane (i.e., $-w_1 = w_2 = \cdots = w_d = 1$), as the training and test sets can always be re-scaled accordingly.

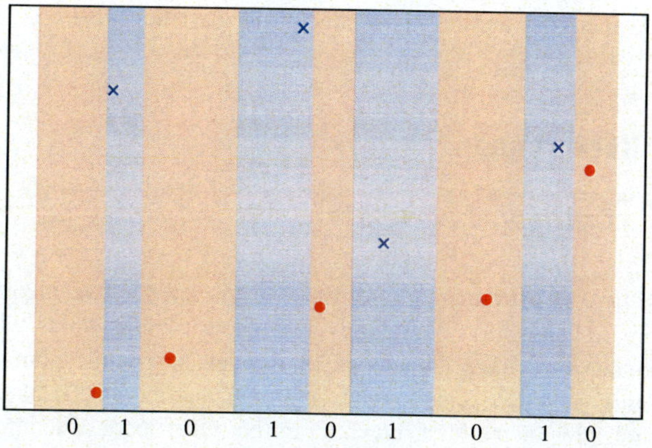

Fig. 1 Example of overfitting inspired by Hanna et al. [1]. Only one encoder is needed. For points $\mathbf{x}$ in the blue region, $\mathcal{E}_1(\mathbf{x}) = y(\mathbf{x}) = 1$. Similarly, for those in the red regions, $\mathcal{E}_1(\mathbf{x}) = y(\mathbf{x}) = 0$. This scheme achieves zero classification error on the training set

3 Related Work

Previous work has investigated distributed quantization schemes for both classification [1] and traditional settings [3]. More recently, [1] introduced the family of *on-the-line* quantizers for linear binary classification in $\mathbb{R}^2$, which uses the same encoder for both dimensions. The core of the algorithm works by noting that there is a finite set of boundaries that affect the quantization error for a fixed training set of size N, which are the $2N$ coordinates of the data points. A greedy optimization strategy is given that finds an optimal solution to problem (1), under the constraint that $\mathcal{E}_1 = \mathcal{E}_2$ and P is the empirical distribution of the training set. It works by exhaustively searching the set of potential boundaries and keeping those that minimize the loss. To avoid overfitting, the authors impose continuity conditions on the quantization regions. The computational complexity is shown to be $\mathcal{O}(N^2 2^R)$, where $R = R_1 = R_2$ is the rate.

This configuration can be viewed as related to distributed estimation and detection with communication-constrained links. See [4, 5] and references therein. In [6], scalar quantization was studied. Many of these works assume that the sensor measurements are independently distributed given the detection hypothesis. Further, it is assumed that these conditional distributions are known. In our framework, neither this conditional independence nor the knowledge of these distributions is assumed.

Several recent works focused on distributed probability estimation, property testing, and simulation [7–9]. However, these works assume that each node observes all features, while we focus on the case where each node observes a single feature.

Applying the pre-trained classifier over the quantized features can be viewed as functional compression problem. Distributed compression for functional computation with distortion has been studied (for example) in [10, 11]. Extending the ideas of [1], our suggested algorithm performs quantization for unknown source distribution and without any a priori knowledge of the classifier function.

4 Background

4.1 Distributed Lloyd-Max

Vanilla non-distributed Lloyd-Max [12] is an iterative algorithm that computes optimal quantization boundaries and reconstruction points with respect to a loss function. It requires knowledge of the data distribution but can be used with a training set by considering its empirical distribution. It alternates between two optimization steps. First, given an initial set of reconstruction points, it computes the bins by assigning each data point to the reconstruction point that minimizes the loss (bin step). Then, using the newly computed bins, it updates the reconstruction points to those that minimize the expected loss for each bin (reconstruction step). This method is known to converge to a local minimum of the loss. A good minima can be found by running the algorithm on random initializations. Note that binning happens at the encoder, and reconstruction at the decoder.

In [2], a distributed version of Lloyd-Max is introduced. They provide conditions under which the loss function converges to a local minimum. The algorithm is similar to vanilla non-distributed Lloyd-Max, in the sense that it also alternates between two steps. However, to account for the distributed nature of the problem, it performs the bin step locally at each encoder while holding all other encoders fixed. The reconstruction step follows as in the non-distributed setting, since it happens at the decoder side.

4.2 Generalization and Regularization

In our context, a learning algorithm $\mathcal{A}$ is a function that receives a training set of i.i.d. points $\mathcal{T} = \{\mathbf{x}^{(i)}\}_{i=1}^N \sim P^N$ and outputs a distributed quantization scheme $\mathcal{A}(\mathcal{T}) = (\mathcal{E}, \mathcal{D})$, where $\mathcal{E} = (\mathcal{E}_1, \ldots, \mathcal{E}_d)$ are the encoders and $\mathcal{D}$ the decoder. A loss function $\mathcal{L}$, such as the reconstruction or 0-1 loss, is used to quantify the performance. Given a fresh sample $\mathcal{T}' \sim P^N$, we define two important quantities,

$$\begin{aligned} \text{Training error} \quad & \mathcal{L}(\mathcal{A}(\mathcal{T}), \mathcal{T}) \\ \text{Test error} \quad & \mathcal{L}(\mathcal{A}(\mathcal{T}), \mathcal{T}'). \end{aligned} \tag{2}$$

Note that in both cases $\mathcal{A}$ is trained on the training set $\mathcal{T}$, what differs is on which dataset the loss is evaluated. An algorithm is said to generalize well if both quantities are always similar in value.

When the training error is low and testing error is high, we say that the algorithm has overfitted. This can happen when $\mathcal{A}(\mathcal{T})$ has enough complexity that it outputs a quantization scheme that memorizes the training set (see Fig. 1) [13]. Regularization can be seen as a way to reduce overfitting in $\mathcal{A}$ by augmenting the loss function such that it penalizes highly complex quantization schemes [13].

In many practical cases, part of the training set $\mathcal{T}$ is held out (i.e., not used during training) and is used to tune *hyperparameters* [13]. This third dataset is called the validation set $\mathcal{T}_V$. Confusingly, the remaining dataset $\mathcal{T}_T = \mathcal{T} \setminus \mathcal{T}_V$ is also called the training set. Hyperparameters, in our context, can be understood as any parameter that is not optimized during training. Thus, the train/test paradigm is augmented to train/validate/test. Multiple quantizers, with different hyperparameter values, are trained on $\mathcal{T}_T$, and the one that performs best on $\mathcal{T}_V$ is chosen. Test loss is evaluated on $\mathcal{T}'$, as before. This setting is employed in RCAQ and is discussed in Sect. 5.

5 Methods

Our algorithm is a direct application of distributed Lloyd-max presented in [2] and discussed in Sect. 4.1. As the name entails, RCAQ works by regularizing the loss function with a term that penalizes bad reconstruction. The intuition stems from viewing this as a continuity condition in the data space. Points that are close together should tend to be quantized similarly. For this to be possible, we must add an extra step in the decoding process that maps the quantization indices $\mathcal{E}(\mathbf{x})$ to a reconstruction point $\mathcal{D}(\mathcal{E}(\mathbf{x})) = \hat{\mathbf{x}} \in \mathbb{R}^d$ before the classification function $\hat{y}$ is applied, as in traditional quantization.

5.1 *Algorithm*

RCAQ is a two-stage quantizer. Encoders perform binning before quantization. Formally, each encoder has a binning function $B_i : \mathbb{R} \to \mathbb{R}$ that maps the data points to evenly spaced bins. The bin centers are then quantized with $\mathcal{E}_i$ before being sent to the decoder. The functions $\{B_i\}_{i=1}^d$ are *not* optimized during training. Instead, an exhaustive search is performed over the number of bins during the validation phase discussed in Sect. 4.2. Thus, each encoder is represented by the pair $(B_i, \mathcal{E}_i)$, and the decoder by $(\mathcal{D}, \hat{y})$. A diagram is shown in Fig. 2.

Our loss function is

$$\mathcal{L}_{\mathcal{T}}(\mathcal{E}, \mathcal{D}, \hat{y}) = \frac{1}{N} \sum_{\mathbf{x} \in \mathcal{T}} \left(\gamma \mathbb{1}[\hat{y}(\hat{\mathbf{x}}) \neq y(\mathbf{x})] + (1 - \gamma) \left\| \hat{\mathbf{x}} - \mathbf{x} \right\|^2 \right), \tag{3}$$

where $\mathcal{T}$ is the training set, $N = |\mathcal{T}|$ is the training set size, $0 \leq \gamma \leq 1$ controls the strength of regularization, $\left\| \hat{\mathbf{x}} - \mathbf{x} \right\|^2$ is the reconstruction penalty, and $\mathbb{1}$ is the indicator function. For $\gamma = 1$, $\mathcal{L}$ becomes the 0-1 loss, while for $\gamma = 0$, it is the original Lloyd-max distortion term (i.e., MSE).

A small number of bins acts as a second regularizer, as it precludes the encoder from having highly discontinuous quantization regions. Larger values for b increase

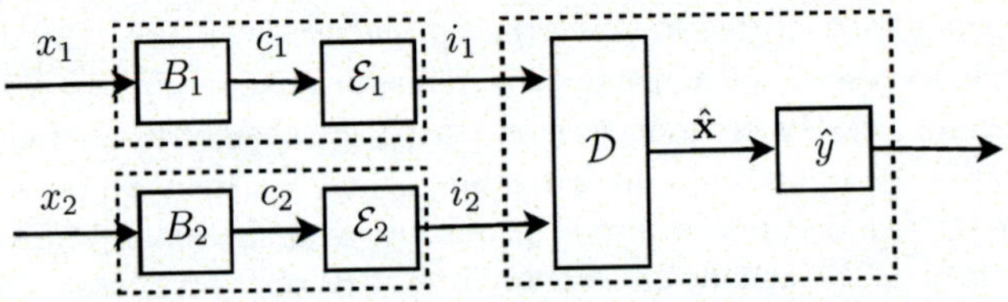

Fig. 2 RCAQ high-level diagram. Data point $\mathbf{x} = (x_1, x_2)$ is mapped to bin centers (c_1, c_2) and then to integer quantization indices $(i_1, i_2) \in [2^{R_1}] \times [2^{R_2}]$. The decoder deterministically selects a reconstruction point $\hat{\mathbf{x}}$ and applies the classifier $\hat{y}$

the resolution at the encoder, making it possible to have discontinuous regions and therefore less distortion. However, in our experiments, we found that the algorithm sometimes opts for discontinuity even for a relatively small $b \approx 16$. The pseudocode for the optimization algorithm is presented and discussed next.

5.2 *Optimizing the Encoder*

Algorithm 1 assumes equal rates and the number of bins in all dimensions, but this can be trivially extended. The inner optimization loop updates $\mathcal{E}_i$ by exhaustively searching which integer should be mapped to each bin. The overall computational complexity, if implemented naively, is $\mathcal{O}(Id\sum_{b=1}^{b_{\max}} 2^{Rb})$, where R is the rate, b is the number of bins, and I is the number of iterations to convergence. However, it is possible to parallelize over the bins by noting that the integer assigned to bin i will not affect the loss incurred at bins $j \neq i$. In other words, the bins are independent during optimization. The outer validation loop can also be done in parallel. Also, we found that convergence is fast and does not depend significantly on other quantities, and hence, I is fixed in our experiments. In practice, this reduces the complexity to

$$\mathcal{O}\left(d2^R\right). \tag{4}$$

When compared to the family of on-the-line quantizers ($d = 2$), there is a reduction in complexity of $\mathcal{O}(N^2)$. The order in which the encoders are optimized matters, as the optimization step for encoder $\mathcal{E}_k$ uses the previous, and already

Algorithm 1 Reg. classification-aware quantization

```
Given rate R and training set 𝒯;
Partition 𝒯 into 𝒯_T and validation set 𝒯_V;
Consider R_i = R and b_i = b for all i ∈ {1, ..., d};
for number of bins b ∈ {1, 2, ..., b_max} do
    Initialize encoders {(B_i, ℰ_i)}_{i=1}^d;
    Initialize decoder (𝒟, ŷ);
    while ℒ_{𝒯_T}(ℰ, 𝒟, ŷ) has not converged do
        for i ∈ {1, ..., d} do
            ℰ_i ← arg min_{ℰ_i} ℒ_{𝒯_T}(ℰ, 𝒟, ŷ);
        end
        (𝒟, ŷ) ← arg min_{(𝒟,ŷ)} ℒ_{𝒯_T}(ℰ, 𝒟, ŷ);
    end
    Evaluate 0-1 loss on validation set 𝒯_V;
end
Output codec that achieved lowest 0-1 loss on 𝒯_V;
```

optimized, encoders $\{\mathcal{E}_i\}_{i=1}^{k-1}$. Therefore, it is not possible to parallelize over the inner optimization loop, unless we relax the algorithm and optimize subsets of $\{\mathcal{E}_1, \ldots, \mathcal{E}_d\}$ simultaneously. We do not investigate this latter scenario in this chapter.

5.3 *Optimizing the Decoder*

Optimizing the decoder $(\mathcal{D}, \hat{y})$ consists of updating the reconstruction point function $\mathcal{D}$ as in regular Lloyd-max, but with the loss function defined in Eq. (3). The classification function $\hat{y}$ is simply the classifier itself. Note that, during minimization, the 0-1 loss component will only be affected if the reconstruction point is pushed to the other side of the decision boundary. This implies that the loss function (3) is discontinuous with respect to the reconstruction point. Surprisingly, this does *not* add complexity to the minimization process, as there are only two possible updates for any reconstruction point. Either $\hat{\mathbf{x}}$ is updated to the average of the data points assigned to it (i.e., as in vanilla Lloyd-max, which considers only MSE), or, if

$$\sum_{\mathbf{x} \in \mathcal{T}_T(\hat{\mathbf{x}})} \gamma \mathbb{1}[\hat{y}(\hat{\mathbf{x}}') \neq y(\mathbf{x})] > \sum_{\mathbf{x} \in \mathcal{T}_T(\hat{\mathbf{x}})} (1 - \gamma) \left\|\hat{\mathbf{x}}' - \mathbf{x}\right\|^2, \tag{5}$$

$\hat{\mathbf{x}}$ must be pushed to the other side of the boundary, where $\mathcal{T}_T(\hat{\mathbf{x}})$ is the set of training points that are currently assigned to reconstruction point $\hat{\mathbf{x}}$. Intuitively, this must be done in a way such that the increase in MSE is minimal. It is easy to show that the optimal pushing direction is exactly that of the normal vector defining the hyperplane, i.e., a straight line to the boundary. Therefore, the update procedure for a single reconstruction point $\hat{\mathbf{x}}$ is

$$\underset{\hat{\mathbf{x}}' \in \{\hat{\mathbf{x}}, \hat{\mathbf{x}} + \boldsymbol{\Delta}\}}{\arg\min} \sum_{\mathbf{x} \in \mathcal{T}_T(\hat{\mathbf{x}})} \left(\gamma \mathbb{1}[\hat{y}(\hat{\mathbf{x}}') \neq y(\mathbf{x})] + (1 - \gamma) \left\|\hat{\mathbf{x}}' - \mathbf{x}\right\|^2\right), \tag{6}$$

where $\boldsymbol{\Delta} \propto \mathbf{w}$ is the vector such that $\hat{\mathbf{x}} + \boldsymbol{\Delta}$ crosses the boundary by some small additive constant (we used 10^{-6}).

6 Experiments

We assess the generalization performance of RCAQ and on-the-line quantizers with bivariate and mixture Gaussian data, as a function of both rate and correlation. Experiments were run for $1 \leq d \leq 5$ to confirm that the number of iterations to convergence I does not scale significantly with d. However, experiments shown here are limited to $d = 2$, as on-the-line is defined only in this setting. Artificial

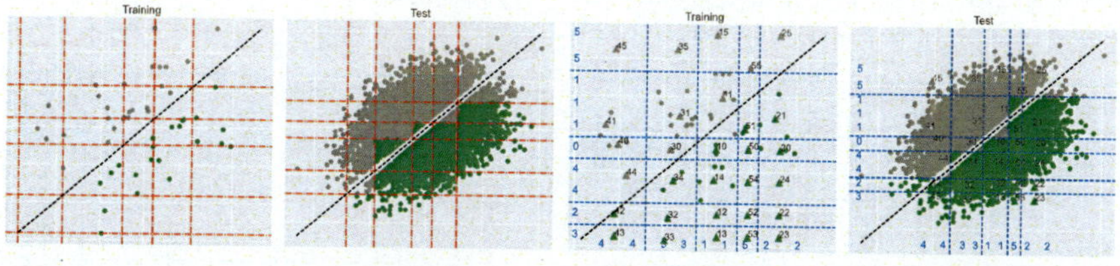

Fig. 3 Quantization bins for on-the-line (left, orange) and RCAQ (right, blue) for both training (50 points) and test sets (10,000 points). Class labels are represented by the color of each point (green or gray). Training losses for on-the-line and RCAQ are 0 and 0.02, respectively, while testing losses are 0.08 and 0.05. Reconstruction points are shown as triangles with their respective class labels corresponding to their colors. The number of bins for RCAQ is 10, and the integer quantization index for each bin is shown in blue for each encoder. On-the-line does not use reconstruction points, as it assigns class labels to each quantization cell directly

margins are created at the separation hyperplane (i.e., classifier) by adding a small perturbation in the direction of the normal vector (see Fig. 3 for a visual). Intuitively, since there are less points concentrating around the decision boundary, larger margins should result in better generalization performance [13]. Indeed, this was observed experimentally for both RCAQ and on-the-line.

Empirically, we found that RCAQ is *not* sensitive to γ, which serves mostly to equalize the scales of the 0-1 and reconstruction losses. In all experiments, $\gamma = 0.95$.

In the following sections, we detail the experimental setup. The experiments presented are a subset of the totality performed and were chosen for being good representatives of the overall performance of both RCAQ and on-the-line.

6.1 Bivariate Gaussians

Bivariate Gaussians with zero mean, equal variances, and correlation coefficient ρ were used, resulting in the following covariance matrix:

$$\mathbf{\Sigma} = \begin{pmatrix} 1 & \rho \\ \rho & 1 \end{pmatrix}. \tag{7}$$

We generated 6 training sets with correlation coefficient $\rho \in \{0, 0.2, 0.4, 0.6, 0.8, 1.0\}$ and training set size $N = 300$. Both on-the-line and RCAQ were trained on all 6, and performance was measured on 6 different test sets, each of size 10,000, with the same values of ρ as the training sets. The rate was fixed at $2^{R_1} = 2^{R_2} = 6$ integers per encoder (i.e., dimension) everywhere. The 0-1 loss for the test set is shown in Fig. 4 for both methods. For RCAQ, the train/validation split is set to 50/250.

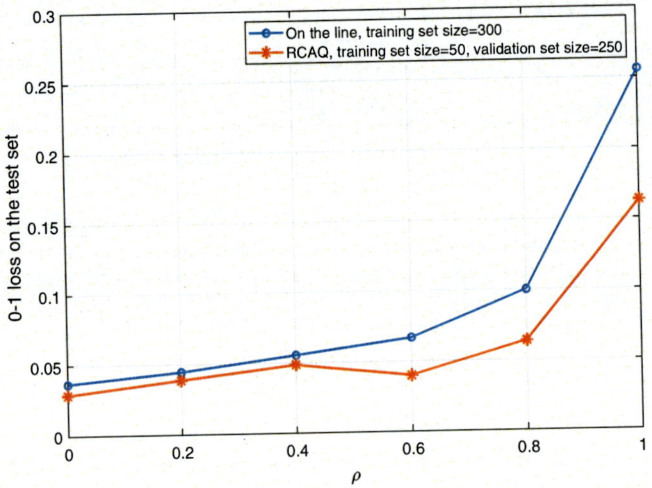

Fig. 4 0-1 loss (lower is better) for on-the-line (blue) and RCAQ (orange) quantizers. Test set of size 10,000 was used. Results are shown as a function of the correlation coefficient for a bivariate Gaussian distribution

Our experiments indicate that RCAQ is, at the very least, competitive with on-the-line, but with a reduction in complexity of $\mathcal{O}(N^2)$. As ρ increases, the loss increases in general, as more points concentrate along the decision plane making it harder to quantize. The effect of the margin is the opposite, and increasing it pushes points away from the boundary, making quantization easier. This is true for both encoders, as previously mentioned.

In Fig. 3, we provide the quantization regions and indices for on-the-line and RCAQ for $\rho = 0.4$ and $N = 50$. Note that on-the-line does not use reconstruction points, as the quantization cells are assigned a class label directly. RCAQ uses reconstruction points and chooses the class label by placing the reconstruction points below or above the hyperplane. The quantization cells for on-the-line that contain the hyperplane are always squares. To see this, note that the hyperplane is a 45° line and serves as the diagonal of the cell. RCAQ does not have this restriction, allowing those cells to be rectangles. This improves the generalization error, as the cells that intersect the hyperplane are the ones that can potentially misclassify points. In this example, on-the-line overfits as it achieves 0 training and 0.08 testing errors. RCAQ has a larger training error (0.02), but smaller testing error (0.05) (Fig. 5).

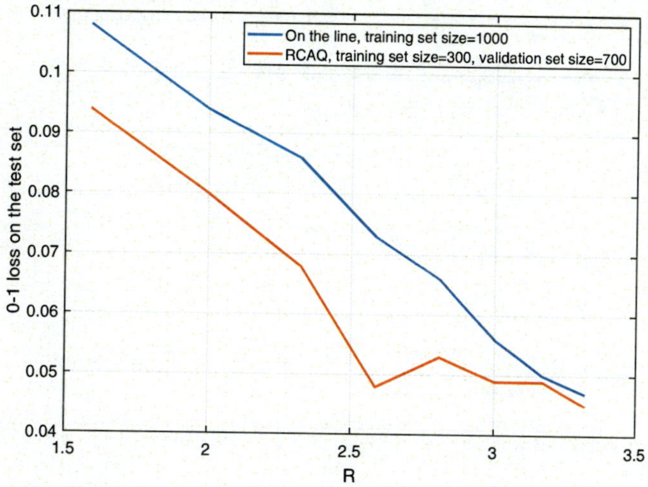

Fig. 5 0-1 loss (lower is better) for on-the-line (blue) and RCAQ (orange) quantizers. Test set of size 10,000 was used. Results are shown as a function of the rates $R_1 = R_2 = R$ (i.e., 2^R bins at each encoder) for a Gaussian mixture distribution

6.2 *Gaussian Mixtures*

We perform an experiment by fixing the rate of both RCAQ and on-the-line with data from a Gaussian mixture. The mean was uniformly sampled between 0 and 1, and a fixed diagonal covariance matrix $10^{-2}\mathbb{I}_d$ was used, where $\mathbb{I}_d$ is the identity matrix with d rows and columns. The dataset sizes were 1000 for training and 10,000 for testing. RCAQ's train/validation split was 300/700.

Caution is needed when applying on-the-line at high rates. On-the-line iterates over the training data and inserts boundaries that minimize the training error. Therefore, if the training error reaches zero, it will stop adding boundaries even if more rate is available. To avoid this, and to guarantee a fair comparison, we evaluated on-the-line only for large dataset sizes ($N = 1000$). Still, RCAQ clearly outperforms on-the-line on the test loss.

6.3 *Gauging the Potential of RCAQ*

As shown in previous sections, RCAQ performs competitively with on-the-line, with significantly less complexity. However, we postulate that RCAQ can perform better for small training set size.

A major component of RCAQ is the validation step that is used to find the number of evenly spaced bins for each encoder, $\{B_i\}_{i=1}^d$. To motivate future work,

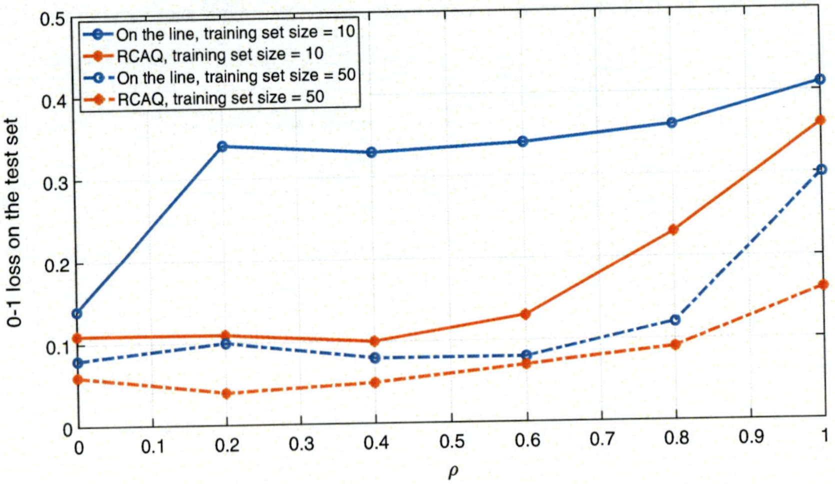

Fig. 6 Testing loss for on-the-line (blue) and RCAQ (orange) quantizers. Two training sets of sizes 10 (solid) and 50 (dashed) were used. Results are shown as a function of the correlation coefficient for a Bivariate Gaussian distribution. Both methods achieved training loss close to zero, but RCAQ was slightly higher

we performed experiments to understand if there exists a good set $\{b_i^*\}_{i=1}^d$ to begin with, i.e., to assess the potential of RCAQ with respect to the number of evenly spaced bins. To do this, we forgo the validation step, fix $b_1 = b_2 = \cdots = b_d = b$, and exhaustively search over $1 \leq b \leq 32$ by *evaluating on the test set*. Clearly, this setup is not feasible in practice, as we are optimizing over the test set. However, it does allow us to gauge the potential performance of RCAQ. On-the-line was trained only with the training set; hence, the results shown are representative of its performance. Results are shown in Fig. 6. Our experiments indicate that RCAQ can potentially outperform on-the-line. With few training points, on-the-line is not able to generalize, as the algorithm works by inserting quantization boundaries on the coordinates of the training points themselves. This is reflected in the gap between the losses of RCAQ and on-the-line, as the potential advantage incurred by RCAQ diminishes as the training set size increases. Also, RCAQ does not limit the solution to $\mathcal{E}_1 = \mathcal{E}_2$, which allows more flexibility.

7 Conclusion

In this chapter, we proposed a new quantization algorithm for distributed classification called *Regularized Classification-Aware Quantization* (RCAQ). We provide experiments that show that RCAQ can generalize to out-of-sample data from few training points. Through vectorization, an implementation is provided with computational complexity $\mathcal{O}(d2^R)$, where d is the number of encoders and R is the rate at each encoder. RCAQ is competitive with current baselines such as on-the-line, but with significantly lower computational complexity, as it does not loop over the dataset to pick boundaries.

We provide experiments that motivate future work by showing that there exist good configurations of evenly spaced boundaries that can outperform the existing methods. Finding them, however, is left as an open problem. More experiments are needed to assess if this scheme will scale to high-dimensional datasets.

Acknowledgement This work was funded by Huawei.

References

1. Hanna, O.A., Ezzeldin, Y.H., Sadjadpour, T., Fragouli, C., & Diggavi, S.: On distributed quantization for classification. IEEE J. Selected Areas Inf. Theory, **1**(1), 237–249 (2020)
2. Rebollo-Monedero, D., Girod, B.: Design of optimal quantizers for distributed coding of noisy sources. In: Proceedings. (ICASSP '05). IEEE International Conference on Acoustics, Speech, and Signal Processing, 2005, vol. 5, pp. v/1097–v/1100 (2005). https://ieeexplore.ieee.org
3. Wang, S., Fang, Y., Cheng, S.: Distributed Source Coding: Theory and Practice. Wiley, Hoboken (2017)
4. Chamberland, J.-F., Veeravalli, V.V.: Wireless sensors in distributed detection applications. IEEE Signal Process. Mag. **24**(3), 16–25 (2007)
5. Luo, Z.-Q.: Universal decentralized estimation in a bandwidth constrained sensor network. IEEE Trans. Inf. Theory **51**(6), 2210–2219 (2005)
6. Longo, M., Lookabaugh, T.D., Gray, R.M.: Quantization for decentralized hypothesis testing under communication constraints. IEEE Trans. Inf. Theory **36**(2), 241–255 (1990)
7. Han, Y., Özgür, A., Weissman, T.: Geometric lower bounds for distributed parameter estimation under communication constraints. In: Conference on Learning Theory, pp. 3163–3188. PMLR (2018).
8. Diakonikolas, I., Grigorescu, E., Li, J., Natarajan, A., Onak, K., Schmidt, L.: Communication-efficient distributed learning of discrete distributions. In: *Advances in Neural Information Processing Systems* (2017)
9. Acharya, J., Canonne, C.L., Tyagi, H.: Distributed simulation and distributed inference (2018). Preprint arXiv:1804.06952
10. Wagner, A.B.: On distributed compression of linear functions. IEEE Trans. Inf. Theory **57**(1), 79–94 (2010)
11. Doshi, V., Shah, D., Médard, M., Effros, M.: Functional compression through graph coloring. IEEE Trans. Inf. Theory **56**(8), 3901–3917 (2010)
12. Lloyd, S.: Least squares quantization in PCM. IEEE Trans. Inf. Theory **28**(2), 129–137 (1982)
13. Shalev-Shwartz, S., Ben-David, S.: Understanding Machine Learning: From Theory to Algorithms. Cambridge University Press, Cambridge (2014)

SVD-Based NFDM with Index Modulation

Nghia H. Nguyen, Ha H. Nguyen, and Brian Berscheid

1 Introduction

Non-orthogonal frequency-division multiplexing (NFDM) is considered as a promising substitute for the conventional orthogonal frequency-division multiplexing (OFDM). The main idea of NFDM is that, by packing subcarriers closer than the minimum spacing required for orthogonality as in OFDM, a smaller bandwidth is consumed while delivering an equal data rate. However, such an advantage is generally challenged by possible performance degradation due to inter-carrier interference (ICI). Such interference becomes more severe as the subcarrier spacing is reduced.

To date, there are a few designs to combat the ICI problem in NFDM. Of course, the best error performance that NFDM[1] can achieve is by employing the maximum likelihood (ML) detector [1]. To avoid the very high (in many cases, impractical) complexity, alternative detection methods have been discussed in [2, 3]. These methods, however, suffer from large performance degradation.

Instead of examining low-complexity alternatives to the ML receiver, a recent work in [4] revisits the design of both the transmitter and the receiver in an NFDM

[1] Another common name for NFDM is spectrally efficient frequency-division multiplexing (SEFDM), which is used in [1].

N. H. Nguyen (✉) · B. Berscheid
Department of Electrical & Computer Engineering, University of Saskatchewan, Saskatoon, SK, Canada
e-mail: nghia.nguyen@usask.ca; brian.berscheid@usask.ca

H. H. Nguyen *(deceased)*
Saskatoon, SK, Canada

H. Nguyen et al. (eds.), *30th Biennial Symposium on Communications 2021*,
Signals and Communication Technology, https://doi.org/10.1007/978-3-031-06947-5_6

system. With the design proposed in [4], the singular-value decomposition (SVD) of the modified Fourier matrix is performed to obtain precoding and decoding matrices for the transmitter and receiver, respectively, which helps to completely eliminate ICI at the receiver in an AWGN channel. In particular, the design in [4] converts the NFDM system into a set of parallel independent subchannels, i.e., similar to OFDM, albeit with the number of effective channels less than the number of subcarriers. Nevertheless, compared to OFDM under the same bit error rate performance and transmit power, the design in [4] shows that NFDM can enjoy spectral efficiency enhancement of 23% for certain system configurations. Such spectral efficiency enhancement comes at very little complexity increase when compared to OFDM.

In a different line of research, the concept of index modulation (IM) has been extensively explored to increase the data rate of a conventional OFDM system [5]. In such a system, called OFDM-IM, instead of sending constellation symbols over all subcarriers at any given time, the patterns of activated subcarriers are determined by the so-called index bits. This means that a portion of information bits can be used to index, and hence be carried by, the subcarrier activation patterns. Although a portion of constellation bits are "lost" in those inactive subcarriers (when compared to OFDM), the loss could be well compensated by the amount of index bits. At the receiver, the ML detection can be employed, but only when the employed signal constellation is small and the number of subcarriers is not very large [6, 7]. Alternatively, the two-stage detection method discussed in [8] is a more practical choice, which sacrifices some performance degradation for a much simpler implementation.

The applicability of the IM concept is not just limited to OFDM. As a matter of fact, the integration of IM with NFDM has been considered in [9, 10]. It is important to point out that the studies in [9, 10] combine IM with the *conventional NFDM* and, as such, require the use of the highly complicated ML detection to adequately handle the inherent ICI. This also means that the systems in [9, 10] are limited to the use of small constellations such as BPSK and QPSK for reasons of computational complexity and therefore are only applicable to the very low spectral efficiency regime.

Given the clear advantages of the SVD-based NFDM design in [4] over the conventional NFDM, it is natural and interesting to combine it with IM. This is precisely the objective of this chapter. Thanks to the nonexistence of ICI in the SVD-based NFDM, we can apply the low-complexity two-stage detection method used in conventional OFDM-IM to the proposed NFDM-IM system. This involves detecting the subcarrier pattern first, followed by the demodulation of constellation symbols. By analyzing the potential sources of bit errors in the resulting system, we also design constellations to optimize the system performance. In particular, results show that the proposed NFDM-IM with an improved 8-QAM constellation enjoys 13% improvement compared to SVD-based NFDM in terms of spectral efficiency.

2 System Model

This part reviews the key signal processing steps of the NFDM design proposed in [4]. At the transmitter, most signal processing steps are similar to that in the conventional OFDM transmitter, except the use of a modified Fourier transform. With N subcarriers, this transform is represented by matrix $\mathbf{F}$, where the n, kth element is defined by $f_{n,k} = \frac{1}{\sqrt{N}} \exp(j2\pi\alpha nk/N)$, $n, k \in \{0, \ldots, N-1\}$ and $\alpha < 1$ is the compression factor. The value of α determines how much the subcarrier spacing in NFDM is compressed as compared to OFDM.

Let $\mathbf{F} = \mathbf{U}\mathbf{D}\mathbf{V}^{\mathrm{H}}$ be the singular-value decomposition of $\mathbf{F}$. In SVD-based NFDM design, only $N_a \leq N$ subchannels are active, where N_a is properly chosen to strike a balance between having well-behaved power spectral density and good spectral efficiency. Let $\mathbf{s} = [s_0, \ldots, s_{N_a-1}, 0, \ldots, 0]$ denote a length-N baseband symbol vector. This signal is precoded with matrix $\mathbf{V}$ and scaled by $\mathbf{G} = \mathrm{diag}\{[g_0, \ldots, g_{N-1}]\} = \mathbf{D}^{-1}$. Then, the transmit NFDM signal can be expressed as

$$\mathbf{y} = \mathbf{FVGs} = \mathbf{Us}. \tag{1}$$

Over an AWGN channel, the received signal is

$$\mathbf{r} = \mathbf{Us} + \mathbf{w}, \tag{2}$$

where $\mathbf{w} \sim \mathcal{CN}(0, N_0\mathbf{I})$. Then, one can apply the transformation $\mathbf{U}^{\mathrm{H}}$ on $\mathbf{r}$ to obtain $\hat{\mathbf{s}} = \mathbf{U}^{\mathrm{H}}\mathbf{r} = \mathbf{s} + \mathbf{U}^{\mathrm{H}}\mathbf{w}$. Since $\mathbf{U}$ is unitary, $\mathbf{U}^{\mathrm{H}}\mathbf{w}$ is also distributed as $\mathcal{CN}(0, N_0\mathbf{I})$. This means that, as far as demodulation of $\hat{\mathbf{s}}$ is concerned, there are N_a parallel subchannels that are independent from each other and have the same SNR. This important property inspires our proposal to combine SVD-based NFDM with index modulation.

Figure 1 shows the block diagram of the proposed combination of NFDM with index modulation, called NFDM-IM. First, λ information bits are separated into λ_1 constellation bits and λ_2 index bits, where $\lambda = \lambda_1 + \lambda_2$. The constellation bits are

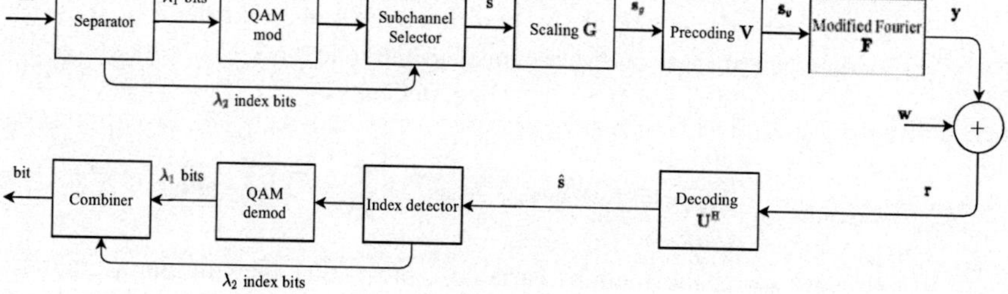

Fig. 1 Block diagram of the proposed NFDM-IM system

mapped to constellation symbols, while the index bits determine the set of K out of N_a subchannels, denoted as $\mathcal{K}$, to carry those constellation symbols. Obviously, the number of index bits is $\lambda_2 = \left\lfloor \log_2 \left(\frac{N_a!}{K!(N_a-K)!} \right) \right\rfloor$. The symbol vector $\mathbf{s}$ is $\mathbf{s} = [s_0, \ldots, s_{N_a-1}, 0, \ldots, 0]$, where s_i is an M-QAM symbol if $i \in \mathcal{K}$, and zero otherwise.

The remaining steps in the transmitter are the same as that in the SVD-based NFDM transmitter, and the output signal is given as $\mathbf{y} = \mathbf{U}\mathbf{s}$. The average NFDM-IM symbol energy can be calculated as

$$\mathbb{E}\{||\mathbf{y}||^2\} = \mathbb{E}\left\{||\mathbf{s}||^2\right\} = \left\{\sum_{i \in \mathcal{K}} \mathbb{E}|s_i|^2\right\} = K\bar{E}, \tag{3}$$

where $\bar{E}$ is the average symbol energy of the constellation.

As pointed out above, by applying the transform operation $\mathbf{U}^{\mathrm{H}}$ on $\mathbf{r}$, one obtains the set of N_a parallel subchannels. It follows that, similar to OFDM-IM [8, 11], the detection process can be carried out in two steps: The first step is to detect the set of active subcarriers, denoted as $\hat{\mathcal{K}}$, and the second step is to demodulate λ_1 constellation bits from these subcarriers. Note that this is markedly different from the complicated ML detection required for the conventional NFDM-IM system as in [10]. Specifically, the K active channels are detected by identifying the K channels having the largest powers among N_a available parallel subchannels. The activating pattern allows us to retrieve index bits, while constellation bits are obtained from the detected symbols on those channels.

In the next section, performance analysis of the proposed NFDM-IM is discussed, and designs of QAM constellations are presented to improve the system's error performance.

3 Performance Analysis and Constellation Design

3.1 Performance Analysis

Similar to the analysis in [11] for fast OFDM-IM, the overall error event is caused by errors made in one of the two detection stages. In the first stage, an error is made when at least one subchannel in the inactive set is marked as "active." The probability of this event can be calculated as

$$P_{\text{idx}} = P(X_{\min} < Z_{\max}) = \int_0^{\infty} F_{X_{\min}}(x) f_{Z_{\max}}(x) dx, \tag{4}$$

where $X_{\min}$ is the random variable representing the minimum magnitude of the received signals in the K active subchannels, while $Z_{\max}$ is the maximum value in the remaining $N_a - K$ inactive subchannels. The functions $F_X(\cdot)$ and $f_X(\cdot)$ generally denote the cdf and pdf of random variable X.

Let X and Z be the magnitudes of the received signals corresponding to any active and inactive subchannels, respectively. They can be shown to have the forms $X = |(a + jb) + w|$ and $Z = |0 + w|$, where $(a + jb)$ represents a constellation symbol and $w \sim \mathcal{CN}(0, N_0)$. The cdf of X can be obtained by averaging over M constellation symbols, which is

$$F_X(x) = \frac{1}{M} \sum_{m=1}^{M} 1 - Q_1(\mu_m, x). \tag{5}$$

In the above expression, $\mu_m = \sqrt{a_m^2 + b_m^2}$ is the Euclidean distance from the mth constellation symbol to the origin, and Q_1 is the Marcum Q-function. On the other hand, Z is Rayleigh distributed, whose pdf and cdf are $F_Z(x) = 1 - \exp(-x^2/N_0)$, and $f_Z(x) = \frac{x}{N_0/2} \exp(-x^2/N_0)$.

Note that

$$F_{X_{\min}}(x) = 1 - (1 - F_X(x))^K, \tag{6}$$

$$\begin{aligned} f_{Z_{\max}}(x) &= \frac{dF_{Z_{\max}}(x)}{dx} = \frac{d(F_Z(x))^{N_a - K}}{dx} \\ &= (N_a - K)(F_Z(x))^{N_a - K - 1} f_Z(x). \end{aligned} \tag{7}$$

Plugging $F_Z(x)$, $f_Z(x)$ and $F_X(x)$ into (4) yields

$$\begin{aligned} P_{\text{idx}} =& (N_a - K) \int_0^{\infty} \left(1 - (1 - \frac{1}{M} \sum_{m=1}^{M} 1 - Q_1(\mu_m, x))^K \right) \\ & (1 - \exp(-x^2/N_0))^{N_a - K - 1} \frac{x}{N_0/2} \exp(-x^2/N_0) dx, \end{aligned}$$

which can be numerically evaluated.

The overall bit error probability can be expressed as

$$\begin{aligned} P_e = & \underbrace{P_{\text{idx}} \left(\frac{\lambda_2}{2\lambda} + \frac{\lambda_1}{\lambda} \left(\left(\frac{K-2}{K} \right) P_{\text{const}} + \frac{1}{K} \right) \right)}_{P_{e,1}} \\ & + \underbrace{(1 - P_{\text{idx}}) \frac{\lambda_1}{\lambda} P_{\text{const}}}_{P_{e,2}}, \end{aligned} \tag{8}$$

where $P_{e,1}$ and $P_{e,2}$ are probabilities of error corresponding to the incorrect and correct index detection events, respectively, whereas P_{const} is the bit error probability of the constellation.

3.2 *Constellation Design*

With the two-step detection process, errors can be made at either or both steps. One may observe that the likelihood of incorrectly detecting which subcarriers are active is strongly influenced by the minimum distance from the origin to any constellation point (denoted as Δ_1), whereas the error in detecting the constellation bits is largely determined by the minimum distance between constellation points (denoted as Δ_2). As such, it makes sense and is beneficial to properly design a constellation to strike an optimum balance among the two types of errors, leading to a minimum value of P_e. In the context of low-rate communications, we shall focus on the design of small constellations[2] namely 8-QAM and 16-QAM.

3.2.1 Hexagonal 8-QAM

The original hexagonal 8-QAM constellation is shown in Fig. 2a, where $\Delta_1 = 0.5\Delta_2$. It is worth noting that this is the best among many 8-ary constellations in terms of error performance [12] when applied in a single-carrier system over an AWGN channel. The exact bit error probability for this constellation is given as [13]

$$P_{\text{const}} = \frac{1}{3}\left(CQ(\Delta_2/(2\sigma)) - C_K C_A\right), \tag{9}$$

where $\sigma = \sqrt{N_0/2}$. The constants $C = 13/4$, $C_K = 9/4$ are the average number of nearest neighbors of all constellation points and the average number of pairs of nearest neighbors, respectively. The coefficient C_A is the correction term to

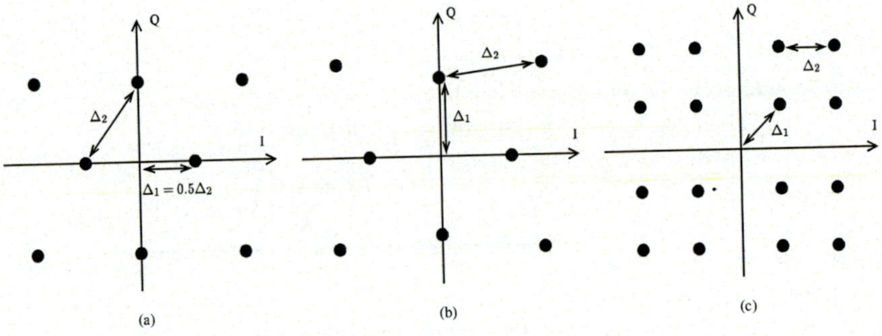

Fig. 2 Constellation designs for 8-QAM and 16-QAM. (**a**) Original hexagonal 8-QAM. (**b**) Improved hexagonal 8-QAM. (**c**) Improved 16-QAM

[2] Due to the perfect symmetry of BPSK and QPSK, there is no flexibility to optimize these two constellations.

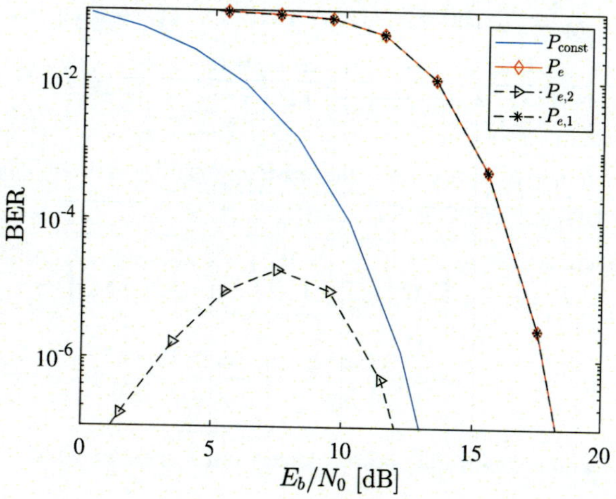

Fig. 3 Components of the error probability for NFDM-IM $\alpha = 0.8$ with conventional hexagonal 8-QAM

compensate for double counting of the overlap of a 3-PSK constellation and is given by

$$C_A = 2Q\left(\frac{\Delta_2}{2\sigma}\right)Q\left(\frac{\Delta_2}{2\sqrt{3}\sigma}\right) - \frac{2}{3}Q^2\left(\frac{\Delta_2}{\sqrt{6}\sigma}\right). \tag{10}$$

Figure 3 shows that there is a significant gap between $P_{e,1}$ and $P_{e,2}$, which implies that the overall error rate is dominated by the scenario where the active subcarriers are incorrectly detected. This result can intuitively be predicted by the short distance from the origin to its nearest points in the conventional hexagonal 8-QAM system. To improve P_e, an improved 8-QAM constellation is suggested as in Fig. 2b. It is designed such that $\Delta_2 \le \sqrt{2}\Delta_1$, which ensures Δ_2 is the smallest distance.

With the proposed 8-QAM design, the bit error probability of the constellation is given as

$$\begin{aligned} P_{\text{const}} &= \frac{1}{3}\left(\frac{1}{8}\sum_{m=1}^{M} P\{s_{m'\neq m}|s_m\} - B_K B_A\right) \\ &= \frac{1}{3}\Bigg(2Q\left(\frac{\Delta_2}{2\sigma}\right) + Q\left(\frac{\Delta_1}{\sqrt{2}\sigma}\right) \\ &\quad + 0.5Q\left(\frac{\Delta_1}{\sigma}\right) - B_K B_A\Bigg), \end{aligned} \tag{11}$$

where $B_K = C_K$ and

$$B_A \approx 2Q\left(\frac{\Delta_2}{2\sigma}\right)Q\left(\frac{\Delta_2}{2\sqrt{3}\sigma}\right) - \frac{2}{3}Q^2\left(\frac{\Delta_2}{\sqrt{6}\sigma}\right). \tag{12}$$

By defining $\beta = \Delta_1/\Delta_2$, the error probability term P_{const} in Eq. (8) can be expressed as a function of β and Δ_1 as

$$\begin{aligned} P_{\text{const}} =& \frac{1}{3}\Bigg(2Q\left(\frac{\Delta_2}{2\sigma}\right) + Q\left(\frac{\beta\Delta_2}{\sqrt{2}\sigma}\right) + 0.5Q\left(\frac{\beta\Delta_2}{\sigma}\right) \\ & - \frac{13}{2}Q\left(\frac{\Delta_2}{2\sigma}\right)Q\left(\frac{\Delta_2}{2\sqrt{3}\sigma}\right) - \frac{13}{6}Q^2\left(\frac{\Delta_2}{\sqrt{6}\sigma}\right)\Bigg). \end{aligned} \tag{13}$$

On the other hand, P_{idx} is calculated from $F_X(x)$, which is

$$\begin{aligned} F_X(x) &= \frac{1}{M}\sum_{m=1}^{M} 1 - Q_1(\mu_m, x) \\ &= 1 - \frac{1}{2}Q_1(\beta\Delta_2, x) - \frac{1}{2}Q_1\left(\frac{\Delta_2\sqrt{(2-\beta^2)} + \beta}{\sqrt{2}}, x\right). \end{aligned} \tag{14}$$

Under the average constellation energy constraint $\bar{E} = 0.5(\beta^2 + 1 + \beta\sqrt{2-\beta^2})\Delta_2^2$, the optimal value of β for each $\bar{E}$ can be found by a numerical search in order to minimize P_e.

3.2.2 Rectangular 16-QAM

The improved rectangular 16-QAM constellation is shown in Fig. 2c. The bit error probability of such a constellation can be calculated as

$$P_{\text{const}} = \frac{1 - (1 - P_{\text{const-1D}})^2}{\log_2 M}, \tag{15}$$

where $P_{\text{const-1D}}$ is the symbol error probability of the corresponding one-dimensional 4-ASK constellation. This error probability can be shown to be

$$P_{\text{const-1D}} = Q\left(\frac{\Delta_2}{2\sqrt{N_0/2}}\right) + \frac{2}{\sqrt{M}}Q\left(\frac{\beta\Delta_2}{\sqrt{N_0}}\right). \tag{16}$$

Under the average symbol energy constraint of $\bar{E}$, Δ_2 and β are related by

$$\Delta_2 = \frac{\bar{E}}{\beta^2 + \beta\sqrt{2} + 1}. \tag{17}$$

Furthermore, the adjustment of the symbol locations in the signal space changes the distribution of X to

$$\begin{aligned} F_X(x) &= \frac{1}{M}\sum_{m=1}^{M} 1 - Q_1(\mu_m, x) \\ &= 1 - \frac{1}{4}\Big[Q_1(\mu_1, x) + Q_1(\mu_2, x) \\ &\quad + Q_1(\mu_3, x) + Q_1(\mu_4, x)\Big], \end{aligned} \tag{18}$$

where

$$\mu_1 = \Delta_1 = \beta\Delta_2, \tag{19}$$

$$\mu_2 = \mu_3 = \Delta_2\sqrt{\frac{3}{2}\beta^2 + \beta\sqrt{2} + 1}, \tag{20}$$

$$\mu_4 = \Delta_2\sqrt{\beta^2 + 2\sqrt{2}\beta + 2}. \tag{21}$$

Again, numerical search is sufficient and simple enough to find the optimal ratio β to achieve the best error performance.

4 Numerical Results

In this section, numerical results for the proposed NFDM-IM with improved constellations are presented. The number of subcarriers is $N = 128$, the compression factor is $\alpha \in \{1, 0.5, 0.2\}$, and the subcarrier spacing is $\Delta_f = 15\,\text{kHz}$. Depending on the compression factor, the number of available subcarriers N_a is determined based on a heuristic rule: $g_i \leq 1$ with $i \in \{1, \ldots, N_a\}$. The rule is set to control the out-of-band emission of the transmitted signal in the frequency domain (see [4, 14] for further discussion). With N_a being the total number of subchannels available for indexing, the optimal number of subchannels that should be activated in a given symbol duration should be $K = \frac{M}{M+1}N_a$ in order to maximize the spectral efficiency [11], which is $\eta = \frac{\lambda}{T_s B}$, where $T_s = 1/\Delta_f$ is the symbol duration. Another important factor is the occupied bandwidth, which is approximated as $B \approx \alpha(N-1)\Delta_f + \Delta_f$.

Table 1 Comparison when $N = 128,\ M = 8$

	α	N_a	K	B (kHz)	λ	η (bits/s/Hz)
NFDM	1	128	n/a	1920	384	3.0
	0.5	64	n/a	967	192	2.97
	0.2	27	n/a	396	81	3.06
NFDM-IM	1	128	113	1920	384	3.14
	0.5	64	56	967	200	3.1
	0.2	27	24	396	92	3.48

Table 2 Comparison when $N = 128,\ M = 16$

	α	N_a	K	B (kHz)	λ	η (bits/s/Hz)
NFDM	1	128	n/a	1920	512	4
	0.5	64	n/a	967	256	3.96
	0.2	27	n/a	396	108	4.1
NFDM-IM	1	128	113	1920	515	4.02
	0.5	64	56	967	256	3.96
	0.2	27	24	396	107	4.05

Table 1 presents the comparison between SVD-based NFDM in [4] and the NFDM-IM (with the improved constellation) proposed in this chapter when $M = 8$. One can see that, with $\alpha = 0.2$, the highest bandwidth efficiency of 3.48 bits/s/Hz is achieved by the proposed NFDM-IM, while NFDM without index modulation yields 3.06 bits/s/Hz. However, when $M = 16$, the results presented in Table 2 show that, even by using the improved 16-QAM constellation, there is no benefit in spectral efficiency by exploiting index modulation. This is because with a higher constellation such as 16-QAM, the amount of index bits is not high enough to compensate for the reduction in constellation bits due to inactive subchannels in NFDM-IM.

Next, comparison of the bit error probabilities is made for the following three SVD-based systems: NFDM (P_{const}), NFDM-IM with conventional constellation (P_e), and NFDM-IM with the improved constellation ($P_{e}\text{–opt}$). For the results presented in Figs. 4 and 5, the compression factor is $\alpha = 0.2$. It can be observed that when $M = 8$, to obtain a BER of 10^{-6}, NFDM-IM with the regular 8-QAM requires $E_b/N_0 = 15.5$ dB, while NFDM-IM with the improved 8-QAM needs only 12.2 dB, i.e., there is 3.3 dB gain. In comparison with SVD-based NFDM, the NFDM-IM with improved 8-QAM basically achieves the same error rate, especially when $E_b/N_0 > 10$ dB. While having the same error performance, NFDM-IM with the improved constellation enjoys a 13% enhancement in spectral efficiency, specifically at 3.48 bits/s/Hz as compared to 3 bits/s/Hz of the SVD-based NFDM.

When $M = 16$, one can also observe a 1.1 dB gain when the improved constellation is used in NFDM-IM instead of the regular 16-QAM constellation (i.e., by comparing P_e and $P_{e}\text{–opt}$). However, there is still a gap of 0.2 dB for NFDM-IM with improved 16-QAM to match the performance of the SVD-based NFDM. Since

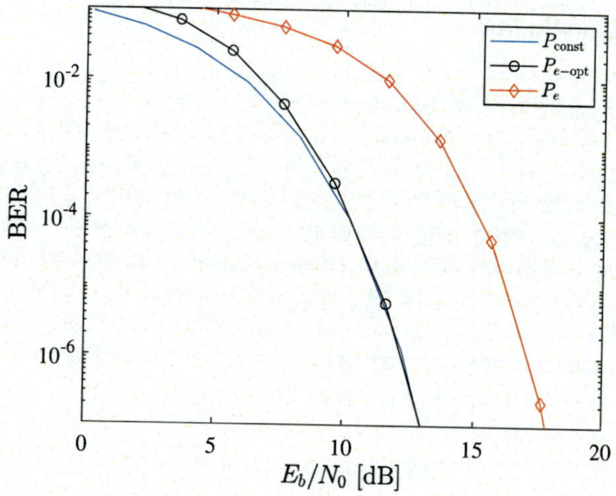

Fig. 4 BER comparison when $M = 8$

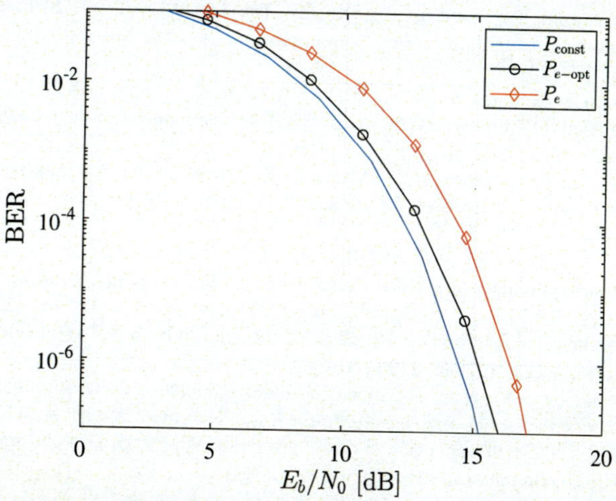

Fig. 5 BER comparison when $M = 16$

both the error performance and spectral efficiency of NFDM-IM are not better than SVD-based NFDM for $M = 16$, it is likely that there is no benefit in designing large constellations ($M \geq 16$) for NFDM-IM. As such, the proposed constellation optimization method is best suited for small constellations, which are often used in low-rate low-power applications.

5 Conclusion

In this chapter, the combination of SVD-based NFDM and index modulation was proposed and analyzed. The ICI that exists in the conventional NFDM is eliminated with SVD-based NFDM design, which allows simple signal detection when combined with index modulation. Furthermore, signal constellations are designed to improve the overall system error performance. The results show that at the target BER of 10^{-6}, a 13% enhancement in spectral efficiency can be obtained over SVD-based NFDM by using the proposed NFDM-IM with the improved 8-QAM constellation. In general, the application of index modulation to the frequency dimension in a multi-carrier system, whether it is OFDM or SVD-based NFDM, is only beneficial in low-rate low-power scenarios.

References

1. Darwazeh, I., Ghannam, H., Xu, T.: The first 15 years of SEFDM: A brief survey. In: International Symposium on Communication Systems, Networks & Digital Signal Processing, pp. 1–7 (2018)
2. Isam, S., Kanaras, I., Darwazeh, I.: A truncated SVD approach for fixed complexity spectrally efficient FDM receivers. In: IEEE Wireless Communications and Networking Conference, pp. 1584–1589 (2011)
3. Xu, T., Grammenos, R.C., Marvasti, F., Darwazeh, I.: An improved fixed sphere decoder employing soft decision for the detection of non-orthogonal signals. IEEE Commun. Lett. **17**, 1964–1967 (2013)
4. Nguyen, N.H., Nguyen, H.H., Berscheid, B.: SVD-based design for non-orthogonal frequency division multiplexing. IEEE Commun. Lett. **25**, 1343–1347. https://doi.org/10.1109/LCOMM.2020.3045719
5. Ishikawa, N., Sugiura, S., Hanzo, L.: Subcarrier-index modulation aided OFDM – Will it work? IEEE Access **4**, 2580–2593 (2016)
6. Abu-alhiga, R., Haas, H.: Subcarrier-index modulation OFDM. In: IEEE International Symposium on Personal, Indoor and Mobile Radio Communications, pp. 177–181 (2009)
7. Ko, Y.: A tight upper bound on bit error rate of joint OFDM and multi-carrier index keying. IEEE Commun. Lett. **18**, 1763–1766 (2014)
8. Frenger, P.K., Svensson, N.A.B.: Parallel combinatory OFDM signaling. IEEE Trans. Commun. **47**, 558–567 (1999)
9. Liu, H., Liu, L., Wang, P.: Spectrally efficient nonorthogonal frequency division multiplexing with index modulation. In: International Conference on Parallel and Distributed Computing, Applications and Technologies, pp. 290–293 (2016)
10. Nakao, M., Sugiura, S.: Spectrally efficient frequency division multiplexing with index-modulated non-orthogonal subcarriers. IEEE Wirel. Commun. Lett. **8**, 233–236 (2019)
11. Nguyen, N.H., Berscheid, B., Nguyen, H.H.: Fast-OFDM with index modulation for NB-IoT. IEEE Commun. Lett. **23**, 1157–1160 (2019)
12. Nguyen, H., Shwedyk, E.: A First Course in Digital Communications. Cambridge University Press, Cambridge (2009)
13. Rugini, L.: Symbol error probability of hexagonal QAM. IEEE Commun. Lett. **20**, 1523–1526 (2016)
14. Ganji, M., Zou, X., Jafarkhani, H.: A block-based non-orthogonal multicarrier scheme. In: 2019 IEEE Global Communications Conference (2019)

Data-Driven Measurement Matrix Design for Sparse Channel Reconstruction in Massive MIMO Systems

Pengxia Wu

1 Introduction

Downlink channel estimation is challenging for massive multiple-input multiple-output (MIMO) systems since the overhead of training pilots is proportional to the number of antennas at the base station (BS). To reduce the training pilot overhead, compressive sensing techniques have been employed to develop various sparse channel estimation schemes [1–6]. Since massive MIMO channels are also known to exhibit sparsity feature in angular domain (beamspace) [7], high-dimensional beamspace channels can be accurately reconstructed from a small number of received pilots based on the compressive sensing theory [8]. In such compressive sensing-aided beamspace channel estimation schemes, the transmit pilot matrix is associated with the measurement matrix, which has an important impact on the channel reconstruction performances. Randomized measurement matrices drawn from Gaussian and Bernoulli distribution are commonly adopted in the existing beamspace channel estimation schemes. However, these random matrices have been criticized for the suboptimal reconstruction performance, especially when the measurements for reconstructions are insufficient. It is undesirable to use more measurements to improve reconstructions since a larger number of measurements correspond to the heavier training pilot overhead. Instead, one can optimize the measurement matrix to improve reconstructions without increasing training pilot overhead.

The measurement matrix performs a linear dimensionality reduction by projecting high-dimensional sparse vectors onto the subspace spanned by the columns of measurement matrix. Real-world sparse datasets often have certain underlying

P. Wu (✉)
School of Engineering, The University of British Columbia, Kelowna, BC, Canada
e-mail: pengxia.wu@ubc.ca

H. Nguyen et al. (eds.), *30th Biennial Symposium on Communications 2021*,
Signals and Communication Technology, https://doi.org/10.1007/978-3-031-06947-5_7

structural features beyond simple sparsity [9]. The random matrices perform random projections, which are incapable of finding a compact subspace that can retain most useful information by exploiting the underlying structural features of the sparse dataset. Fortunately, given enough training data, deep learning techniques have the advantage in exploring the hidden structural features. The data-driven measurement matrices have been investigated for real-valued nonnegative data such as the image datasets [9–11] and have better performances than the widely adopted random measurement matrices. However, the proposed approaches cannot be applied to the general channel datasets for massive MIMO systems in a straightforward way.

In this chapter, we propose a general deep learning model to design the data-driven measurement matrix. More specifically, we propose a generic framework of deep compressive sensing autoencoder, which mimics the compressive sensing process including the linear dimensionality reduction and sparse reconstruction. We use the deep unfolding method to transform an iterative, model-based sparse recovery algorithm into a deep network whose parameters can be learned from training data as the decoder. The measurement matrix is assigned to be the trainable parameters of the whole autoencoder, such that the measurement matrix can be automatically learned from the training dataset of beamspace channel vectors. Moreover, we introduce the residual learning module to improve the training performance of the autoencoder. The acquired data-driven measurement matrix has meaningful applications in mmWave massive MIMO systems since the measurement matrix is associated with the pilot matrix for downlink channel estimations. Simulation results show the acquired data-driven measurement matrix performs well with the original model-based sparse recovery algorithm and has better reconstruction performances than benchmark random matrices.

2 System Model

We consider a downlink massive MIMO system, where the BS has N antennas and each user equipment (UE) has a single antenna. We use $\mathbf{h}^s \in \mathbb{C}^N$ to denote the spatial-domain channel between the BS and a UE, and let $\mathbf{h} \in \mathbb{C}^N$ denote the channel in virtual angular domain. Assuming a narrowband block-fading channel, the spatial-domain channel vector $\mathbf{h}^s$ is given by Tse and Pramod [12]

$$\mathbf{h}^s = \sqrt{\frac{N}{N_p}} \sum_{l=1}^{N_p} \beta^{(l)} \boldsymbol{\alpha}(\phi^{(l)}), \tag{1}$$

where N_p is the number of paths; $l = 1$ is the index for the line-of-sight path; $2 \leq l \leq N_p$ is the index for non-line-of-sight paths; $\beta^{(l)}$ is the complex path gain; $\boldsymbol{\alpha}(\phi^{(l)})$ is the corresponding array steering vector that contains a list of complex spatial sinusoids to represent the relative phase shifts for the incident far-field waveform across the array elements. For the N-element uniform linear array, the array steering vector $\boldsymbol{\alpha}(\phi^{(l)})$ is given by

$$\boldsymbol{\alpha}(\phi^{(l)}) = \frac{1}{\sqrt{N}}[1, e^{-j2\pi\phi^{(l)}}, \ldots, e^{-j2\pi\phi^{(l)}(N-1)}]^T, \tag{2}$$

where $\phi^{(l)}$ denotes the spatial direction of the lth path, and it is related to the physical angle $\theta^{(l)}$ by $\phi^{(l)} = \frac{d}{\lambda}\sin\theta^{(l)}$ for $-\frac{1}{2} \le \phi^{(l)} \le \frac{1}{2}$ and $-\frac{\pi}{2} \le \theta^{(l)} \le \frac{\pi}{2}$, where λ is the wavelength, and $d = \frac{\lambda}{2}$ is the antenna spacing.

The spatial-domain channel vector $\mathbf{h}^s$ in (1) can be transformed into the virtual angular domain by Tse and Pramod [12]

$$\mathbf{h} = \mathbf{U}\mathbf{h}^s, \tag{3}$$

where $\mathbf{U}$ denotes the discrete Fourier transform (DFT) matrix having the size of $N \times N$, and it can be expressed using a set of orthogonal array steering vectors as

$$\mathbf{U} = [\boldsymbol{\alpha}(\phi_1), \boldsymbol{\alpha}(\phi_2), \ldots, \boldsymbol{\alpha}(\phi_N)]^H, \tag{4}$$

where $\phi_i = \frac{1}{N}(i - \frac{N+1}{2})$ for $i = 1, 2, \ldots, N$ is the spatial direction predefined by the array having half-wavelength spaced antennas. The beamspace channel vector $\mathbf{h}$ exhibits sparsity feature. Most of the channel coefficients in the beamspace channel vector $\mathbf{h}$ are zero or nearly zero.

For the pilot-aided downlink channel estimation schemes, the BS transmits the known pilots $\mathbf{P}$ to the UEs. The received pilot symbols at the UE can be expressed as [13]

$$\mathbf{r} = \mathbf{P}\mathbf{h}^s + \mathbf{w}, \tag{5}$$

where $\mathbf{r} \in \mathbb{C}^M$ is the received pilot vector; $\mathbf{P} \in \mathbb{C}^{M\times N}$ is the downlink pilot matrix transmitted over M time slots; $\mathbf{h}^s \in \mathbb{C}^N$ is the spatial-domain channel vector; $\mathbf{w}$ is the received noise vector and $\mathbf{w} \sim \mathcal{CN}(0, \sigma_n^2\mathbf{I})$. According to (3), the channel vector $\mathbf{h}^s$ in (5) can be replaced by $\mathbf{h}^s = \mathbf{U}^H\mathbf{h}$. Thus, the received pilot symbols in (5) can be rewritten as

$$\begin{aligned}\mathbf{r} &= \underbrace{\mathbf{P}\mathbf{U}^H}_{\boldsymbol{\Phi}}\mathbf{h} + \mathbf{w} \\ &= \boldsymbol{\Phi}\mathbf{h} + \mathbf{w},\end{aligned} \tag{6}$$

where $\boldsymbol{\Phi}$ is the measurement matrix, which relates to the pilot matrix $\mathbf{P}$ by $\boldsymbol{\Phi} = \mathbf{P}\mathbf{U}^H$. Once we have a well-designed measurement matrix $\boldsymbol{\Phi}$, the pilot matrix $\mathbf{P}$ can be simply obtained by multiplexing a DFT matrix $\mathbf{U}$, i.e., $\mathbf{P} = \boldsymbol{\Phi}\mathbf{U}$.

The goal of the beamspace channel estimation is to reconstruct the sparse channel vector $\mathbf{h} \in \mathbb{C}^N$ from the lower-dimensional measurements $\mathbf{r} \in \mathbb{C}^M$ for $M \ll N$ with the known $\boldsymbol{\Phi}$. To this end, we can adopt the real-valued matrix $\boldsymbol{\Phi} \in \mathbb{R}^{M\times N}$ by forcing the imaginary part to be zeros, i.e., $\Im(\mathbf{\Phi}) = \mathbf{0}$, $\mathbf{\Phi} = \Re(\mathbf{\Phi})$, where $\Re(\cdot)$ and $\Im(\cdot)$, respectively, denote the real and imaginary parts of a complex vector. Thus, the computation complexity can be reduced to a half of the original complex-valued matrix-vector multiplication $\mathbf{\Phi h}$. Equation (6) can be expressed as

$$\begin{bmatrix} \Re(\mathbf{r}) \\ \Im(\mathbf{r}) \end{bmatrix} = \begin{bmatrix} \mathbf{\Phi} & \mathbf{0} \\ \mathbf{0} & \mathbf{\Phi} \end{bmatrix} \begin{bmatrix} \Re(\mathbf{h}) \\ \Im(\mathbf{h}) \end{bmatrix} + \begin{bmatrix} \Re(\mathbf{w}) \\ \Im(\mathbf{w}) \end{bmatrix}. \tag{7}$$

Equation (7) can be expressed as two independent real-valued linear equations as

$$\begin{aligned} \Re(\mathbf{r}) &= \mathbf{\Phi}\Re(\mathbf{h}) + \Re(\mathbf{w}) \\ \Im(\mathbf{r}) &= \mathbf{\Phi}\Im(\mathbf{h}) + \Im(\mathbf{w}). \end{aligned} \tag{8}$$

Equation (8) implies that we can treat a complex-valued channel sample $\mathbf{h}$ as two independent real-valued vectors $\Re(\mathbf{h})$ and $\Im(\mathbf{h})$. In the remainder of this chapter, we do not make particular distinctions for the real or imaginary part of $\mathbf{h}$, and we uniquely refer $\mathbf{x}$ as a sparse channel vector instead. Therefore, we can express the two equations in (8) as

$$\mathbf{y} = \mathbf{\Phi x} + \mathbf{n}, \tag{9}$$

where $\mathbf{y} \in \mathbb{R}^M$ is the measurements; $\mathbf{\Phi} \in \mathbb{R}^{M \times N}$ is the measurement matrix; $\mathbf{x} \in \mathbb{R}^N$ represents the sparse channel vector to be estimated; $\mathbf{n} \in \mathbb{R}^M$ is the Gaussian noise vector and $\mathbf{n} \sim \mathcal{N}(0, \sigma_n^2\mathbf{I})$.

3 Learning a Measurement Matrix via Deep Compressive Sensing Autoencoder

In this section, we first propose a generic framework of deep compressive sensing autoencoder. Then, following this framework, we propose two specific autoencoder models constructed by the deep unfolding of the iterations of ℓ_1-minimization sparse reconstruction. The measurement matrix is assigned as the trainable parameters of the constructed deep compressive autoencoders, such that the data-driven measurement matrix can be acquired by training the autoencoders on given dataset.

3.1 *Framework of Deep Compressive Sensing Autoencoder*

Both a compressive sensing process and an autoencoder have the similar "encoding-decoding" structure such that these two techniques can be incorporated to design a particular kind of autoencoder, namely the deep compressive sensing autoencoder.

A complete compressive sensing process involves two phases: the linear encoding or compression and the nonlinear decoding or sparse recovery. In the phase of linear compression, the measurement matrix is used to project sparse vectors onto a compact subspace and get the measurements that can best represent the sparse vectors. In the phase of sparse recovery, the measurement matrix participates in computations of the sparse reconstruction algorithms. Therefore, a complete compressive sensing process can be interpreted as a special kind of autoencoder that consists of a linear encoder and a sparse reconstruction decoder, and the "encoder" and the "decoder" can be explicitly defined as

$$
\begin{aligned}
g(\cdot) &:= \mathbf{\Phi x}, \\
f(\cdot) &:= \underset{\mathbf{x}}{\operatorname{argmin}} \|\mathbf{x}\|_1 \quad \text{s.t.} \quad \mathbf{\Phi x} = \mathbf{y}.
\end{aligned}
$$

Based on this observation, we propose a framework of deep compressive sensing autoencoder, which mimics the process of compressive sensing under the autoencoder architecture. The encoder performs linear compression, and the decoder solves the ℓ_1-minimization sparse reconstruction. More importantly, we parameterize the deep compressive sensing autoencoder with measurement matrix $\mathbf{\Phi}$ as the trainable parameters for optimizing the measurement matrix. Similar to a standard autoencoder that automatically optimizes the trainable parameters through training with given dataset by backpropagation, the compressive sensing autoencoder also optimizes the measurement matrix $\mathbf{\Phi}$ similarly. The optimization objective is to minimize the average input–output reconstruction errors, and the training optimizer can adopt the standard deep learning techniques such as the stochastic gradient descent (SGD) algorithm based on backpropagation [14]. The proposed deep compressive sensing autoencoder has an interpretable network structure. More specifically, from the viewpoint of forward computation, the compressive sensing autoencoder performs the linear dimensionality reduction jointly with the unfolded sparse reconstruction, such that it can be regarded as a complete compressive sensing process. From the viewpoint of backward propagation, the autoencoder back propagates and minimizes the reconstruction errors with the measurement matrix as customized weights, such that it can be regarded as an optimizer of the measurement matrix.

3.2 Architecture of Deep Compressive Sensing Autoencoders

The overall network architecture of deep compressed sensing autoencoder is shown in Fig. 1. The encoder is simply a linear fully connected layer $\mathbf{y} = \mathbf{\Phi x}$ without an activation function. The compressed measurement $\mathbf{y}$ is the input of the decoder, and the first-layer decoder is set as $\mathbf{x}^{(1)} = \mathbf{\Phi}^T\mathbf{y}$. From the first-layer decoder $\mathbf{x}^{(1)}$ to the output layer of the reconstruction $\hat{\mathbf{x}}$, the network can be regarded as a stack of iterative updates $\mathbf{x}^{(t+1)} = f_{\mathbf{\Phi}}(\mathbf{x}^{(t)}, \mathbf{y})$ for $1 \leq t \leq L-1$. This iterative

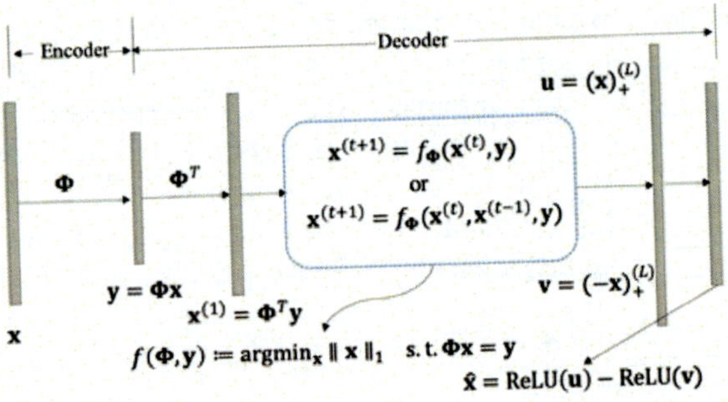

Fig. 1 Structure diagram of deep compressed sensing autoencoder

update is defined according to the iterative solution of sparse reconstruction. This method of transforming an iterative solution to a deep neural network is called deep unfolding [15]. Deep unfolding regards each step of iterations as a single layer of neural network to transform a model-based iterative algorithm into a stack layer of neural networks. Unlike the original iterative algorithms having predefined stopping conditions and manually tuning parameters, the unfolded neural networks have a fixed number of iterative layers, and the trainable parameters are automatically learned based on the training data. To obtain an unfolding architecture of the decoder, we need to derive an iterative solution of sparse reconstruction.

The sparse reconstruction is an ℓ_0-norm minimization optimization problem [1]

$$\min_{\mathbf{x}} \|\mathbf{x}\|_0 \qquad \text{s.t.} \quad \mathbf{\Phi x} = \mathbf{y}, \tag{10}$$

where $\|\cdot\|_0$ represents the zero norm, which is defined as the number of nonzero elements in a vector. The combinatorial optimization problem (10) is NP-hard[16]. A common method is to relax the problem (10) into a convex ℓ_1-minimization optimization problem as

$$\min_{\mathbf{x}} \|\mathbf{x}\|_1 \qquad \text{s.t.} \quad \mathbf{\Phi x} = \mathbf{y}, \tag{11}$$

where $\|\cdot\|_1$ represents the one norm, which is defined as the sum of the absolute values of all elements in a vector. The projected subgradient descent can be used to solve the problem (11), and the update is given by Boyd et al. [17]

$$\mathbf{x}^{(t+1)} = \mathcal{P}\big(\mathbf{x}^{(t)} - \alpha_t \cdot \text{sign}(\mathbf{x}^{(t)})\big), \tag{12}$$

where t indicates the tth update for $t > 0$; α_t is the step size; sign$(\cdot)$ represents the Signum function, which is the subgradient of $\|\cdot\|_1$; $\mathcal{P}(\cdot)$ represents the projection operation onto the convex set $\{\mathbf{x}' : \mathbf{\Phi x}' = \mathbf{y}\}$. The projection of a vector $\mathbf{z}$ onto the set $\{\mathbf{x}' : \mathbf{\Phi x}' = \mathbf{y}\}$ has the following closed-form solution:

[1] We consider the noiseless case, and the extension to noisy case is left for future work.

$$\mathcal{P}(\mathbf{z}) = \mathbf{z} + \mathbf{\Phi}^{\dagger}(\mathbf{y} - \mathbf{\Phi}\mathbf{z}), \tag{13}$$

where $\mathbf{\Phi}^{\dagger} = \mathbf{\Phi}^T(\mathbf{\Phi}\mathbf{\Phi}^T)^{-1}$ is the Moore–Penrose pseudo-inverse of $\mathbf{\Phi}$. We substitute $\mathbf{z} = \mathbf{x}^{(t)} - \alpha_t \cdot \text{sign}(\mathbf{x}^{(t)})$ from (12) into the projection operation (13) and obtain the tth-step update as

$$\mathbf{x}^{(t+1)} = \mathbf{x}^{(t)} + \mathbf{\Phi}^{\dagger}(\mathbf{y} - \mathbf{\Phi}\mathbf{x}^{(t)}) - \alpha_t(\mathbf{I} - \mathbf{\Phi}^{\dagger}\mathbf{\Phi}) \cdot \text{sign}(\mathbf{x}^{(t)}). \tag{14}$$

The update (14) can be regarded a simple iterative sparse reconstruction algorithm for reconstructing $\hat{\mathbf{x}}$, given a proper starting point $\mathbf{x}^{(1)}$, step size α_t, and stop condition. We unfold this iteration algorithm for constructing the decoder of compressed sensing autoencoder in Fig. 1. More precisely, we use the update (14) to define the forward computations of the $(t+1)$th-layer decoder. Before doing so, we make the following adjustments:

- The pseudo-inverse $\mathbf{\Phi}^{\dagger}$ is replaced by the transpose $\mathbf{\Phi}^T$ to release intensive computations.
- The step size parameter is set as $\alpha_t = \alpha/t$ according to the diminishing step size rule [17].

In this way, the update (14) becomes

$$\begin{aligned}\mathbf{x}^{(t+1)} &= \mathbf{x}^{(t)} + \mathbf{\Phi}^T\mathbf{y} - \mathbf{\Phi}^T\mathbf{\Phi}\mathbf{x}^{(t)} \\ &\quad -(\alpha/t)(\mathbf{I} - \mathbf{\Phi}^T\mathbf{\Phi}) \cdot \text{sign}(\mathbf{x}^{(t)}).\end{aligned} \tag{15}$$

Thus, the computational graph[2] of the $(t+1)$th-layer decoder $\mathbf{x}^{(t+1)} = f_{\mathbf{\Phi}}(\mathbf{x}^{(t)}, \mathbf{y})$ for $1 \leq t \leq L-1$ would be defined basically based on the update (15). We note that the measurements $\mathbf{y}$ participates in the computations of update (15). By treating the measurements $\mathbf{y}$ in (15) in different ways, we can define two kinds of decoders with different computation graphs. Implementing these two distinct decoder structures in the deep compressed sensing autoencoder in Fig. 1, we can construct two autoencoder models, namely the ℓ_1-minimization simplified autoencoder (ℓ_1-SAE) and the ℓ_1-minimization generic autoencoder (ℓ_1-GAE).

ℓ_1-SAE Decoder By substituting $\mathbf{y} = \mathbf{\Phi}\mathbf{x}^{(t)}$ in update (15), we obtain

$$\mathbf{x}^{(t+1)} = \mathbf{x}^{(t)} - (\alpha/t)(\mathbf{I} - \mathbf{\Phi}^T\mathbf{\Phi}) \cdot \text{sign}(\mathbf{x}^{(t)}), \tag{16}$$

which represents the tth-layer computation of the ℓ_1-SAE decoder. The computation graph of iteration (16) for the tth-layer decoder of ℓ_1-SAE is shown in Fig. 2.

ℓ_1-GAE Decoder By treating $\mathbf{y} = \mathbf{\Phi}\mathbf{x}^{(t-1)}$, the update (15) becomes

[2] In deep learning, a computation graph is often used to represent a math function in the language of graph theory.

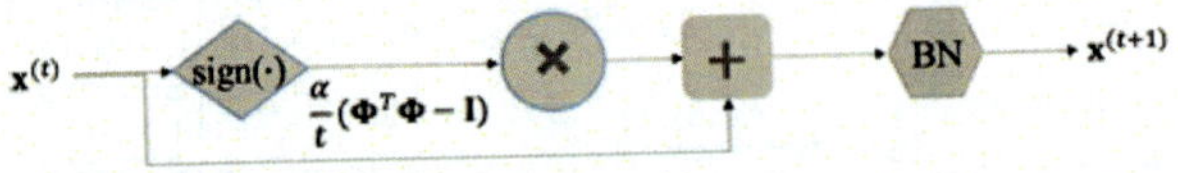

Fig. 2 Computation graph structure of ℓ_1-SAE decoder for $\mathbf{x}^{(t+1)} = f_{\Phi}(\mathbf{x}^{(t)}, \mathbf{y})$ for $0 \leq t \leq L-1$; the module BN represents batch normalization

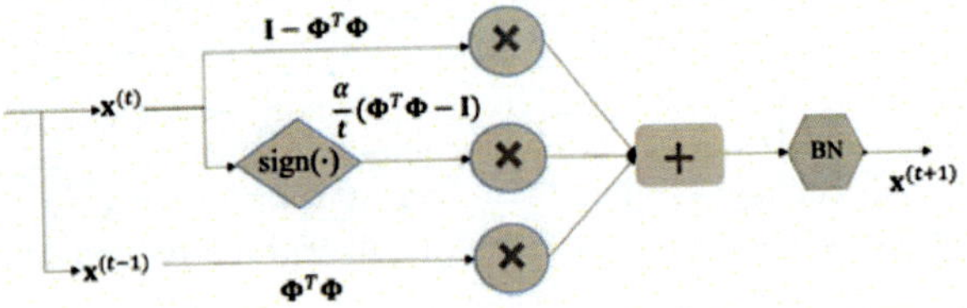

Fig. 3 Computation graph structure of ℓ_1-GAE decoder for $\mathbf{x}^{(t+1)} = f_{\Phi}(\mathbf{x}^{(t)}, \mathbf{y})$ for $0 \leq t \leq L-1$; the module BN represents batch normalization

$$\begin{aligned} \mathbf{x}^{(t+1)} = & \ \mathbf{x}^{(t)} + \Phi^T \Phi \mathbf{x}^{(t-1)} - \Phi^T \Phi \mathbf{x}^{(t)} \\ & -(\alpha/t)(\mathbf{I} - \Phi^T \Phi) \cdot \text{sign}(\mathbf{x}^{(t)}). \end{aligned} \tag{17}$$

The iteration (17) represents the computation of the tth-layer decoder of the ℓ_1-GAE. The computation graph of iteration (17) for the tth-layer decoder of ℓ_1-GAE is shown in Fig. 3. The benefit of treating $\mathbf{y}$ as $\Phi\mathbf{x}^{(t-1)}$ comes from the role of $\mathbf{x}^{(t-1)}$, which can be understood as introducing the residual learning modules [18] to improve the autoencoder performance.

Through $L-1$ layers of decoding computations, we obtain the channel vector $\mathbf{x}^{(L)}$ at the Lth-layer decoder. We represent $\mathbf{x}^{(L)}$ using the difference of its positive and negative parts, that is

$$\begin{aligned} \mathbf{x}^{(L)} &= \mathbf{u} - \mathbf{v} \\ \mathbf{u} &= \mathbf{x}_+^{(L)}, \mathbf{v} = (-\mathbf{x}^{(L)})_+, \end{aligned} \tag{18}$$

where $(\cdot)_+$ denotes the positive-taking operator that keeps the positive components of a vector and sets other elements be zeros. More specifically, $\mathbf{x}_+^{(L)}$ indicates $(\mathbf{x}_+^{(L)})_i = \max\{\mathbf{x}_i, 0\}$ for $i = 1, \cdots, N$, and $(-\mathbf{x}^{(L)})_+$ indicates $((-\mathbf{x}^{(L)})_+)_i = \max\{-\mathbf{x}_i^{(L)}, 0\}$ for $i = 1, \cdots, N$, where $(\cdot)_+$ denotes the ith element of a vector. The ReLU [3] activation function is added on the nonnegative vectors $\mathbf{u}$ and $\mathbf{v}$. Then we can obtain the output of the decoder, which is also the finally reconstructed channel vector and is expressed as

[3] ReLU stands for rectified linear unit and is a type of activation function. Mathematically, it is defined as $\text{ReLU}(x) = \max(0, x)$ when x is a scale. When the input is a vector, $\text{ReLU}(\cdot)$ is applied in an element-wise manner.

$$\hat{\mathbf{x}} = \mathrm{ReLU}\,(\mathbf{u}) - \mathrm{ReLU}\,(\mathbf{v})\,. \tag{19}$$

It is worth to mention that the ℓ_1-AE proposed in [9, 11] can be regarded as the special case of ℓ_1-SAE when the data $\mathbf{x}$ is nonnegative, so that the variable splitting (18) is not required.

The loss function of the autoencoders is defined as the mean square ℓ_2-error (MSE) between input samples $\mathbf{x}$ and reconstructed output vectors $\hat{\mathbf{x}}$

$$L(\mathbf{x}, \hat{\mathbf{x}}) = \frac{1}{n}\sum_{i=1}^{n} \|\mathbf{x} - \hat{\mathbf{x}}\|_2^2, \tag{20}$$

where n is the number of training samples. The encoder and decoder of the autoencoder are jointly optimized on given dataset to minimize the reconstruction error (20). The trainable variables are $\{\boldsymbol{\Phi}, \alpha\}$. During training, the measurement matrix is $\boldsymbol{\Phi}$ as the trainable variables are automatically tuned to be optimal based on the given dataset. Therefore, the obtained measurement matrix is data-driven and can adapt to the dataset features. After training, the optimal data-driven measurement matrix $\boldsymbol{\Phi} \in \mathbb{R}^{M\times N}$ can be extracted from the trained autoencoder models.

4 Numerical Results

We consider a massive MIMO system having 256 antennas at the BS and a single antenna at the UE. The 20,000 spatial-domain channel realizations are randomly generated according to the channel model (1), where the number of scattering clusters is set as $N_p = 3$. We neglect the power leakages and transform the generated channel vectors into the s-sparse beamspace channel vectors with sparsity level $s = 3$, which means we only care about the nonzero elements with largest three magnitudes in the beamspace channel vectors. We split the real part and imaginary part of the complex beamspace channel vectors and obtain a dataset consisting of 40,000 real-form sparse channel vectors. Then we split the dataset into training, validation, and test dataset by the ratio of 0.8/0.1/0.1. The depth of decoder is set to be 10 layers, i.e., $L = 10$. The trainable measurement matrix $\boldsymbol{\Phi}$ is initialized by the truncated normal distribution with standard deviation $\sigma = 1/\sqrt{256}$. The step size is α as another trainable parameter is initialized by $\alpha = 1.0$, and its value will be automatically updated to an appropriate value during training. The SGD is used as the training optimizer. The experiments are repeated with different compressed dimensions M, which is taken from 6 to 30 with the interval 3.

We use $\boldsymbol{\Phi}_{\ell_1 s} \in \mathbb{R}^{M\times N}$ and $\boldsymbol{\Phi}_{\ell_1 g} \in \mathbb{R}^{M\times N}$ to denote the learned matrices by ℓ_1-SAE and ℓ_1-GAE. Four random matrices are adopted as baselines, including the random Gaussian matrix $\mathbf{G}$, the random Bernoulli matrix $\mathbf{B}$,[4] the partial Fourier

[4] For random Bernoulli matrix, entries are -1 or 1 with equal probability.

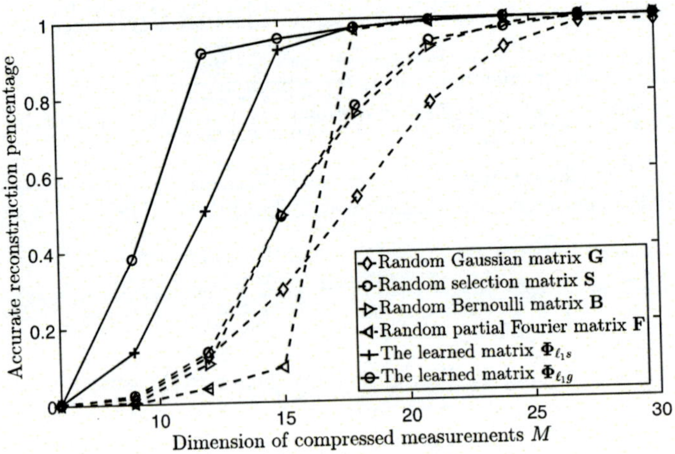

Fig. 4 Accurate reconstruction percentages of various measurement matrices

matrix **F**, and the random selection matrix **S**.[5] Reconstruction performances of the data-driven measurement matrices are evaluated and compared with random matrices for different compressed dimensions M. To evaluate the reconstruction performance of measurement matrices, we adopt the linear programming to solve the ℓ_1-minimization sparse reconstructions over test dataset. Metrics for evaluating the reconstruction performances include the root mean square error (RMSE) and the accurate reconstruction percentage. The RMSE is defined as the square root of mean squared error $\mathbb{E}[\|\mathbf{x} - \hat{\mathbf{x}}\|_2^2]$. The accurate reconstruction percentage is calculated as the ratio of the number of accurate reconstructed samples over the total number of samples in test dataset. A sample is counted to be reconstructed accurately if its reconstruction ℓ_2-norm error is not more than 10^{-8}, i.e., $\|\mathbf{x} - \hat{\mathbf{x}}\|_2 \leq 10^{-8}$. The results of accurate reconstruction percentages are shown in Fig. 4. The results of average reconstruction RMSE are shown in Fig. 5.

We first compare the performances of data-driven measurement matrices $\mathbf{\Phi}_{\ell_1 s} \in \mathbb{R}^{M \times N}$ and $\mathbf{\Phi}_{\ell_1 g} \in \mathbb{R}^{M \times N}$, which are acquired by training ℓ_1-SAE and ℓ_1-GAE. We can see in Fig. 4 the matrix $\mathbf{\Phi}_{\ell_1 g} \in \mathbb{R}^{M \times N}$ has a higher accurate reconstruction percentage than the matrix $\mathbf{\Phi}_{\ell_1 s} \in \mathbb{R}^{M \times N}$; in Fig. 5, the matrix $\mathbf{\Phi}_{\ell_1 g} \in \mathbb{R}^{M \times N}$ has a lower RMSE compared with matrix $\mathbf{\Phi}_{\ell_1 s} \in \mathbb{R}^{M \times N}$. This result confirms the proposed residual learning module in ℓ_1-GAE, namely the term $\mathbf{\Phi}^T \mathbf{\Phi} \mathbf{x}^{(t-1)}$ in (17), can significantly improve the reconstruction performance of the autoencoder.

Then, we compare the reconstruction performances of the data-driven measurement matrices with the random matrices. From Fig. 4, we can see the data-driven measurement matrices exhibit obvious higher accurate reconstruction percentages, especially for small compressed dimensions. For example, when $M = 12$, the

[5] For random selection matrix, entries are 0 or 1 with equal probability.

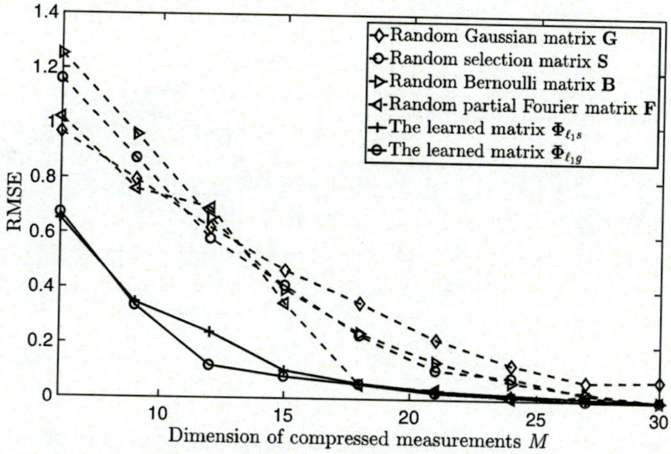

Fig. 5 Reconstruction RMSE of various measurement matrices

data-driven matrices $\mathbf{\Phi}_{\ell_1 g}$ achieve over 90% accurate reconstruction percentage, while the accurate reconstruction percentages of random matrices are below 15%. From Fig. 5, we can see the reconstruction errors of the data-driven matrices are overall lower than the random matrices. We can also infer the data-driven matrix can achieve accurate reconstructions using fewer measurements. For example, when achieving the 95% accurate reconstruction percentage, the least number of required measurements is $M = 15$ by the matrices $\mathbf{\Phi}_{\ell_1 g}$, while the least number of required measurements is $M = 21$ by the random partial Fourier matrix and $M = 24$ by the other random matrices. We can see using data-driven matrix can reduce the number of measurements by more than two-fold of sparsity levels.

Increasing the compressed dimension M will improve the reconstruction accuracy but will lead to a lower spectrum efficiency. To analyze the trade-off between the spectrum occupation and the reconstruction accuracy, following [6], we define the effective achievable rate as $R_e = R_0(1 - \frac{M}{B})P$, where R_0 denotes the maximal achievable rate for a user, $\frac{M}{B}$ denotes the pilot occupation ratio in transmission block, B is the block length that is set as 50 symbols, and P denotes the probability of successful recoveries. The effective achievable rates for various measurement matrices are compared and shown in Fig. 6. We can see the data-driven matrices have obviously higher effective achievable rates than the random matrices. The maximum effective achievable rates of data-driven matrices can be achieved at the lower compressed dimension M. The matrix $\mathbf{\Phi}_{\ell_1 g}$ achieves its maximal effective achievable rate at $M = 12$, and the matrix $\mathbf{\Phi}_{\ell_1 s}$ achieves its maximal effective achievable rate at $M = 15$. For random matrices, the maximal effective achievable rates occur around $M = 21$. Moreover, the maximum effective achievable rates of random matrices are lower than the maximum effective achievable rates achieved by the data-driven matrices.

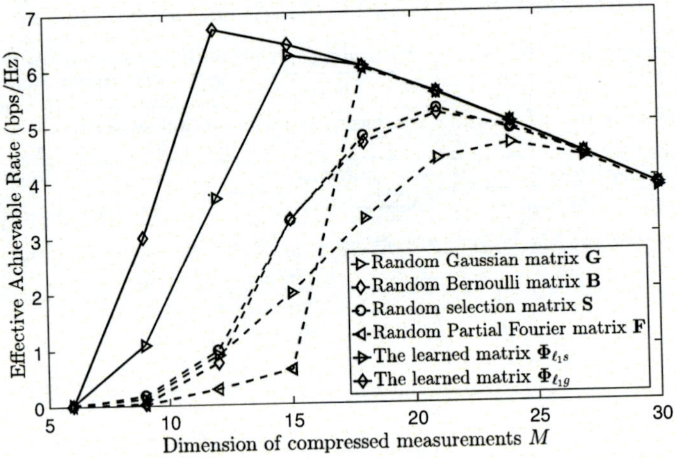

Fig. 6 Effective achievable rate using different measurement matrices

In summary, based on the comparisons of reconstruction performances including the accurate reconstruction percentages in Fig. 4 and the RMSE results in Fig. 5, we claim the data-driven matrices can achieve more accurate reconstructions using fewer measurements compared with the random matrices. This is meaningful for overhead reduction of transmit pilots for channel estimations in mmWave massive MIMO systems. As shown in the results of effective achievable rates in Fig. 6, using data-driven measurement matrices can significantly improve the effective achievable rate.

5 Conclusion

A generic framework of deep compressive sensing autoencoder was proposed to design the data-driven measurement matrix. The deep compressive sensing autoencoder consists of an encoder that is a fully connected layer for linear compression and a decoder that is a stack of unfolded iterative update of the ℓ_1-minimization sparse recovery. More importantly, we treat the measurement matrix as the trainable weights of both the encoder and decoder, such that the measurement matrix can be optimized based on the sparse channel dataset by back propagating the reconstruction errors. The acquired data-driven measurement matrix outperforms the random matrices in terms of reconstruction accuracy with the original ℓ_1-minimization sparse recovery algorithm. Since the measurement matrix is associated with the pilot matrix for channel estimation, the proposed data-driven approach provides valuable insights for designing the environment-adaptive pilots by deep learning with the assistance of large dataset in this big-data era. In this chapter, we considered the noiseless case when constructing the autoencoder, and in the future work, the noisy case can be extended to deal with the approximately sparse channel dataset.

References

1. Li, X., Fang, J., Li, H., Wang, P.: Millimeter wave channel estimation via exploiting joint sparse and low-rank structures. IEEE Trans. Wirel. Commun. **17**(2), 1123–1133 (2018)
2. Lin, X., Wu, S., Jiang, C., Kuang, L., Yan, J., Hanzo, L.: Estimation of broadband multiuser millimeter wave massive MIMO-OFDM channels by exploiting their sparse structure. IEEE Trans. Wirel. Commun. **17**(6), 3959–3973 (2018)
3. Gao, X., Dai, L., Han, S., Chih-Lin, I, Wang, X.: Reliable beamspace channel estimation for millimeter-wave massive MIMO systems with lens antenna array. IEEE Trans. Wirel. Commun. **16**(9), 6010–6021 (2017)
4. Gao, Z., Hu, C., Dai, L., Wang, Z.: Channel estimation for millimeter-wave massive MIMO with hybrid precoding over frequency-selective fading channels. IEEE Commun. Lett. **20**(6), 1259–1262 (2016)
5. Gao, Z., Dai, L., Wang, Z.: Channel estimation for mmWave massive MIMO based access and backhaul in ultra-dense network. In: IEEE International Conference on Communications (ICC), pp. 1–6 (2016)
6. Alkhateeb, A., Leus, G., Heath, R.W.: Compressed sensing based multi-user millimeter wave systems: How many measurements are needed? In: Proceedings of the IEEE International Conference on Acoustics, Speech, and Signal Processing, pp. 2909–2913 (2015)
7. Swindlehurst, A.L., Ayanoglu, E., Heydari, P., Capolino, F.: Millimeter-wave massive MIMO: the next wireless revolution? IEEE Commun. Mag. **52**(9), 56–62 (2014)
8. Bajwa, W.U., Haupt, J., Sayeed, A.M., Nowak, R.: Compressed channel sensing: a new approach to estimating sparse multipath channels. Proc. IEEE **98**(6), 1058–1076 (2010)
9. Wu, S., Dimakis, A.G., Sanghavi, S., Yu, F.X., Holtmann-Rice, D., Storcheus, D., Rostamizadeh, A., Kumar, S.: Learning a compressed sensing measurement matrix via gradient unrolling. In: Proceedings of the 36th International Conference on Machine Learning (PMLR), pp. 6828–6839 (2019)
10. Arora, S., Khodak, M., Saunshi, N., Vodrahalli, K.: A compressed sensing view of unsupervised text embeddings, bag-of-n-grams, and LSTMs. In: Proceedings of the International Conference on Learning Representations (2018)
11. Wu, P., Liu, Z., Cheng, J.: Learning a measurement matrix in compressed CSI feedback for millimeter wave massive MIMO systems (2019). Preprint, arXiv:1903.02127
12. Tse, D., Pramod, V.: Fundamentals of Wireless Communication. Cambridge University Press, Cambridge (2005)
13. Choi, J.W., Shim, B., Ding, Y., Rao, B., Kim, D.I.: Compressed sensing for wireless communications: Useful tips and tricks. IEEE Commun. Surveys Tut. **19**(3), 1527–1550 (2017)
14. Qin, Z., Ye, H., Li, G.Y., Juang, B.F.: Deep learning in physical layer communications. IEEE Wirel. Commun. **26**(2), 93–99 (2019)
15. Hershey, J.L.R.J., Weninger, F.: Deep unfolding: Model-based inspiration of novel deep architectures (2014). Preprint, arXiv:1409.2574
16. Amaldi, E., Kann, V.: On the approximability of minimizing nonzero variables or unsatisfied relations in linear systems. Theoret. Comput. Sci. **209**, 237–260 (1998)
17. Boyd, S., Xiao, L., Mutapcic, A.: Subgradient methods. Notes for EE392o Stanford University Autumn (2003–2004)
18. He, K., Zhang, X., Ren, S., Sun, J.: Deep residual learning for image recognition. In: Proceedings of the IEEE Conference on Computer Vision and Pattern Recognition (CVPR), pp. 770–778 (2016)

A Low-Complexity Splitting Receiver for Ultrasonic Index Modulation-Based Intra-Body Communications

Qianqian Wang, Quansheng Guan, and Fei Ji

1 Introduction

In healthcare systems, there are several revolutionary and promising applications, such as brain–computer interface, cardiac pacemaker, capsule endoscope, and nanorobot [1–3]. These applications all demand advanced intra-body communications (IBCs). Several technologies have been extensively investigated for IBCs, including radio frequency (RF) waves, optical communication, molecular communication, capacities coupling, and resonant coupling. Unfortunately, these technologies are limited by transmission distance and power, and thus they cannot achieve IBCs with high data rate at meaningful distances [1]. For instance, the widely used RF waves have severe attenuation because the water content of the human body is up to 65%, and the continuous RF radiation from implants imposes potential risks for the human body [2]; the transmission distance of optical and molecular communications for IBCs is limited to several millimeters [1, 3].

Q. Wang
School of Electronic and Information Engineering, South China University of Technology, Guangzhou, China

Key Laboratory of Marine Environmental Survey Technology and Application, Ministry of Natural Resources, Guangzhou, China

School of Engineering, The University of British Columbia, Kelowna, BC, Canada
e-mail: eeqqw@mail.scut.edu.cn

Q. Guan (✉) · F. Ji
School of Electronic and Information Engineering, South China University of Technology, Guangzhou, China

Key Laboratory of Marine Environmental Survey Technology and Application, Ministry of Natural Resources, Guangzhou, China
e-mail: eeqshguan@scut.edu.cn; eefeiji@scut.edu.cn

H. Nguyen et al. (eds.), *30th Biennial Symposium on Communications 2021*,
Signals and Communication Technology, https://doi.org/10.1007/978-3-031-06947-5_8

Compared with the aforementioned methods, the ultrasonic wave is promising to enable high-data-rate IBCs, up to several MHz at a much longer distance (about tens of centimeters) [2]. Ultrasonic wave, owing to off-the-shelf medical transceivers, has been applied to the diagnostic and therapeutic purposes for decades, showing clinical safety. It allows a much higher transmission power approved by FDA and lower signal attenuation than RF waves [4]. However, to implement ultrasonic IBCs, there are several challenges to overcome. First, the ubiquitous tissues and organs can cause reflection and refraction of the transmitted signals, resulting in severe multipath effects. Second, IBC has strict constraints on transmission power, implant size, and energy consumption since organs are vulnerable and battery replacement in the human body is costly.

Motivated by the fact that medical ultrasonic applications usually adopt ultrasonic pulses [2], ultrasonic pulse-based IBC provides an effective solution to the aforementioned challenges. To be specific, the multipath effect can be alleviated by properly setting the duty cycle of the pulse, and in this case, the impact of thermal and mechanical effects can be reduced [5]. Although the ultrasonic pulse-based IBC systems, including time-hopping ultrasonic wideband (TH-UsWB) [5] and direct-sequence UsWB [6], are safe and energy-efficient, they only achieve a data rate of hundreds of kbps. High data rates are required to enable versatile intra-body applications, such as real-time and high-definition video, or control small implanted devices.

By using the pulse index to convey additional information bits without consuming additional energy, ultrasonic index modulation (UsIM) can achieve IBCs with high data rate and high energy efficiency [7]. Unfortunately, the maximum likelihood (ML) receiver has high computational complexity, and thus, it is unsuitable for size- and energy-constrained IBCs. Although the maximum ratio combination (MRC) and the energy detection (ED) receivers are investigated with lower complexity than the ML receiver, both two receivers sacrifice bit-error rate (BER) performance [7].

Focusing on the detection of the ML, MRC, and ED receivers of UsIM, we find that the ML and MRC receivers employ coherent detection (CD) to detect the pulse amplitude and phase, and the ED receiver adopts non-coherent ED to detect the energy of the received signals. In general, the ultrasonic pulse signal space contains the degrees of freedom (DoFs) in amplitude, phase, and energy. These receivers exploit only partial dimensions of the received signal space [8]. Thus, the underexploitation of DoFs in detection will incur a loss in signal-to-noise ratio (SNR) for the received signal, and even the ML decision cannot achieve the optimal BER.

In this chapter, we introduce a splitting-detection (SD) structure [9, 10] for UsIM to exploit all DoFs and thus improve the BER performance. Specifically, the received signal is split into two streams that can be separately processed by CD and ED and then combined as joint decision variables to achieve diversity combining. Since integrating SD with the ML receiver will lead to higher hardware and computational complexity than the ML receiver, we consider integrating SD with the MRC receiver. Besides, we find that the absolution operation on the detected signal in the MRC receiver is unnecessary and even causes high BER when adopting pulse position modulation (PPM). Therefore, we propose to integrate the SD structure with a joint decision by the simplified-MRC (sMRC) receiver, which removes the absolution operation from MRC, namely SD-JsMRC. The main contributions are summarized as follows:

- An SD-JsMRC receiver is proposed for the UsIM-based IBCs. The integrated UsIM-SD-JsMRC system can achieve IBCs with high data rate, low BER, and low complexity, providing a promising scheme for future healthcare applications.
- The theoretical BER expression of UsIM-SD-JsMRC is derived. Extensive Monte Carlo simulations verify the theoretical analysis and show the superiority of SD-JsMRC over the ML, MRC, ED, and sMRC receivers in terms of lower BER.

The remainder of this chapter is organized as follows. Section 2 proposes the SD-JsMRC receiver for UsIM-based IBCs. Section 3 derives the theoretical BER and analyzes the system performance. The simulation results are explored in Sect. 4, and conclusions are drawn in Sect. 5.

2 SD-JsMRC

The UsIM-SD-JsMRC system, as shown in Fig. 1, contains three parts: transmitter, channel, and receiver. The transmitter employs UsIM to modulate the information bits and then generates ultrasonic pulses that propagate via the intra-body channel. The received signal is detected by SD and is then demodulated by the JsMRC receiver. The detailed signal processing in each block is illustrated in the sequel.

2.1 Transmitted Signal of UsIM

Suppose that the channel is organized by frames, and a frame consists of N_h chips. The information bits are divided into blocks with p bits, including p_1 *modulation*

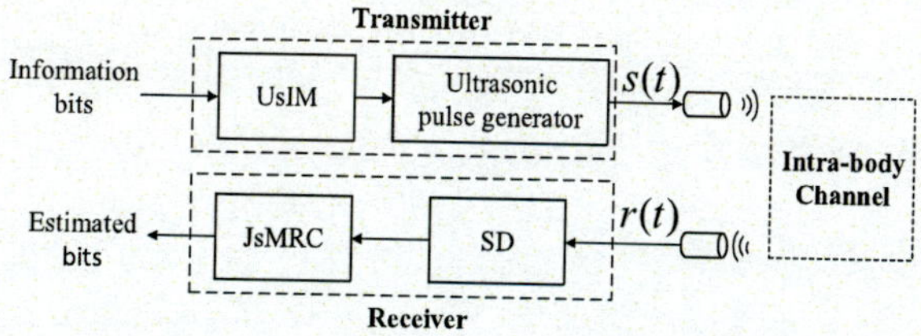

Fig. 1 System structure of the UsIM-SD-JsMRC for ultrasonic IBCs

bits and p_2 *index bits*. The index bits are conveyed by the chip indices. Hence, we have $p = p_1+p_2$. Assuming that one of N_h chips is activated to transmit modulation bits, the number of index bits and the indices of active chips have the relationship of $p_2 = \lfloor \log_2 \mathbb{C}(N_h, 1) \rfloor$, where $\lfloor \cdot \rfloor$ is the floor function, and $\mathbb{C}(\cdot, \cdot)$ denotes the binomial coefficient. For brevity, let $N_h = 2^{p_2}$.

The active chip uses orthogonal binary PPM (i.e., $p_1 = 1$) to modulate the modulation bit x_s ($x_s \in \{0, 1\}$) due to the fact that PPM is widely used in pulse-based communication systems and can be received by both the CD and ED receivers [11]. Other inactive chips in a frame should have symbols of zeros. Hence, the transmitted symbols are created as $\mathbf{x} = [x_1, \cdots, x_{N_h}]$. The transmitted signal waveform of UsIM in one frame is expressed as [7]

$$s(t) = \sum_{k=0}^{N_h-1} \sqrt{E_p} p(t - x_k T_d - kT_c)\delta(k - \Omega_i{}^{-1}(1)), \tag{1}$$

where the parameters are described as follows:

- $p(t)$ represents the transmitted monocycle waveform of one pulse with normalized energy of $\int_{-\infty}^{\infty} p^2(t)dt = 1$.
- E_p denotes the energy of one transmitted pulse.
- T_d is the pulse displacement within one chip, and the pulse width is T_p.[1] For brevity, we assume $T_d = T_p$.
- T_c is the time duration of one chip.
- $\delta(\cdot)$ represents the Dirac delta function.
- Ω_i is the ith index of active chips.

Note that $\Omega_j(r) = i$ ($i \in \{0, 1\}, r \in \{0, \cdots, N_h - 1\}$); then $\Omega_j^{-1}(i) = r$, where r and i denote the chip index and state, respectively. An example of indices and bits mapping of UsIM for $N_h = 4$ is shown in Table 1.

Table 1 Example of indices and bits mapping of UsIM for $N_h = 4$

Indices	Index bits	Chip states
Ω_1	[0 0]	{1 0 0 0}
Ω_2	[0 1]	{0 1 0 0}
Ω_3	[1 0]	{0 0 1 0}
Ω_4	[1 1]	{0 0 0 1}

[1] The relationship between T_d and T_p should be $T_d \gg T_p$ to avoid inter-chip interference caused by the multipath effect [6, 7].

2.2 Splitting Detection

Figure 2 depicts the structure of the proposed SD-JsMRC receiver. The received signal is expressed as

$$r(t) = h(t)s(t) + n(t), \tag{2}$$

where $h(t)$ is the channel coefficient, which is assumed to be invariant within one chip duration, and $n(t)$ is the additive white Gaussian noise having the first-order distribution of $\mathcal{N}\left(0, \frac{N_0}{2}\right)$. The channel fading coefficient $h(t)$ follows the generalized Nakagami distribution [12]

$$f(h) = \frac{2zm^m h^{2zm-1}}{\Gamma(m)\Omega^m} e^{-\frac{m}{\Omega}h^{2z}}, \tag{3}$$

where m, Ω, and z are the shaping, spreading, and generalization parameters of the generalized Nakagami distribution function; $\Gamma(\cdot)$ denotes the gamma function.

The received signal is first split into the CD and ED branches using a splitting ratio ρ, as shown in Fig. 2. The signal in the CD branch is multiplied by the template signal $m(t)$, i.e.,

$$y(t) = \sqrt{\rho}r(t)m(t), \tag{4}$$

where

$$m(t) = \sum_{k=0}^{N_h-1} \sqrt{E_p}\left[p(t - kT_c) + p(t - kT_c - T_p)\right]. \tag{5}$$

The ED branch processes the received signal by the square operation to generate the ED signal, which is given by

$$z(t) = \left(\sqrt{1-\rho}r(t)\right)^2. \tag{6}$$

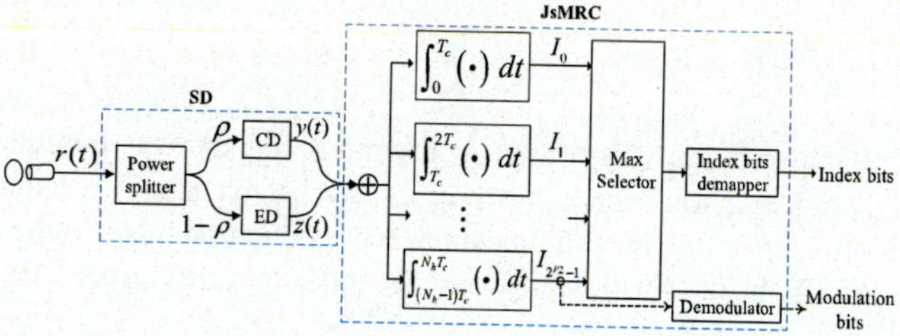

Fig. 2 Structure of the SD-JsMRC receiver

Then, the CD and ED signals are combined and integrated within each chip duration. The ith integral is expressed as

$$I_i = \int_{\tau_i}^{\tau_i+T_c} [y(t) + z(t)]\, dt, \tag{7}$$

where $i = 0, \ldots, 2^{p_2} - 1$, and τ_i denotes the start time of the ith chip within one frame. Please note that compared with MRC, which requires additional absolute operations on I_i [7], the structure of sMRC is simplified.

2.3 *Joint Decision of JsMRC*

After the SD, the integral outputs I_i, $i=0, \cdots, 2^{p_2}-1$, correspond to 2^{p_2} possible indices of active chips, are used to demap the index bits. The decision criterion follows:

$$\hat{i} = \mathop{\arg\max}_{i\in\{0,1,\ldots,2^{p_2}-1\}} \{I_i\}. \tag{8}$$

The index bits, accordingly, can be recovered from $\hat{i}$.

By comparing the integral values in two parts of one demapped chip, i.e., $T_1 \in [\tau_{\hat{i}}, \ \tau_{\hat{i}} + T_p]$, and $T_2 \in [\tau_{\hat{i}} + T_p, \ \tau_i + T_c]$, the modulation bits can be recovered based on a zero threshold [11].

3 Performance Analysis

In this section, we derive and analyze the theoretical BER and the complexity of the SD-JsMRC receiver for UsIM.

3.1 *BER*

The transmission bits consist of the index bits and the modulation bits, and thus the error decision occurs in two cases: (*i*) the erroneous detection of index bits will result in error bits in both index bits and modulation bits; (*ii*) the error decision only occurs in the modulation bits when the index bits are correctly detected. The average BER is calculated as [13]

$$P_b = P_{ed} P_{b,i} + (1 - P_{ed})\, P_{b,ii}, \tag{9}$$

where P_{ed} is the probability of error detection of the index bits, and $P_{b,i}$ and $P_{b,ii}$ represent the average bit-error probability for Case i and Case ii, respectively.

In Case i, on the one hand, the active chip may be erroneously detected as the remaining $2^{p_2} - 1$ chips. Different detected active chips correspond to the different number of erroneous index bits. It can be calculated that there are $\frac{p_2!}{t!(p_2-t)!}$ error events that correspond to t erroneous index bits. On the other hand, in the case of erroneous detection of active chips, the error probability of the modulation bits equals $\frac{1}{2}$ [14]. Therefore, we have

$$P_{b,i} = \frac{1}{p}\left(\frac{1}{2^{p_2}-1}\sum_{t=1}^{p_2}\frac{tp_2!}{t!\,(p_2-t)!} + \frac{1}{2}\right). \tag{10}$$

In Case ii, $P_{b,ii}$ is given by

$$P_{b,ii} = \frac{1}{p}P_{ss}, \tag{11}$$

where P_{ss} is the error probability of the modulation bits.

From (9)–(11), it can be found that as long as we calculate P_{ed} and P_{ss}, the average BER can be deduced. From Fig. 2, we have

$$\begin{aligned} P_{ed} &= 1 - P_r\left(I_{\hat{i}} > \max_{\substack{i\in\{0,1,\ldots,2^{p_2}-1\}\\ i\neq\hat{i}}}\{I_i\}\right) \\ &= 1 - \left(1 - P_r\left(I_{\hat{i}} < \breve{I}_i\right)\right)^{2^{p_2}-1}, \end{aligned} \tag{12}$$

where $\breve{I}_i = \min\{I_i\}$, and $i \in \{0, 1, \ldots, 2^{p_2} - 1\}$, $i \neq \hat{i}$. Let us calculate the probability of $I_{\hat{i}} < \breve{I}_i$, which is expressed as

$$P_e = P_r\left(I_{\hat{i}} < \breve{I}_i\right). \tag{13}$$

If the ith chip is active and the modulation bit "0" is transmitted, the detected CD signal is given by

$$\begin{aligned} y_i &= \int_{\tau_i}^{\tau_i+T_c} y(t)dt \\ &= \int_{\tau_i}^{\tau_i+T_p} \sqrt{\rho}r(t)m(t)dt + \int_{\tau_i+T_p}^{\tau_i+T_c} \sqrt{\rho}r(t)m(t)dt \\ &= h\sqrt{\rho}E_p + \sqrt{\rho E_p}n_{1i} + \sqrt{\rho E_p}n_{2i}. \end{aligned} \tag{14}$$

The detected ED signal is expressed as

$$z_i = \int_{\tau_i}^{\tau_i + T_c} z(t)dt = (1-\rho)\left(\left(h\sqrt{E_p} + \eta_{1i}\right)^2 + \eta_{2i}^2\right). \quad (15)$$

In (14) and (15), $n_{1i}, n_{2i}, \eta_{1i}$, and η_{2i} are independent identically distributed Gaussian random variables according to $\mathcal{N}\left(0, \frac{N_0}{2}\right)$; the integral of $n^2(t)$ from ED in $t \in [\tau_i, \tau_i + T_p]$ can be decomposed in a sum of approximately $2T_p W_{rx}$ independent random variables, where W_{rx} is the noise bandwidth that equals the bandwidth of the transmitted signal [11]. Since ultrasonic pulses are wideband signals, it is easy to achieve $2T_p W_{rx} > 40$. In this case, a central limit theorem applies to the sum so that η_{1i}^2 and η_{2i}^2 are approximately Gaussian distributed [11], having a mean of $N_0 T_p W_{rx}$ and a variance of $N_0^2 T_p W_{rx}$.

Suppose the jth chip is inactive and $I_j = \min\{I_i\}$, and we have $\breve{I}_i = y_j + z_j$, where $j \in \{0, \ldots, 2^{p_2} - 1\}$ and $j \neq i$. The detected CD and ED signals are, respectively, given by

$$y_j = \int_{\tau_j}^{\tau_j + T_c} y(t)dt = \sqrt{\rho E_p} n_{1j} + \sqrt{\rho E_p} n_{2j}, \quad (16)$$

$$z_j = \int_{\tau_j}^{\tau_j + T_c} z(t)dt = (1-\rho)\left(\eta_{1j}^2 + \eta_{2j}^2\right). \quad (17)$$

Then, $I_{\hat{i}} < \breve{I}_i$ is equivalent to

$$\begin{aligned} & h\sqrt{\rho}E_p + \sqrt{\rho E_p} n_{1i} + \sqrt{\rho E_p} n_{2i} + \\ & \quad (1-\rho)\left(h^2 E_p + 2h\sqrt{E_p}\eta_{1i} + \eta_{1i}^2 + \eta_{2i}^2\right) \\ & \quad < \sqrt{\rho E_p} n_{1j} + \sqrt{\rho E_p} n_{2j} + (1-\rho)\left(\eta_{1j}^2 + \eta_{2j}^2\right) \\ & \Rightarrow \left((1-\rho)h^2 + h\sqrt{\rho}\right) E_p < n_x, \end{aligned} \quad (18)$$

where $n_x = \sqrt{\rho E_p}\left(n_{1j} + n_{2j} - (n_{1i} + n_{2i})\right) + (1-\rho)\left(\eta_{1j}^2 + \eta_{2j}^2 - \left(\eta_{1i}^2 + \eta_{2i}^2\right)\right)$ $-2h\sqrt{E_p}(1-\rho)\eta_{1i}$, following

$$\mathcal{N}\left(0, 2\rho E_p N_0 + 2(1-\rho)^2 h^2 E_p N_0 + 4(1-\rho)^2 N_0^2 T_p W_{rx}\right). \quad (19)$$

Therefore, we have

$$P_e = E_h \left\{ Q\left(\frac{(1-\rho)h + \sqrt{\rho}}{\sqrt{\frac{2\rho}{\gamma} + \frac{2(1-\rho)^2 h^2}{\gamma} + \frac{4(1-\rho)^2 h^2 T_p W_{rx}}{\gamma^2}}} \right) \right\}, \tag{20}$$

where $\gamma = \frac{h^2 E_p}{N_0}$ is the SNR in the receiver, $E_h\{\cdot\}$ denotes the expectation with respect to h, and $Q(\cdot)$ represents the Gaussian Q-function. Substituting (20) and (12) into (10), we can obtain $P_{b,i}$.

Next, the modulation bit within the ith chip can be demodulated according to

$$h\sqrt{\rho E_p} + \sqrt{\rho E_p} n_{1i} + (1-\rho)\left(h\sqrt{E_p} + \eta_{1i}\right)^2 \underset{\text{“1”}}{\overset{\text{“0”}}{\gtrless}} \sqrt{\rho E_p} n_{2i} + (1-\rho)\eta_{2i}^2. \tag{21}$$

The error probability P_{ss} conditioned on h is given by

$$\begin{aligned} P_{ss|h} &= \frac{1}{2} P_r \left(h\sqrt{\rho E_p} + \sqrt{\rho E_p} n_{1i} + (1-\rho)\left(h\sqrt{E_p} + \eta_{1i}\right)^2 \right. \\ &\left. < \sqrt{\rho E_p} n_{2i} + (1-\rho)\eta_{2i}^2 \,|\, \text{“0” sent} \right) \\ &+ \frac{1}{2} P_r \left(h\sqrt{\rho E_p} + \sqrt{\rho E_p} n_{2i} + (1-\rho)\left(h\sqrt{E_p} + \eta_{2i}\right)^2 \right. \\ &\left. < \sqrt{\rho E_p} n_{1i} + (1-\rho)\eta_{1i}^2 \,|\, \text{“1” sent} \right). \end{aligned} \tag{22}$$

The average error probability P_{ss} is calculated as

$$P_{ss} = E_h \left\{ Q\left(\frac{(1-\rho)h + \sqrt{\rho}}{\sqrt{\frac{\rho}{\gamma} + \frac{2(1-\rho)^2 h^2}{\gamma} + \frac{2(1-\rho)^2 T_p W_{rx}}{\gamma^2}}} \right) \right\}. \tag{23}$$

Finally, substituting (23) into (11), we can calculate $P_{b,ii}$.

3.2 Complexity

For p bits, the ML receiver requires search 2^p possible $\mathbf{x}$, and the multiplications of order $\sim \mathcal{O}(2^p)$ [7]. For MRC-based receivers, including MRC, sMRC, and JsMRC, they first search 2^{p_2} possible indices of active chips and then search two possible modulation bits within the active chip. Therefore, the multiplications of order $\sim \mathcal{O}(2^{p_2}+2)$ for MRC-based receivers. Obviously, MRC-based receivers can significantly reduce complexity compared with the ML receiver.

Regarding the hardware complexity, as shown in Fig. 2, replacing SD by CD and adding absolute operations on I_i, the structure becomes the MRC receiver [7]. Only replacing SD by CD can generate the sMRC receiver. Thus, sMRC has the lowest complexity compared with other MRC-based receivers. Although the complexity of SD-JsMRC is slightly increased compared with sMRC due to the joint use of CD and ED, introducing SD will significantly improve the BER performance compared with the ML, MRC, and sMRC receivers, which will be verified in Sect. 4.

4 Simulation Results

In this section, we compare the performance of ML, MRC, ED, sMRC, and SD-JsMRC receivers for UsIM via Monte Carlo simulations. The maximum transmission power is set to 16 μW (i.e., SNR = 12 dB) over a transducer area of 1 cm^2, which is below the maximum limit (720 $\mathrm{mW/cm}^2$) imposed by the FDA [4, 12]. In the transmitter, the parameters are set as $N_h = 4$, $p_1 = 1$, and $p_2 = 2$. Based on the existing experimental results of intra-body channel modeling, the parameters in (3) are $m = 0.59$, $\Omega = 0.05$, and $z = 1.12$ [12]. The channel state information is perfectly known in the receivers.

Figure 3 depicts the theoretical and simulation BER of the SD-JsMRC receiver versus the splitting ratio ρ in the cases of SNR = 6, 9 dB over the Gaussian (i.e., $|h| = 1$) and generalized Nakagami channels. It can be observed that the theoretical curves agree with the simulation counterparts well, which validates the theoretical BER in Sect. 3.1. We can also find that the lowest BER is achieved at a value of $0 < \rho < 1$, neither at $\rho = 0$ (i.e., the ED receiver) nor at $\rho = 1$ (i.e., the CD receiver), which proves that using SD can reduce BER due to the receive diversity in the signal space. Moreover, it is shown in Fig. 3 that a much lower BER can be achieved by increasing SNR, which means that a greater SNR for SD can provide a higher BER gain compared with both the CD and ED receivers. Additionally, although the channel fading affects the value of the optimal splitting ratio for the lowest BER, the BER gain still exists, showing the superiority of SD in both Gaussian and fading channels.

Figure 4 compares the BERs for the ML, MRC, ED, sMRC, and SD-JsMRC receivers. The BER of the SD-JsMRC receiver is measured at the optimal splitting ratio, which achieves the lowest BER. We can see that the SD-JsMRC receiver

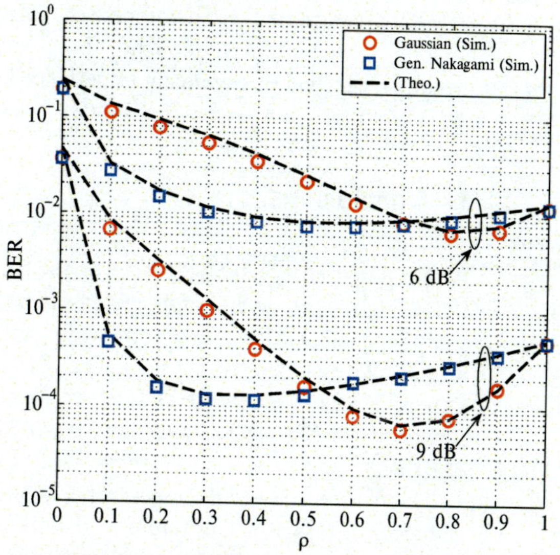

Fig. 3 Theoretical and simulation BER versus the splitting ratio ρ in the cases of SNR $= 6, 9$ dB over the Gaussian and generalized Nakagami channels

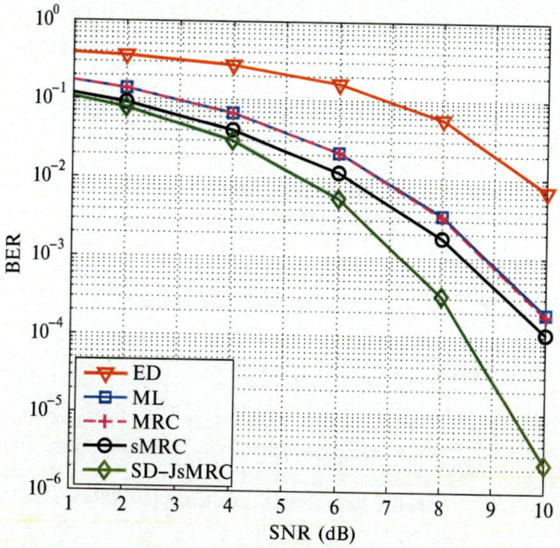

Fig. 4 BER comparisons among the ML, MRC, ED, sMRC, and SD-JsMRC receivers

achieves the lowest BERs compared with the other receivers, benefiting from the diversity combining from CD and ED. The ML and MRC receivers have similar BER curves due to the independent channel coefficients within chips [7]. The sMRC receiver achieves lower BER than the MRC receiver because the absolute operation on the received signal results in SNR loss in the MRC receiver. The ED receiver has the highest BER, which is inevitable due to the severe non-linear noise.

It is worth noting that UsIM conveys two index bits and one modulation bit, achieving a data rate, which is three times of that of TH-UsWB that transmits one bit within one frame [7]. Therefore, SD-JsMRC and UsIM can be integrated into a comprehensive system to implement high-data-rate and reliable reception communications, which is of significant importance for future healthcare applications.

5 Conclusions

We proposed an SD-JsMRC receiver for the UsIM-based IBC system. By using UsIM, high-data-rate and high-energy-efficiency transmission can be achieved. By employing SD-JsMRC, low complexity and low BER reception can be implemented. We also derived the theoretical BER expression of SD-JsMRC for USIM. Extensive Monte Carlo simulations verify the theoretical analysis and show the superiority of SD-JsMRC over the ML, MRC, ED, and sMRC receivers in terms of BER.

This chapter showed the superiority of comprehensively designing the transmitter and receiver for IBCs. Our future work will focus on designing coding schemes and splitting schemes for the SD-JsMRC receiver to improve the performance for future healthcare systems. At the moment, we are prototyping the proposed ultrasonic IBC system to evaluate its empirical performance.

Acknowledgments This work was supported in part by the National Natural Science Foundation of China (under Grants U1801261 and 62001128, 61671208, U1701265, U1809211, and 61971198) and in part by the Key Program of Marine Economy Development (Six Marine Industries) Special Foundation of Department of Natural Resources of Guangdong Province (GDNRC [2020]009). The work of Q. Wang is funded by China Scholarship Council No. 202006150069. The corresponding author is Quansheng Guan.

References

1. Tomlinson, W.J., et al.: Comprehensive survey of galvanic coupling and alternative intra-body communication technologies. IEEE Commun. Surveys Tuts **21**(2), 1145–1164 (2019)
2. Santagati, G.E., Melodia, T., Galluccio, L., Palazzo, S.: Ultrasonic networking for e-health applications. IEEE Wirel. Commun. **20**(4), 74–81 (2013)
3. Trevlakis, S.E., Boulogeorgos, A.A., Chatzidiamantis, N.D., Karagiannidis, G.K.: All-optical cochlear implants. IEEE Trans. Mol., Biol. Multi-Scale Commun. **6**(1), 13–24 (2020)
4. U.S. food drug administration (FDA), Marketing clearance of diagnostic ultrasound systems and transducers (2019). [Online]. Available: https://www.fda.gov/media/71100/download
5. Santagati, G.E., Melodia, T., Galluccio, L., Palazzo, S.: Medium access control and rate adaptation for ultrasonic intrabody sensor networks. IEEE/ACM Trans. Netw. **23**(4), 1121–1134 (2015)
6. Wang, Q., Guan, Q., Ma, B., Liu, J.: Direct-sequence ultrasonic wideband technology for intra-body communications. IEEE Commun. Lett. **23**(10), 1744–1747 (2019)
7. Wang, Q., Guan, Q., Cheng, J., Ma, B.: Ultrasonic indexed modulation and multiple access for intra-body networks. IEEE Trans. Commun. **69**(1), 108–120 (2021)

8. Tse, D., Viswanath, P.: Fundamentals of Wireless Communication. Cambridge Univ. Press, Cambridge (2005)
9. Liu, W., Zhou, X., Durrani, S., Popovski, P.: A novel receiver design with joint coherent and non-coherent processing. IEEE Trans. Commun. **65**(8), 3479–3493 (2017)
10. Wang, Q., Guan, Q., Cheng, J., Ji, F.: A splitting-detection joint-decision receiver for ultrasonic intra-body communications. IEEE Trans. Commun. **69**(6), 3586–3597 (2021)
11. Witrisal, K., et al.: Noncoherent ultra-wideband systems. IEEE Signal Process. Mag. **26**(4), 48–66 (2009)
12. Santagati, G.E., Melodia, T.: Experimental evaluation of impulsive ultrasonic intra-body communications for implantable biomedical devices. IEEE Trans. Mobile Comput. **16**(2), 367–380 (2017)
13. Li, Q., Wen, M., Basar, E., Chen, F.: Index modulated OFDM spread spectrum. IEEE Trans. Wirel. Commun. **17**(4), 2360–2374 (2018)
14. Kaddoum, G., Nijsure, Y., Tran, H.: Generalized code index modulation technique for high-data-rate communication systems. IEEE Trans. Veh. Technol. **65**(9), 7000–7009 (2016)

Modular Spectrum Utilization for Next-Generation Fixed Transmission Networks

Jordan Naterer, Elena-Iulia Dobre, Ayman Mostafa, and Lutz Lampe

1 Introduction

The insistent rise of mobile data service demands continuously increases the data requirements for global communication networks. Furthermore, the fifth generation of wireless communications places greater emphasis on data requirements for use cases such as ultra-reliable low-latency communications and massive machine type communications in Internet-of-Things devices. Gradually over the years, the bottleneck of radio access network capacity has shifted from the radio interface to the backhaul network.

Backhaul links are commonly established either by high-capacity optical fibre lines or low-cost fixed microwave wireless links. While optical fibre can achieve significantly higher data rates and larger capacities, the high cost of deployment and the geographical installation constraints are restrictive. On the other hand, microwave links can be quickly deployed and require a lower capital expenditure compared to fibre. Also, microwave transmission is the preferred medium for the backhaul of the next generation of cellular networks in cases where optical fibre links cannot be deployed [1]. However, microwave transmission requires spectrum leasing, which raises the operating expense of the network.

The co-authors dedicate this paper to the main contributor Jordan Naterer, who passed away in October 2020. Our thoughts are with him.

J. Naterer · E.-I. Dobre (✉) · A. Mostafa · L. Lampe
Department of Electrical and Computer Engineering, The University of British Columbia, Vancouver, BC, Canada
e-mail: elenad@ece.ubc.ca

H. Nguyen et al. (eds.), *30th Biennial Symposium on Communications 2021*,
Signals and Communication Technology, https://doi.org/10.1007/978-3-031-06947-5_9

Frequency spectrum regulations in microwave transmission bands ensure that systems operating in adjacent frequency channels do not interfere with one another. However, it is observed in many higher-frequency microwave links that channels that are used in the same transmission hop have unoccupied neighbouring channel slots. For example, our analysis of the latest spectrum allocation database in the 18 GHz band in the USA [2] showed that only 1.2% of fixed point-to-point transmit channels have active neighbours. Similar results have been found for microwave bands in Canada and the UK, for example [3, Fig. 1]. Hence, from the perspective of a communication link, there is no interference from a neighbouring channel in the spectrum skirt regions. At the same time, the signal-to-noise ratio (SNR) levels experienced by fixed microwave links are, for the majority of time, sufficiently high to support reliable communication in spectrum skirts. For example, fixed microwave links experience SNRs of around 45–50 dB and higher for 99% of the time in a year [4].

We propose using the spectrum skirts for data transmission to increase the data rate achievable within a single microwave transmission channel without additional costs for licenses. In previous work [3], some of us have developed single-carrier pulse-shaped transmission to leverage this opportunity. In this chapter, we introduce two alternative transmission methods that, different from [3], consist of the conventional microwave transmission signal and additional signal components (or data streams) that can be placed in spectrum skirts. Compared to the solution in [3], the new methods have the advantage that the use of spectrum skirts can be turned on or off by adding additional data streams in a modular fashion. The proposed methods are complementary to other approaches for improving spectral efficiency such as dual-polarization or multiple-antenna transmission.

The first new method, which we refer to as superposition transmission, overlaps the strong primary signal with a weak secondary signal whose power spectral density (PSD) extends to the spectrum skirts. We present an interference cancellation method for the detection of these two signal streams. The second method uses a multi-carrier approach and adds two low-power sideband signals, which are orthogonal to the main signal. For the multi-carrier method, we apply the approach from [3] to design custom pulse shapes that fit under the spectrum mask to best exploit the available sidebands. In all cases, we consider transmission impairments such as a dispersive channel and local-oscillator phase noise, which are experienced by practical microwave links. It should be noted that phase noise is limiting the performance of microwave communication links operating at high SNRs. In particular, the maximum constellation size suitable for reliable communication is limited by phase-noise impairments. Our numerical performance results show that both schemes offer data rate improvements on the order of 30 and 40% over conventional microwave transmission.

2 Transmission Schemes for Modular Spectrum Utilization

In this section, we introduce the two new methods that use the spectrum skirts. Figure 1 shows a spectrum mask for microwave transmission in the 17–30 GHz band according to [5, Table 3e] together with the PSDs for: (1) a conventional microwave system employing a root-raised cosine (RRC) pulse shape, (2) the superposition transmission method, and (3) the multi-carrier transmission method. In the following, we will provide the details for the latter two.

2.1 *Superposition Transmission*

The first spectrum utilization method overlaps two carrier signals at the transmitter and sequentially detects them at the receiver. The transmitter architecture in Fig. 2

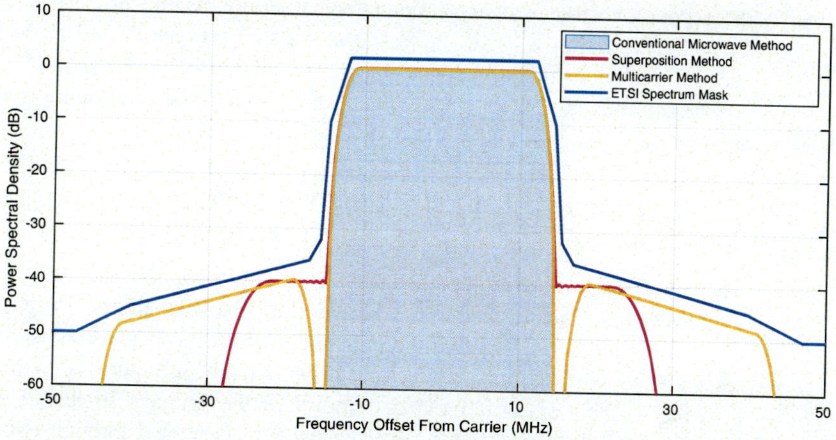

Fig. 1 Illustration of the use of spectrum skirts for a spectrum mask according to [5]. The conventional microwave system uses a root-raised cosine pulse, which can occupy a bandwidth of 29.4 MHz for the chosen emission mask. The power spectral densities for the superposition and multi-carrier transmission schemes show the use of spectrum skirts

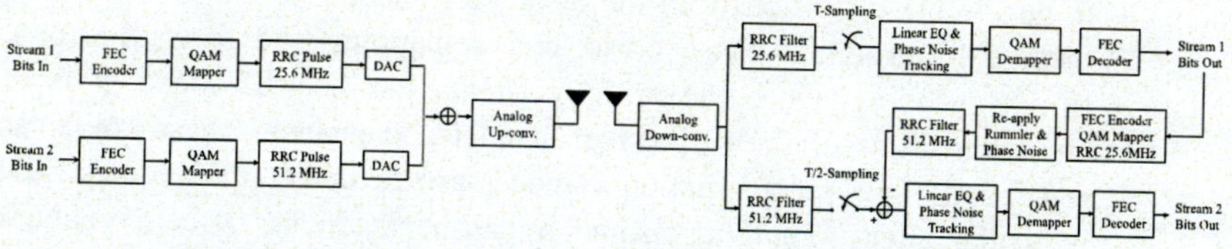

Fig. 2 Superposition transmission scheme architecture. The displayed baud-rate figures are example settings, applying the spectrum mask for the 17–30 GHz band according to [5, Table 3e] and being consistent with the PSD for superposition transmission shown in Fig. 1

(left subfigure) is similar to a standard microwave link, with a signal chain added to the primary stream before analog upconversion and antenna transmission. The continuous-time representation of the superposition transmission signal is

$$s(t) = \sqrt{P_1} \sum_k I_1[k] p_1(t - kT_1) + \sqrt{P_2} \sum_k I_2[k] p_2(t - kT_2), \tag{1}$$

where $I_1[k]$ and $I_2[k]$ are the quadrature amplitude modulation (QAM) symbols of the primary and secondary data streams, respectively. The pulse shapes $p_1(t)$ and $p_2(t)$ represent the narrowband and wideband pulses used to shape our primary and secondary pulses, while T_1 and T_2 denote each stream's symbol interval. Furthermore, P_1 and P_2 are the powers for the two streams. The QAM orders of each stream may vary based on channel conditions.

Regarding pulse shapes, the superposition scheme allows the second stream to incorporate a larger bandwidth to effectively utilize the spectrum skirts. For example, the PSD for the combined RRC pulse shapes with Nyquist rates of 25.6 and 51.2 MHz and a roll-off factor of 0.15 are displayed in Fig. 1. Transmitting the secondary stream at a symbol rate that is double the primary stream, i.e., $T_2 = T_1/2$, implies a 40 dB difference in transmit powers P_1 and P_2 to meet the spectral mask constraints.

The continuous-time representation of the received baseband signal is expressed as

$$r(t) = e^{j\phi_r(t)} \int_{-\infty}^{\infty} s(\tau) e^{j\phi_t(\tau)} h(t - \tau) d\tau + n(t), \tag{2}$$

where $h(t)$ denotes a multipath channel, usually modelled according to [6] for microwave links, and $n(t)$ denotes additive white Gaussian noise (AWGN). Furthermore, $\phi_t(t)$ and $\phi_r(t)$ represent the transmitter and receiver phase-noise processes, respectively. The receiver architecture in Fig. 2 (right subfigure) is more complex than the transmitter, as it requires removal of the primary stream to detect the secondary stream. First, the received signal $r(t)$ is passed through two matched filters with bandwidths according to the transmitter pulse shapes for the two data streams. Figure 2 provides specific values for the pulse shapes and bandwidths considering the 17–30 GHz transmission band with the spectrum mask from [5, Table 3e]. Then, the primary data stream is detected by applying a linear equalizer to handle the distortion from the dispersive channel, a phase-noise compensation method to eliminate phase noise, and demapping and forward-error correction (FEC) decoding. Once the primary stream has been decoded, it is re-encoded and estimations of the dispersive channel and channel phase-noise samples are re-applied. This reconstruction is necessary to detect the secondary stream. In particular, after re-applying channel impairments, the first stream is subtracted from the received signal at a larger bandwidth, so that the secondary stream can be decoded with ideally no interference from the primary stream. The main bottleneck for this receiver architecture is the accuracy of reconstruction of the primary stream.

While equalization and phase-noise removal are important for the secondary stream operating at a lower signal power, the residual interference from the primary stream is the limiting factor when detecting the secondary stream due to the large power difference.

2.2 *Multi-carrier Transmission*

The second spectrum utilization method transmits three pulses simultaneously while remaining orthogonal in the frequency domain. The two secondary pulses are sent adjacent to the primary pulse via an offset frequency and rely on well-designed transmitter and receiver filters to neglect interference.

The transmitter architecture is shown in Fig. 3 (left subfigure). The continuous-time baseband representation of the multi-carrier signal is given by

$$
\begin{aligned}
s(t) = \sqrt{P_1} \sum_k I_1[k] p_1(t - kT) + \sqrt{P_2} \sum_k I_2[k] p_2(t - kT) \\
+ \sqrt{P_3} \sum_k I_3[k] p_3(t - kT),
\end{aligned} \tag{3}
$$

where $I_1[k]$, $I_2[k]$, and $I_3[k]$ are the QAM symbols of the primary and two secondary streams and P_1, P_2, and P_3 are the respective signal powers. The pulse shapes $p_1(t)$, $p_2(t)$, and $p_3(t)$ represent the pulse shapes of the primary and two secondary streams transmitted at symbol interval T. The frequency offsets for the secondary streams are incorporated into each pulse shape.

Since the three pulses are orthogonal in the frequency domain, we have the ability to use custom pulse shapes to fill more of the spectrum mask for the side streams. We may formulate this objective as an optimization problem to match the spectrum mask as closely as possible. We consider a linear-phase discrete-time filter of length $2N + 1$ and organize the real and imaginary parts of the filter coefficients in the vectors $\mathbf{p}_{\text{re}}$ and $\mathbf{p}_{\text{im}}$, respectively. Then, the discrete-time Fourier transform is given by

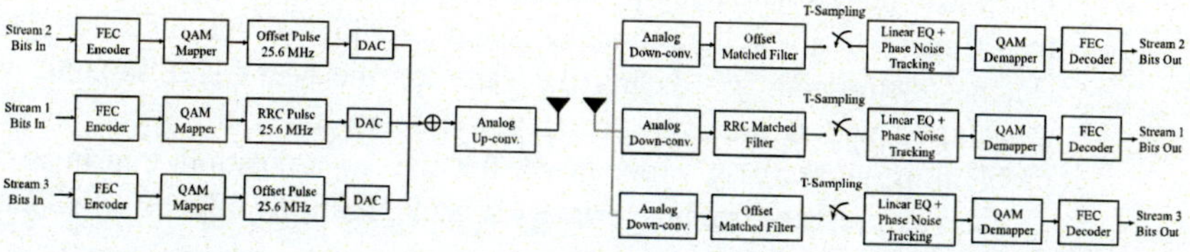

Fig. 3 Multi-carrier transmission scheme architecture. The example of 25.6 MHz wide pulse shapes is illustrated at the transmitter side, consistent with the PSD for multi-carrier transmission shown in Fig. 1

$$P(f) = \mathrm{e}^{-\mathrm{j}2\pi Nf} \left(\mathbf{v}_{\mathrm{re}}(f)^T \mathbf{p}_{\mathrm{re}} + \mathbf{v}_{\mathrm{im}}(f)^T \mathbf{p}_{\mathrm{im}} \right), \tag{4}$$

where

$$\mathbf{v}_{\mathrm{re}}(f) = [1, 2\cos(2\pi f), \ldots, 2\cos(2\pi Nf)]^T \tag{5}$$

$$\mathbf{v}_{\mathrm{im}}(f) = [0, 2\sin(2\pi f), \ldots, 2\sin(2\pi Nf)]^T. \tag{6}$$

Applying the framework from [7], we then formulate the filter design optimization problem as the minimization of a squared error between the desired pulse spectrum mask $D(f)$ and the actual pulse spectrum amplitude $P(f)$, i.e., we solve

$$\min_{\mathbf{p}_{\mathrm{re}}, \mathbf{p}_{\mathrm{im}}} \int_{-F_s/2}^{F_s/2} \left(D(f) - \left| \mathbf{v}_{\mathrm{re}}(f)^T \mathbf{p}_{\mathrm{re}} + \mathbf{v}_{\mathrm{im}}(f)^T \mathbf{p}_{\mathrm{im}} \right| \right)^2 df \tag{7}$$

$$\text{subject to} \left| \mathbf{v}_{\mathrm{re}}(f)^T \mathbf{p}_{\mathrm{re}} + \mathbf{v}_{\mathrm{im}}(f)^T \mathbf{p}_{\mathrm{im}} \right| \leq D(f), \quad \forall f \in [-F_s/2, F_s/2], \tag{8}$$

where F_s denotes the sampling frequency. The optimization problem becomes a convex quadratic problem through the discretization of the frequency axis and can efficiently be solved using interior point algorithms. The optimized pulse shape selected for the numerical results in Sect. 4 is illustrated in Fig. 1. Relative to the centre frequency, the amplitude frequency response of the custom pulse varies by ± 5 dB in its transmission band.

The receiver architecture for the multi-carrier scheme is shown in Fig. 3 (right subfigure). It is more similar to the receiver processing of a conventional microwave transmission system than that for the superposition scheme, as each data stream can be detected in parallel due to the negligible cross interference.

3 Channel Extrapolation

One challenge we encounter when demodulating the sideband signals in both modular schemes is obtaining a channel estimate required for equalization. While the primary stream in both modular schemes enjoys a high SNR and thus reliable channel estimation, this is not the case for the secondary streams, which are transmitted with a relatively lower power to fit under the spectrum skirts. This problem is particularly pronounced for the superposition transmission, where residual interference from the primary scheme limits the effective SNR experienced by the secondary stream.

To deal with this as well as to avoid extra pilot signal overhead for channel estimation for the secondary streams, we make use of the fact that the primary and secondary stream signals pass through the same channel, albeit with different bandwidths in the case of superposition transmission or in different frequency bands in the case of multi-carrier transmission. In particular, since the impulse-response representation of typical microwave channels typically allows for a description with very few parameters (see [6]), we can estimate those parameters from pilot signals in the primary stream and use the corresponding channel estimate for the secondary streams.

Considering a channel impulse response $h(t)$ with L significant taps, e.g., $L = 2$ for the commonly used Rummler channel model [6], and denoting the convolution of the transmitter RRC pulse and its receiver matched filter as $p'(t)$, we can express the overall impulse response as

$$c(t) = \sum_{l=1}^{L} c_l p'(t - t_l) \, , \tag{9}$$

where $\{t_l, c_l\}_{l=1}^{L}$ correspond to the unknown time delays and amplitudes of the transmission channel. The signal $c(t)$ fits the model of a signal with finite rate of innovation and we thus follow the methods from [8, 9] to estimate the desired parameters $\{t_l, c_l\}_{l=1}^{L}$. For this, we choose a T_c such that $c(t > T_c) \approx 0$ and write the Fourier series expression

$$c(t) = \sum_{k=-\infty}^{\infty} C[k] \mathrm{e}^{\mathrm{j}\frac{2\pi}{T_c}kt} \tag{10}$$

with the Fourier series coefficients

$$C[k] = \frac{1}{T_c} P'\left(\frac{2\pi}{T_c}k\right) \sum_{l=1}^{L} c_l \mathrm{e}^{-\mathrm{j}\frac{2\pi}{T_c}kt_l}, \tag{11}$$

where $P'(f)$ represents the continuous-time Fourier transform of $p'(t)$. Removing the effect from the pulse shapes and defining $z_l = \mathrm{e}^{-\mathrm{j}\frac{2\pi}{T_c}t_l}$, we get

$$X[k] = \frac{1}{T_c} \sum_{l=1}^{L} c_l \mathrm{e}^{-\mathrm{j}\frac{2\pi}{T_c}kt_l} = \frac{1}{T_c} \sum_{l=1}^{L} c_l z_l^k, \tag{12}$$

which is solely determined by the transmission channel. To obtain an estimate of z_l in (12), we apply the subspace method from [9], which is summarized in the following algorithm.

Algorithm: Subspace method for microwave channel estimation:
1: Using $M + K - 1$ samples of $X[k]$, construct a $M \times K$ Hankel Matrix

$$\mathbf{X} = \begin{bmatrix} X[0] & X[1] & \cdots & X[K-1] \\ X[1] & X[2] & \cdots & X[K] \\ \vdots & & & \\ X[M-1] & X[M] & \cdots & X[M+K-2] \end{bmatrix}, \tag{13}$$

where $M, K > L$.
2: Compute the singular-value decomposition $\mathbf{X} = \mathbf{U}\mathbf{S}\mathbf{V}^H$.
3: Find the L largest singular values and the corresponding left singular vectors $\mathbf{U}_L$.
4: Compute estimates of $\hat{z}_l$ from (12) as the eigenvalues of the matrix $\underline{\mathbf{U}}_L^+ \cdot \bar{\mathbf{U}}_L$, where $(\cdot)^+$ represents the pseudo-inverse and the overbar and underbar symbols represent omitting the first and last rows of $\mathbf{U}_L$, respectively.

Given $\hat{z}_l$ from the subspace method, we estimate the channel time delays as

$$t_l = \frac{T_c \angle \hat{z}_l}{2\pi}. \tag{14}$$

Finally, to obtain the channel coefficients, we transform (12) into a Vandermonde matrix

$$\mathbf{P} = \begin{bmatrix} 1 & \hat{z}_1 & \hat{z}_1^2 & \cdots & \hat{z}_1^{M+K-1} \\ \cdots & \cdots & & \cdots & \\ 1 & \hat{z}_L & \hat{z}_L^2 & \cdots & \hat{z}_L^{M+K-1} \end{bmatrix} \tag{15}$$

and solve the system of equations as

$$[\hat{c}_1 \ldots, \hat{c_L}] = [X[0], \ldots, X[M+K-2]]\mathbf{P}^+ . \tag{16}$$

One modification we make to the original subspace algorithm is iterating over a number of choices for M to generate various matrices $\mathbf{X}$ (13). We then select a final set of $\{\hat{t}_l, \hat{c}_l\}_{l=1}^{L}$ coefficients that minimize the mean-squared error between the estimated coefficients for the narrowband channel $c[k]$ using conventional pilot-based estimation and a reconstruction of the narrowband channel $\hat{c}[k]$ using the subspace method.

4 Simulations and Results

In this section, we demonstrate the benefits of the proposed transmission schemes through numerical results for achievable data rates and bit-error rate (BER) simulations. We adopt the dispersive Rummler channel model from [6] with a 5 dB notch and transmitter and receiver phase noise at an aggregate level of −90 dBc/Hz at an offset of 100 kHz, which is a significant yet typical phase-noise level for microwave systems. We apply a linear minimum mean-square error (MMSE) equalizer to compensate for the Rummler transmission channel. We estimate and compensate the phase noise via the forward–backward algorithm described in [10] with pilots every 50 data symbols (see [3] for more details of the compensation methods). The signal pulse shapes and thus PSDs are set according to the spectrum mask shown in Fig. 1, where $1/T_1 = 25.6\,\text{MHz}$, $1/T_2 = 51.2\,\text{MHz}$, and $1/T = 25.6\,\text{MHz}$ in (1) and (3), respectively. For the following, we show results as a function of the SNR defined as the symbol energy divided by the noise power per bandwidth unit (or PSD) of the baseline RRC-shaped transmission. This means we consider a scenario where the conventional RRC-based microwave system is operated at a certain SNR, and then the secondary data streams are being added by using the spectrum skirts.

First, Fig. 4 shows the BER results for the superposition and multi-carrier architectures. We use 2^{12} and 2^{14}-QAM for the primary streams of the superposition and multi-carrier schemes, respectively, and 2^m-QAM with $m = 2, 4$ for their secondary streams. A low-density parity-check (LDPC) code with rate 0.75 and codeword length 64,800 bits is used for channel coding. We focus on the BER for the secondary streams in Fig. 4 to illustrate the effect of constellation size and phase noise. Considering the idealized case of transmission over a Rummler channel

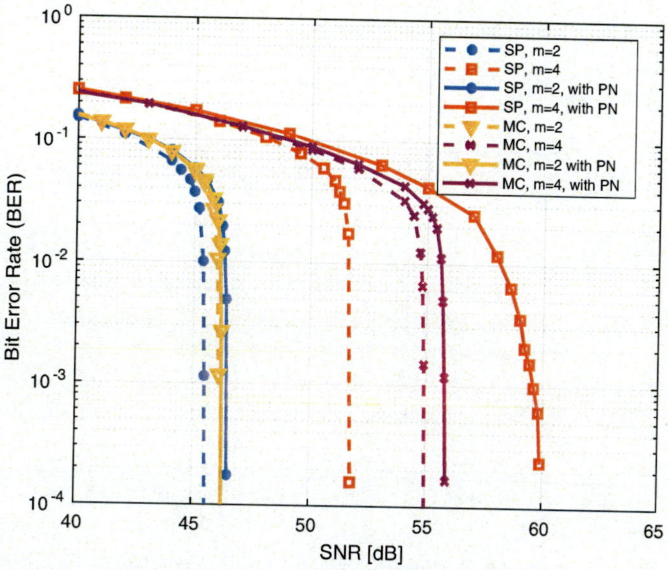

Fig. 4 BER for the secondary streams for: (1) superposition (SP) transmission with 2^{12}-QAM for the primary data stream and 2^m-QAM for the secondary data stream; (2) multi-carrier transmission with 2^{14}-QAM for the primary data stream and 2^m-QAM for the two secondary streams. A Rummler channel without and with phase noise (PN) and FEC coding with a rate-0.75 LDPC code is considered

without phase noise first (dashed lines), we observe quite similar BER curves for the secondary streams with $m = 2$ for superposition and multi-carrier transmission. Increasing the modulation order to $m = 4$, i.e., 16QAM, we notice a significant SNR shift of the BER curves, which is due to the linear equalization that operates at a low SNR for the secondary streams. If phase noise is present (solid lines), we notice the higher sensitivity of the superposition compared to the multi-carrier scheme. This is due to the imperfect interference cancelation of the main data stream as a result of residual phase noise.

Next, Fig. 5 shows the comparison between conventional RRC-based microwave transmission and the two proposed schemes in terms of the achievable information rate (AIR) [11, 12]. The AIR is measured empirically using the decision metrics (log-likelihood ratios) at the receiver, and we consider it as a practical estimate for the achievable rate with coded transmission [12]. The AIR curves in Fig. 5 correspond to the reference case of an AWGN channel (dashed lines) and the

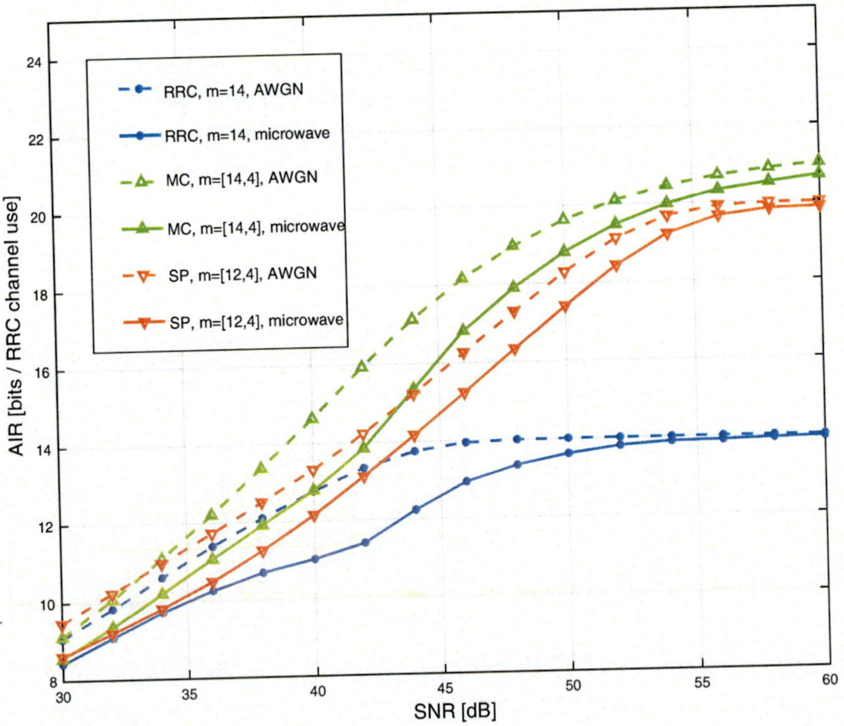

Fig. 5 AIR for different transmission systems: (1) conventional RRC-based transmission with 2^{14}-QAM; (2) multi-carrier (MC) transmission with 2^{14}-QAM in the primary data stream and 2^{4}-QAM in the two secondary data streams; (3) superposition (SP) transmission with 2^{12}-QAM in the primary data stream and 2^{4}-QAM in the secondary data stream. AWGN channel and microwave channel (Rummler channel and phase noise) are considered. AIR is presented in terms of bits per RRC channel use

microwave transmission scenario of dispersive Rummler channel and phase noise (solid lines). The AIR is presented in terms of bits per RRC channel use, which translates into bits per second by multiplying it with the RRC baud rate of $1/T_1 = 25.6\,\text{MHz}$. We observe a notable rate loss between the AWGN and microwave channel scenarios for all schemes. The main limiting factor for the performance of the latter is the phase noise. In particular, this prevents the use of constellations larger than 2^{14}-QAM for the conventional transmission systems. For example, the AIR curve for 2^{16}-QAM (not shown) falls below the one for 2^{14}-QAM in the microwave channel with phase noise. On the other hand, it can be seen that the proposed modular schemes, due to transmission of data in the spectrum skirts, yield significant rate gains over a wide range of practical SNRs for microwave links. For example, the data rate is improved by 28–38% at an SNR of 50 dB. We selected configurations for the primary and secondary data streams that demonstrated preferable performance, but further optimization, in particular adaption as a function of SNR, is possible. The multi-carrier scheme outperforms the superposition transmission, mainly due to the challenge of sufficiently accurate regeneration of the primary stream for interference cancellation in the latter method. Overall, we conclude that the modular spectrum utilization is highly beneficial to improve transmission rates in microwave links.

5 Conclusions

We have proposed the use of spectrum skirts for data transmission in microwave links by adding secondary data streams. The motivation for this approach is the frequent vacancy of adjacent frequency channels and the high operating SNR of microwave links, as well as the practical limitation for the use of higher-order modulations due to phase noise. In the superposition transmission scheme, the secondary data stream is added to the primary data stream, occupying a larger bandwidth and operating at very low power so as to respect the spectrum mask. In the multi-carrier transmission scheme, two sideband signals are added, with custom-made pulse shapes and again very low power to fit under the mask. We have developed a channel estimation method for the proposed modular transmission systems that enables reliable detection of the sideband signals in relatively low SNR. Numerical results for practical channel and phase-noise scenarios indicate significant rate improvements of more than 30% at typical operating SNRs. These benefits are achieved without additional licensing costs and are complementary to gains obtained from other methods for improved spectral efficiency such as multi-antenna transmission.

Acknowledgments This work was supported by the Natural Sciences and Engineering Research Council of Canada (NSERC) and Huawei Technologies Canada.

References

1. Saunders, J.: Mobile backhaul options - spectrum analysis and recommendations, GSM Association, ABI Research, GSMA Head Office, Floor 2, The Walbrook Building, 25 Walbrook, London EC4N 8AF, United Kingdom, Tech. Rep. (September 2018). [Online]. Available: https://www.gsma.com/spectrum/wp-content/uploads/2019/04/Mobile-Backhaul-Options.pdf
2. Universal licensing system. https://wireless2.fcc.gov/UlsApp/UlsSearch/searchAdvanced.jsp. Accessed 20 March 2021
3. Dobre, E., Mostafa, A., Lampe, L., ShahMohammadian, M., Jian, M.: Efficient spectrum utilization via pulse shape design for fixed transmission networks. IEEE Trans. Commun. **69**, 1510–1523 (2021)
4. Huang, X., Guo, Y.J., Zhang, A., Dyadyuk, V.: A multi-gigabit microwave backhaul. IEEE Commun. Mag. **50**(3), 122–129 (2012)
5. ETSI, Characteristics and requirements for point-to-point equipment and antennas; Part 2: Digital systems operating in frequency bands from 1.3 GHz to 86 GHz, Technical Report (2016)
6. Rummler, W.: More on the multipath fading channel model. IEEE Trans. Commun. **29**(3), 346–352 (1981)
7. Davidson, T.N.: Enriching the art of FIR filter design via convex optimization. IEEE Signal Process. Mag. **27**(3), 89–101 (2010)
8. Gedalyahu, K., Tur, R., Eldar, Y.C.: Sampling of pulse streams: Achieving the rate of innovation. In: IEEE Sensor Array and Multichannel Signal Processing Workshop, Israel, pp. 133–136 (2010)
9. Maravic, I., Vetterli, M.: Sampling and reconstruction of signals with finite rate of innovation in the presence of noise. IEEE Trans. Signal Process. **53**(8), 2788–2901 (2005)
10. Mohammadian, H.S., Aharony, A.: Accurate BCJR-based synchronization algorithm for single carrier channels with extremely high order modulations. In: International Conference on Signal Processing and Communication Systems (ICSPCS), Australia, pp. 1–6 (2016)
11. Arnold, D.M., Loeliger, H.-A., Vontobel, P.O., Kavcic, A., Zeng, W., Simulation-based computation of information rates for channels with memory. IEEE Trans. Inf. Theory **52**(8), 3498–3508 (2006)
12. Alvarado, A., Fehenberger, T., Chen, B., Willems, F.M.J.: Achievable information rates for fiber optics: applications and computations. J. Lightw. Technol. **36**(2), 424–439 (2018)

Multi-UAV Trajectory, Resource Allocation Design for UAV-Based Wireless Networks with Dynamic Data Demand for Consecutive Service Periods

Minh Tri Nguyen and Long Bao Le

1 Introduction

Unmanned aerial vehicle (UAV) communications have been emerging as one of the key components for future wireless networks due to their advantages in terms of deployment and flexibility [1]. In particular, by dynamically and cooperatively adjusting the UAV's positions, UAV-based wireless networks offer another dimensions to configure and optimize the wireless networks in several aspects such as coverage, throughput, and spectrum efficiency. By leveraging UAV movements, we can provide more flexible wireless services and make the UAV-based wireless networks more adaptive to highly dynamic communication scenarios. Therefore, the UAV-based cellular wireless networks are expected to play an important role in 5G and beyond-5G wireless systems with many applications[2].

Existing studies of UAV deployment in UAV-based wireless networks consider two important design scenarios. The first type of design is related to UAV position optimization, where UAVs' hovering positions are optimized according to certain design objectives over a flying period, as in [3–5]. Most works in this direction do not consider UAVs' energy constraints and they do not consider transient states when UAVs move from hovering points of the current service duration to the hovering points of the next duration or when UAVs depart from and arrive to a charging station. The second type of design focuses on UAV trajectory optimization, where UAVs' trajectories are optimized over a flying period where it is assumed that UAVs depart from and arrive to the charging station(s) at the beginning and at the end of the flying duration, respectively [6, 7]. Though these designs provide more flexible trajectories than those under the first type, the objectives usually consider

M. T. Nguyen (✉) · L. B. Le
INRS, University of Quebec, Montreal, QC, Canada
e-mail: minh.tri.nguyen@emt.inrs.ca; long.le@emt.inrs.ca

H. Nguyen et al. (eds.), *30th Biennial Symposium on Communications 2021*,
Signals and Communication Technology, https://doi.org/10.1007/978-3-031-06947-5_10

individual flying periods, and hence there is no guarantee that they can satisfy users' high data demand which may/must be served over several consecutive flying periods.

In general, UAV trajectory designs must consider the practical constraints on UAV's limited energy [6, 8]. Additionally, the practical scenarios with dynamic user arrival/requests over consecutive periods are rarely studied in the literature where it is important to engineer a reasonable objective and an efficient UAV trajectory control strategy serving (dynamic) users' demand with low service delay. To this end, our work aims to provide a framework/design that addresses these issues, which remain under-explored in the literature. Specifically, our work considers two novel aspects compared to existing UAV designs. First, while still considering practical UAV's energy constraints, our design does not assume that the UAV's starting and final positions must be the charging station, i.e., by allowing different UAVs to depart from and arrive to their charging station at their desired time instants, we can better utilize the UAVs' energy and network resources. Second, we aim to maximize the weighted demand-aware downlink throughput, by jointly optimizing the resource allocation and UAVs' trajectories.

The formulated problem is non-convex. So, the alternating optimization method is employed to solve the underlying problem for each service period where the bandwidth allocation, power allocation, and UAV trajectory control subproblems are solved iteratively until convergence. And the successive convex approximation (SCA) method is used to convexify and solve the UAV trajectory control subproblem. The proposed algorithm is then employed to support dynamic data transmission demands of users over multiple service periods where we show the significant throughput and access delay improvements of our proposed algorithm over the conventional scheme in one and over multiple service periods.

2 System Models

2.1 Network and Service Models

We consider a UAV-based wireless network where UAVs fly and serve ground users in downlink communication. Time is sliced into small time slots, and the length of each time slot is δ. UAV trajectory and resource allocation optimization is performed in each service period of T time slots. We consider a scenario when users' data transmission requests arrive over time. The users arriving in a particular service period will be served in the very next service period. From now on, all optimization variables correspond to a particular service period.

The new incoming users are known prior to the start of the service period. We assume that each incoming user k demands an amount of data D_k that the UAVs will transmit to them. The number of users in a particular service period is denoted as K. In the following, the terms users and data transmission requests are used interchangeably. The arrival time slot of user k is denoted as η_k, where $\eta_k = \{0, \ldots, T-1\}$.[1] Hence, its (waiting) delay prior to the start of the next service duration is $T - \eta_k$. If user k is served in the following service duration,[2] their access delay (in terms of the number of time slots) is $\tau_k = 2T - \eta_k$.

2.2 System Models

At time slot t, there are N UAVs flying and transmitting data to K ground users. We assume that UAV n flies at the fixed altitude h and its 2D coordinate[3] at t is denoted as $\mathbf{c}_n[t]$. The 2D coordinate of ground user k is denoted as $\mathbf{u}_k$. We consider the practical UAVs' energy constraint, so UAVs must depart from and return to the charging station to recharge their energy before their energy is depleted where the coordinate of the charging station is denoted as $\mathbf{c}_\mathrm{O}$. For brevity, we call the charging station as station.

2.2.1 UAV Energy and Flight Models

We consider rotary-wing UAVs in this work. UAVs are dispatched to transmit data to users and fly back to the station before they run out of energy. The flying energy consumed in time slot t can be computed as follows [6]:

$$E_n^{\mathrm{C}}[t] = \delta \left(P_\mathrm{a} + \frac{3P_\mathrm{a}\|\mathbf{v}_n[t]\|^2}{U^2} + \frac{1}{2}\beta\|\mathbf{v}_n[t]\|^3 + P_\mathrm{b}\left(\sqrt{1 + \frac{\|\mathbf{v}_n[t]\|^4}{4v_0^4}} - \frac{\|\mathbf{v}_n[t]\|^2}{2v_o^2}\right)^{1/2}\right), \tag{1}$$

where $\mathbf{v}_n[t] = (\mathbf{c}_n[t] - \mathbf{c}_n[t-1])/\delta$ denotes the velocity of UAV n during time slot t, and P_a, U, P_b, β, and v_o are the constants that characterize the UAVs' energy consumption model. Please refer to [6] for more details and explanations about the model.

---

[1] Note that users arriving in a particular service duration will be served in the next service duration; thus, these arrival times are relative and their time origin is the start of the current service duration.

[2] In this work, we do not consider the network provision and assume that the demanded data is available in the UAVs [9] or is transmitted to the UAV from ground stations via wireless backhaul [4].

[3] Since UAVs fly continuously during the time slot, as long as their movement in one time slot is not very large compared to their altitude, the position of them during the time slot can be approximated by their positions at the end of that time slot. This will be addressed by a set of constraints later in the problem formulation.

2.2.2 Communication Models

We assume that ground users receive data from UAVs on orthogonal channels where the total system bandwidth is B (Hz). The bandwidth to support the communication between UAV n and user k at time slot t is denoted as $b_{n,k}[t]$. We assume that the Frequency Division Multiple Access (FDMA) technique is employed with continuous allocated bandwidths $b_{n,k}[t]$ as considered in previous works [7]. Therefore, we have the following bandwidth constraint:

$$\sum_{n=1}^{N}\sum_{k=1}^{K} b_{n,k}[t] \leq B, \forall t. \tag{2}$$

Over the allocated bandwidth of $b_{n,k}[t]$, UAV n transmits the power of $b_{n,k}[t]p_{n,k}[t]$ to user k, where $p_{n,k}[t]$ describes the transmit power density (W/Hz). We also consider the practical power constraint expressed as follows, where P_{max} is the maximum transmit power of each UAV:

$$\sum_{k=1}^{K} b_{n,k}[t]p_{n,k}[t] \leq P_{\mathrm{max}}, \forall n, t. \tag{3}$$

The channels between UAVs and users are assumed to be dominated by the Line of Sight (LoS) component, which is reasonable when the UAV flies sufficiently high, as reported from previous field tests [10]. Then, the channel power gain between UAV n and user k at t is $\rho_0/(h^2+\|\mathbf{c}_n[t]-\mathbf{u}_k\|^2)$, where ρ_0 is the power gain at the reference distance of 1 m from the UAV. The total data received by user k at t can be computed as

$$d_k[t]=\delta\sum_{n=1}^{N} b_{n,k}[t]\log\left(1+\gamma\frac{p_{n,k}[t]}{h^2+\|\mathbf{c}_n[t]-\mathbf{u}_k\|^2}\right), \tag{4}$$

where $\gamma = \rho_0/\sigma^2$ is the normalized signal-to-noise ratio (SNR), and σ^2 is the white noise power density (W/Hz) of users' devices.

3 Problem Formulation

Recall that each user k has its data transmission demand/request expressed by the amount of data D_k that must be received through downlink communications from the UAVs. As a result, it is wasteful to transmit data to users beyond what they demand, as in oblivious designs [7]. Therefore, we are interested in demand-aware throughput which is defined for user k as

$$d_k^{\mathsf{a}} = \min\left(D_k, \sum_{t=1}^{t=T} d_k[t]\right). \tag{5}$$

In this work, we propose to maximize the sum of weighted demand-aware throughput for a given service duration from $t=1$ to $t=T$. The weights can be chosen to maintain a certain level of fairness among users in delay-aware applications, i.e., users with longer delay have larger weights and thus they are more prioritized to serve compared to users with shorter delays. The underlying optimization problem is as follows:

$$\mathcal{P}_0 : \max_{\boldsymbol{\Theta}} \sum_{k=1}^{K} \omega_k d_k^{\mathsf{a}},$$

$$\text{s.t. } d_k^{\mathsf{a}} \leq D_k, \forall k, \tag{6a}$$

$$d_k^{\mathsf{a}} \leq \sum_{t=1}^{T} d_k[t], \forall k, \tag{6b}$$

$$E_n^{\mathsf{nec}} + \sum_{t=1}^{T} E_n^{\mathsf{c}}[t] \leq E_n^{\mathsf{o}}, \forall n, \tag{6c}$$

$$\|\mathbf{c}_n[t] - \mathbf{c}_n[t-1]\| \leq \min\left(V_{\mathsf{max}}\delta, D_{\mathsf{max}}\right), \forall n, t, \tag{6d}$$

$$\|\mathbf{c}_n[t] - \mathbf{c}_m[t]\| \geq D_{\mathsf{o}}, \forall n \neq m, t, \tag{6e}$$

constraints (2) and (3),

where $\boldsymbol{\Theta} = \left\{\left\{b_{n,k}[t]\right\}, \left\{p_{n,k}[t]\right\}, \left\{\mathbf{c}_n[t]\right\}, \left\{d_k^{\mathsf{a}}\right\}\right\}$ is the set of optimization variables, ω_k is the weight of user k which is chosen to capture users' priority, and (2), (3), and (6c) capture the transmit power constraints, bandwidth constraints, and total energy constraints, respectively. Constraint (6d) captures the maximum distance that a particular UAV can travel in one time slot, where V_{max} is the maximum speed and D_{max} is the maximum displacement to ensure the LoS channel conditions stay approximately the same. Constraint (6e) is to avoid collision. In (6c), E_n^{o} is the energy of UAV n at the beginning of the service period, and E_n^{nec} is the minimal energy necessary for the UAV to fly from $\mathbf{c}_n[T]$ to the station at $\mathbf{c}_{\mathsf{o}}$ which can be computed as

$$E_n^{\mathsf{nec}} = \frac{E^{\mathsf{min}}}{\delta} \frac{\|\mathbf{c}_n[T] - \mathbf{c}_{\mathsf{o}}\|}{v_{\mathsf{min}}}, \tag{7}$$

where E^{min} is the minimum energy consumption given in (1) with v_{min} being the corresponding minimal energy consuming speed.[4] Note that E_n^{nec} is the function of the UAV's final position $\mathbf{c}_n[T]$. (6c) constrains the position of the UAV n at the end of the service period in such a way that it has sufficient energy to fly back to the station.[5] Thus, problem $\mathcal{P}_0$ does not force the UAV to be at any specific locations at the start and end of the service period.

Problem $\mathcal{P}_0$ has a non-concave objective function and non-convex constraints (6b), (6c), and (6e). Thus, it is a non-convex problem which is difficult to solve optimally.

4 Proposed Design

In this section, we present an iterative algorithm to solve $\mathcal{P}_0$ by jointly applying alternating optimization and the SCA techniques. Specifically, the alternating optimization method is applied to optimize the objective function of $\mathcal{P}_0$ with respect to one set of variables given the others, while the SCA is applied to approximate and convexify the trajectory control subproblem.

4.1 *Optimization of Bandwidth Allocation, Given UAV Trajectories and Transmit Powers*

Given the positions of UAVs and the transmit powers, the bandwidth optimization subproblem can be expressed as

$$\mathcal{P}^{\mathsf{B}}: \max_{\{b_{n,k}[t], d_k^{\mathsf{a}}\}} \sum_{k=1}^{K} \omega_k d_k^{\mathsf{a}},$$

$$\text{s.t. } (2), (3), (6a) \text{ and } (6b).$$

Problem $\mathcal{P}^{\mathsf{B}}$ is linear and can be solved optimally using optimization solvers such as CVX. Then, the communication scheme with obtained bandwidth allocation $\left\{b_{n,k}^*[t]\right\}$ can be realized by using FDMA [7].

---

4 E^{min} and v_{min} can be numerically computed from (1), given the numerical values of all necessary constants.

5 We note here that the proposed energy condition (6c) can adopt any UAV energy consumption model. In this work, we use one in (1) from [6] since it is widely adopted in the community.

4.2 *Optimization of Transmit Power, Given Bandwidth Allocation and UAV Trajectories*

Given the trajectories of UAVs and the bandwidth allocation, the power allocation subproblem can be expressed as

$$\mathcal{P}^{\mathsf{P}}: \max_{\{p_{n,k}[t], d_k^{\mathsf{a}}\}} \sum_{k=1}^{K} \omega_k d_k^{\mathsf{a}},$$
$$\text{s.t. } (3), (6a) \text{ and } (6b).$$

Problem $\mathcal{P}^{\mathsf{P}}$ is a convex optimization problem. So, it can be solved optimally by optimization solvers such as CVX.

4.3 *Optimization of UAV Trajectories, Given Transmit Powers and Bandwidth Allocation*

Given the transmit powers and the bandwidth allocation, the UAV trajectory control subproblem can be stated as follows:

$$\mathcal{P}^{\mathsf{C}}: \max_{\{\mathbf{c}_{n,k}[t], d_k^{\mathsf{a}}\}} \sum_{k=1}^{K} \omega_k d_k^{\mathsf{a}},$$
$$\text{s.t. } (6a), (6b), (6c), (6e) \text{ and } (6d).$$

Problem $\mathcal{P}^{\mathsf{C}}$ is non-convex due to constrains (6b), (6c), and (6e). So, we apply the SCA method to convexity and solve the problem iteratively until convergence. Let the solution obtained from the previous iteration be $\bar{\mathbf{c}}_n[t]$, and the constraint (6e)[6] can be approximated as follows:

$$2(\bar{\mathbf{c}}_n[t]-\bar{\mathbf{c}}_m[t])^T(\mathbf{c}_n[t]-\mathbf{c}_m[t])-\|\bar{\mathbf{c}}_n[t]-\bar{\mathbf{c}}_m[t]\|^2 \geq D_{\mathsf{o}}^2. \quad (8)$$

To tackle the non-convex constraint (6c), we introduce an auxiliary nonnegative variable $\alpha_n[t]$ as follows[11]:

$$\alpha_n^2[t] \geq \left(1+\frac{\mathbf{v}_n[t]^4}{4v_0^4}\right)^{1/2} - \frac{\mathbf{v}_n[t]^2}{2v_0^2}. \quad (9)$$

[6] Actually, we have equivalently squared both sides of (6e).

Using these results, we come up with the following problem:

$$\mathcal{P}^{\mathsf{C1}}: \max_{\{\mathbf{c}_n[t], d_k^{\mathsf{a}}, \alpha_n[t]\}} \sum_{k=1}^{K} \omega_k d_k^{\mathsf{a}},$$

$$\text{s.t. } \delta \sum_{t=1}^{T} \left(P_{\mathsf{a}} + \frac{3P_{\mathsf{a}} \|\mathbf{v}_n[t]\|^2}{U^2} + \frac{1}{2}\beta \|\mathbf{v}_n[t]\|^3 \right)$$

$$+ \delta P_{\mathsf{b}} \sum_{t=1}^{T} \alpha_n[t] + E_n^{\mathsf{nec}} \leq E_n^{\mathsf{o}}, \tag{10a}$$

(6a), (6b), (6d), (8) and (9).

It can be verified that for the optimal solution of $\mathcal{P}^{\mathsf{C1}}$, the equality of (9) must hold; otherwise, we can always decrease $\alpha_n[t]$ to meet the equality of (9). However, (9) is still non-convex. So, we transform it into the following inequality:

$$\alpha_n^2[t] + \frac{\|\mathbf{v}_n[t]\|^2}{v_0^2} \geq \frac{1}{\alpha_n^2[t]}. \tag{11}$$

The constraint (11) is still non-convex; thus, we apply the SCA method again. Let the local values be $\bar{\alpha}_n[t]$, and (11) can be approximated by the following constraint:

$$2\bar{\alpha}_n[t]\alpha_n[t] - \bar{\alpha}_n[t]^2 + \frac{2\bar{\mathbf{v}}_n^T[t]\mathbf{v}_n[t] - \|\bar{\mathbf{v}}_n[t]\|^2}{v_0^2} \geq \frac{1}{\alpha_n^2[t]}, \tag{12}$$

where $\bar{\mathbf{v}}_n[t] = (\bar{\mathbf{c}}_n[t] - \mathbf{c}_n[t-1])/\delta$. Finally, the logarithm terms in (6b) can be approximated as follows [12]:

$$\begin{aligned} \log\left(1 + \frac{\tilde{\gamma}_{n,k}[t]}{h^2 + \|\mathbf{c}_n[t] - \mathbf{u}_k\|^2}\right) \geq \log\left(1 + \frac{\tilde{\gamma}_{n,k}[t]}{h^2 + \|\bar{\mathbf{c}}_n[t] - \mathbf{u}_k\|^2}\right) \\ - \left(\|\mathbf{c}_n[t] - \mathbf{u}_k\|^2 - \|\bar{\mathbf{c}}_n[t] - \mathbf{u}_k\|^2\right) \bar{Z}_{n,k}[t], \end{aligned} \tag{13}$$

where $\tilde{\gamma}_{n,k}[t] = \gamma p_{n,k}[t]$ and

$$Z_{n,k}[t] = \frac{\log_2(e)\tilde{\gamma}_{n,k}[t]}{\left(h^2 + \|\bar{\mathbf{c}}_n[t] - \mathbf{u}_k\|^2\right)\left(\tilde{\gamma}_{n,k}[t] + \left(h^2 + \|\bar{\mathbf{c}}_n[t] - \mathbf{u}_k\|^2\right)\right)}.$$

Using all these results and approximations, problem $\mathcal{P}^{\mathsf{C}}$ can be transformed and solved by iteratively solving the following convex optimization problem:

$$\mathcal{P}^{\mathsf{C2}}: \max_{\{\mathbf{c}_n[t], d_k^{\mathsf{a}}, \alpha_n[t]\}} \sum_{k=1}^{K} \omega_k d_k^{\mathsf{a}},$$

$$\text{s.t. (6a), (6b), (6d), (8), (10a) and (12).}$$

Problem $\mathcal{P}^{\mathsf{C2}}$ is convex,[7] and it can be solved by using CVX.

4.4 Feasible Point for UAV Trajectory Optimization Subproblem

To apply the SCA method, one must find a local feasible point for initial $\bar{\mathbf{c}}_n[t]$, $t = 1, \ldots, T$, which is not trivial, especially when the energy constraint (10a) becomes more and more stringent over time. We describe how to find a particular feasible trajectory for the UAVs in the following.

4.4.1 Hovering Trajectories

The straightforward choice of $\bar{\mathbf{c}}_n[t]$ is their position at the beginning of the considered service period $\mathbf{c}_n[0]$, i.e., initially we can set $\bar{\mathbf{c}}_n[t]=\mathbf{c}_n[0]$, $t = 1, \ldots, T$. From (1), the hovering energy consumed in one time slot is $E^{\mathsf{h}}=\delta(P_{\mathsf{a}}+P_{\mathsf{b}})$. Hence, we can find the maximum number of time slots that UAV n can hover on $\mathbf{c}_n[0]$ with their initial energy E_n^{o} as $t_n^{\mathsf{h}} = \lfloor (E_n^{\mathsf{o}} - E_n^{\mathsf{nec}}) / E^{\mathsf{h}} \rfloor$. If $t_n^{\mathsf{h}} \geq T$, we can let the UAV hover on $\mathbf{c}_n[0]$ for the whole service period.[8] It is also possible that $t_n^{\mathsf{h}} < T$. In that case, the UAV will hover on $\bar{\mathbf{c}}_n[0]$ for t_n^{h} time slots, and then it will fly back to the station at velocity of v_{min} on a straight trajectory in the remaining time slots.

4.4.2 Arrival–Recharge–Departure Trajectories

In case the UAV's energy is nearly depleted and its energy is not sufficient to maintain its flight over the whole service period, the UAV needs to fly back to the station, and then it gets recharged and departs again. It is verified that this happens only when $t_n^{\mathsf{h}} = 0$. In this case, the UAV starts to fly back to the station with

[7] Note that the logarithm terms in (6b) are replaced as described in (13).

[8] Note that it is only the initially feasible trajectory for the optimization problem $\mathcal{P}^{\mathsf{C2}}$. Solving $\mathcal{P}^{\mathsf{C2}}$ may result in a different trajectory.

velocity v_{min} from time slot 1. Then, the number of flying-back time slots for UAV n is $t_n^{\text{back}} = \lfloor E_n^{\text{o}}/E^{\text{min}} \rfloor$. In case $t_n^{\text{back}} < T$, the UAV gets recharged immediately upon its arrival to the station. Then, it departs following a certain trajectory.[9] If this case happens, we remove the energy constraint (6c) for UAV n. We would like to mention three important points here. First, due to the energy constraints of the previous duration, it is guaranteed that the UAVs have enough energy to fly back to the station. Second, we assume that the recharge process takes a negligible amount of time. We could prepare some replacement UAVs that will depart immediately when the energy-depleted UAVs arrive or we can simply swap the fully charged batteries for the energy-depleted UAVs. Nevertheless, these will not change the nature of our framework. Third, we assume that the service period is sufficiently short so that UAVs need to recharge at most once within a single flying duration.

4.5 *Proposed Persistent Design*

It is possible that the system cannot serve all users within a single service period. Therefore, we now describe how the proposed design presented above can be used to serve users' requests over multiple consecutive service periods. This is called persistent design in the following. If the requested data demand cannot be met in the first service period, the leftover data amount will be transmitted to unsatisfied users in the next periods. Note that if a particular data transmission demand is serviced in multiple service period, the service delay, called access delay in the following, will increase accordingly (e.g., we add T to users' access delay if one more service period is required to serve their demands). We summarize our proposed persistent design in Algorithm 2, where ϵ is a number that defines the desired accuracy level.

5 Numerical Results

5.1 *Simulation Setting*

In our considered simulation, there are 4 UAVs that fly at the fixed altitude of $h = 100$ m. The maximum transmit power of each UAV is set equal to 0.1 W. The UAV energy consumption parameters[10] are chosen as follows: $P_{\text{a}} = 79.86$, $U = 120$, $\beta = 0.0185$, $P_{\text{b}} = 88.63$, and $v_0 = 4.03$. Initially, each UAV has a flying

[9] It could be any trajectory. In the simulation, the station is at (0, 0), and we let the UAV depart and fly on a straight trajectory away from the station along a random direction.

[10] The parameters are chosen/computed according to typically used drones and in typical flying conditions. Please refer to [6] for more numerical details about energy models of rotary-wing UAVs.

Algorithm 2 Persistent design for multiple service periods

INPUT: All network parameters.
while 1 **do**
Gather users' information: delay and transmission demand (both uncompleted users from previous periods and new-coming users), determine users' weights.
Find the initially feasible trajectories for the UAVs according to Sect. 4.4.
Solve $\mathcal{P}^0$ by iteratively solving $\mathcal{P}^{\mathsf{C}}$, $\mathcal{P}^{\mathsf{B}}[t]$, $\mathcal{P}^{\mathsf{P}}$ until convergence. Denote the solution as $\{\{b_{n,k}[t]\}, \{p_{n,k}[t]\}, \{\mathbf{c}_n[t]\}, \{d_k^{\mathsf{a}}\}\}$.
for all users k **do**
if $d_k^{\mathsf{a}} \geq D_k - \epsilon$ **then**
User k is satisfied.
else
User k is not satisfied, the user will be served in the next service period.
Let $D_k = D_k - d_k^{\mathsf{a}}$ and $\tau_k = \tau_k + T$ (for the next service period).
end if
end for
Update UAVs' remaining energy (for the next period).
end while
OUTPUT: UAV trajectories and resource allocation solutions for consecutive service periods.

energy capacity of 12×10^4 Joules. The 2D coordinate of the charging station is $(0, 0)$. We assume that the energy for landing and taking off of UAVs is reserved separately from its flying energy. We also assume that the energy budget for circuit and wireless communications is reserved and separated from the flying energy budget. Each user is placed randomly over the network area where its coordinate follows uniform distribution on a disk area with a radius of 1 km. For the sake of simplicity, we consider that all users have an identical data transmission demand of $D = 5$ Mbits unless specified otherwise. The total system bandwidth available for downlink communications is $B = 1$ MHz. The service duration is $T = 10$ s unless specified otherwise. Each time slot has the length of $\delta = 1$ s.

5.2 *Baseline Scheme*

We compare the performances of our proposed persistent design with a baseline (cyclic) scheme in the following. In the baseline scheme, we also perform optimization of the UAV trajectory, power, and bandwidth allocation which follows standard approaches in existing works [6–8]. However, in the cyclic scheme, UAVs have to depart from and arrive to the charging station at $t = 1$ and $t = T$, respectively, as similarly considered in [6–8].

5.3 Simulation Results

First, we study the behaviors of UAVs under our proposed design where we plot in Fig. 1 the varying energy levels of UAVs over a time window of 840 s.[11] A jump in energy level indicates that a UAV has just been recharged/replaced at the station. As can be observed, it is not necessary that UAVs arrive to the station at the same time, and UAVs can fly as long as they have enough energy. The fact that not all UAVs come to the charging station at the same time is beneficial for the system, since it increases the spatial diversity. The improved performances due to this increased diversity will be confirmed by results presented in the next figures.

In Fig. 2, we fix the number of users in each duration and investigate the average percentage of total delivered data to total user demands for several periods. When the number of users increases, the average percentages of total delivered data to total user demand decrease for both designs. However, the proposed design can always deliver more data compared to that due to the cyclic scheme. This confirms the superior performance of our design compared to the baseline scheme.

We now study the practical dynamic scenario with stochastic user arrivals over time. For simulation, user (or data requests) arrivals follow a Poisson distribution

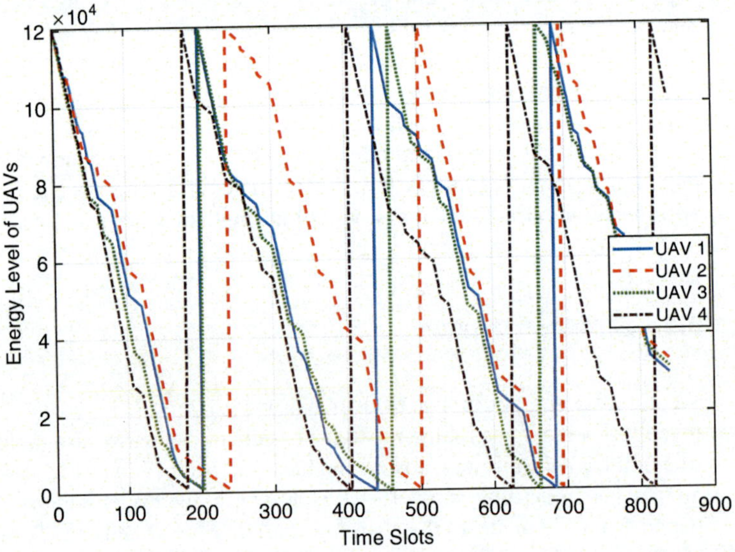

Fig. 1 Energy level of UAVs over time, $T = 10$

[11] The trajectories of multiple UAVs over a long time frame is not intuitive to show. As we can hardly observe anything insightful from these overlapping and complicated trajectories. So we plot the energy levels instead.

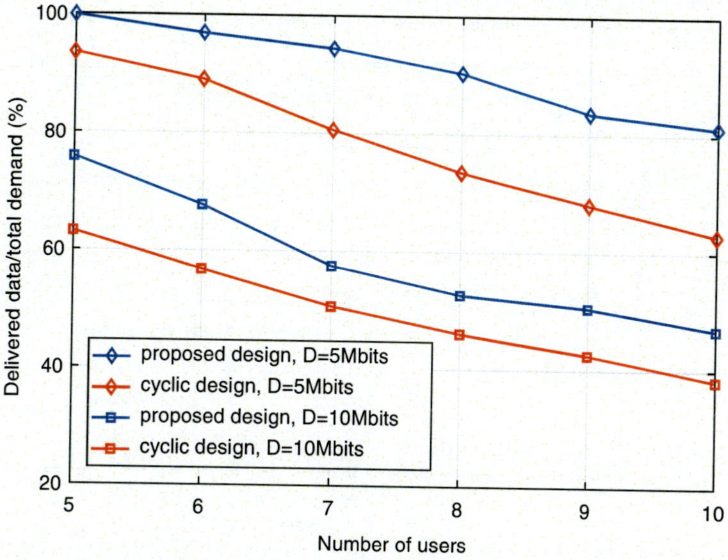

Fig. 2 Delivered data over total demand, $T = 10$

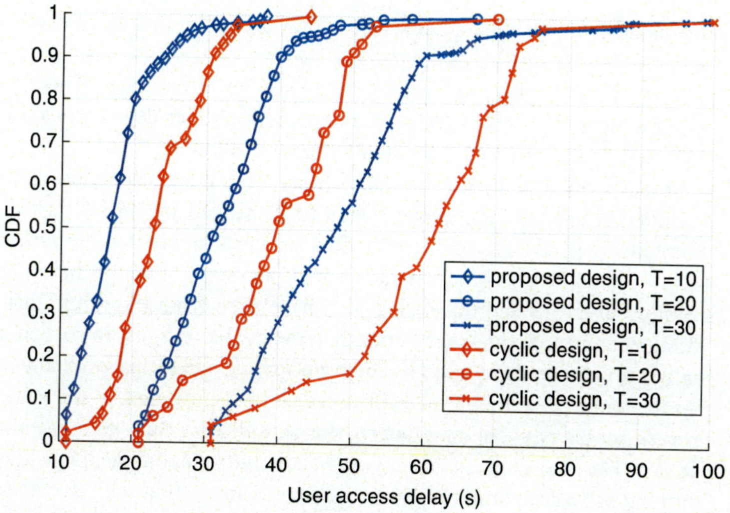

Fig. 3 CDF of user access delay for different duration T

with an expected arrival rate of 0.5 user/second. We set the user's weight equal to its delay ($\omega_k = \tau_k$). In Fig. 3, we show the empirical cumulative distribution function (CDF) of user access delay for different values of service duration (10, 20, 30 s) and compare the performance of our proposed design with the baseline scheme.

Several interesting observations can be drawn from this figure. First, the proposed design provides much shorter access delays compared to those provided by the

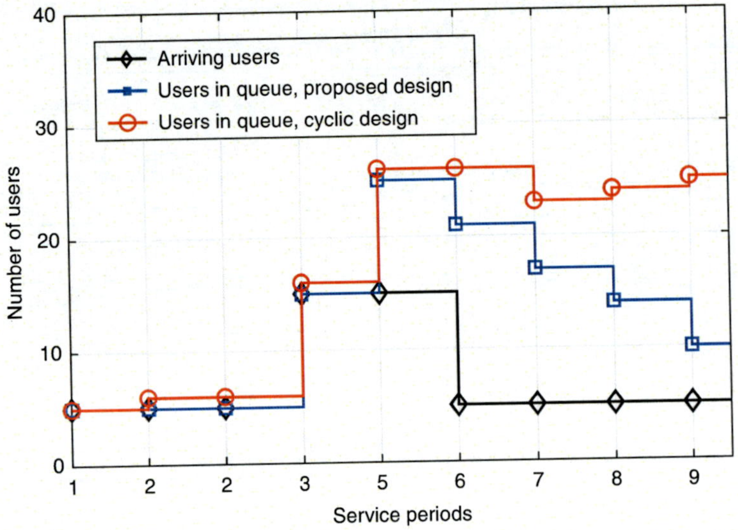

Fig. 4 Number of arriving users and number of users in the queue

cyclic design. Second, higher service durations result in longer access delay. Finally, the smaller the service durations, the steeper the CDF curves, which indicates that the user access delays vary less dramatically with smaller T. Therefore, the system is more adaptive to user requests and provides better degrees of fairness when the optimization is performed for smaller service duration T.

Finally, to investigate the adaptability of our persistent design, we show the number of users in the queue (users to be served in the following periods) over time in Fig. 4. We create a user arrival scenario in which there are 5 users coming every period (10 s). We create a surge at periods 3, 4, and 5 in which there are 15 coming users instead of 5 users. It can be seen from the figure that the proposed design can handle the surge better than the cyclic designs. In particular, the number of users in the queue served by the proposed design drops gradually after the surge. At the same time, the number of users stays approximately the same in the queue served by the conventional design, which indicates that the conventional design is easier to be overloaded. Hence, we can conclude that our design is less vulnerable to surging scenarios, and it is more sustainable in dealing with dynamic user arrivals, compared to the cyclic design.

6 Conclusion

In this chapter, we have proposed a novel persistent design for UAV-based wireless networks to efficiently serve dynamic data transmission demands of wireless users. Toward this end, we have developed an optimization framework to optimize UAV trajectory, power, and bandwidth allocation in each service period and apply this framework to devise a persistent design algorithm. Simulation results have demonstrated the great benefits of our proposed design in efficiently serving users' data demands and achieving better access delay compared to the baseline cyclic scheme.

References

1. Fotouhi, A., Qiang, H., Ding, M., Hassan, M., Giordano, L.G., Garcia-Rodriguez, A., Yuan, J.: Survey on UAV cellular communications: practical aspects, standardization advancements, regulation, and security challenges. IEEE Commun. Surv. Tut. **21**(4), 3417–3442 (2019)
2. Sekander, S., Tabassum, H., Hossain, E.: Multi-tier drone architecture for 5G/B5G cellular networks: challenges, trends, and prospects. IEEE Commun. Mag. **56**(3), 96–103 (2018)
3. Alzenad, M., El-Keyi, A., Lagum, F., Yanikomeroglu, H.: 3-D placement of an unmanned aerial vehicle base station (UAV-BS) for energy-efficient maximal coverage. IEEE Commun. Lett. **6**(4), 434–437 (2017)
4. Nguyen, M.D., Ho, T.M., Le, L.B., Girard, A.: UAV placement and bandwidth allocation for UAV based wireless networks. In: Proceedings of the IEEE Global Communications Conference (GLOBECOM), pp. 1–6. IEEE, Piscataway (2019)
5. Mozaffari, M., Saad, W., Bennis, M., Debbah, M.: Wireless communication using unmanned aerial vehicles (UAVs): optimal transport theory for hover time optimization. IEEE Trans. Wirel. Commun. **16**12, 8052–8066 (2017)
6. Zeng, Y., Xu, J., Zhang, R.: Energy minimization for wireless communication with rotary-wing UAV. IEEE Trans. Wirel. Commun. **18**(4), 2329–2345 (2019)
7. Wu, Q., Zeng, Y., Zhang, R.: Joint trajectory and communication design for multi-UAV enabled wireless networks. IEEE Trans. Wirel. Commun. **17**(3), 2109–2121 (2018)
8. Nguyen, M.T., Le, L.B.: Flight scheduling and trajectory control in UAV-based wireless networks. In: Proceedings of the IEEE Wireless Communications and Networking Conference (WCNC), pp. 1–6. IEEE, Piscataway (2020)
9. Ji, J., Zhu, K., Niyato, D., Wang, R.: Joint cache placement, flight trajectory and transmission power optimization for multi-UAV assisted wireless networks. IEEE Trans. Wirel. Commun. **19**, 5389–5403 (2020)
10. Qualcomm Technologies: LTE unmanned aircraft systems. Available at https://www.qualcomm.com/media/documents/files/lte-unmanned-aircraft-systems-trial-report.pdf (2019/10/07), May 2017
11. Zhan, C., Huang, R.: Energy efficient adaptive video streaming with rotary-wing UAV. IEEE Trans. Veh. Technol. **69**(7), 8040–8044 (2020)
12. Bertsekas, D.P.: Nonlinear programming. J. Oper. Res. Soc. **48**(3), 334–334 (1997)

Intelligent Cognition in an Integrated Satellite–Aerial–Terrestrial Network for Connected Vehicles

Invited Chapter

Farhan Pervez, Lian Zhao, and Cungang Yang

1 Introduction

With the advancement in the interconnectivity of network objects, using cutting-edge communication protocols for a variety of Internet-of-Things (IoT) applications and services, including smart cities, connected cars, and mission-critical support, an unprecedented number of people and devices are able to exchange information [1]. To enable the delivery of various segments of these services, the deployment of 5th generation (5G) terrestrial networks has played a vital role, especially in urban areas [2]. In addition, a satellite network that can reach remote locations is becoming an important medium for IoT applications such as animal monitoring and disaster management. One proposal is to deploy a two-tier satellite–terrestrial network to deliver ubiquitous service that caters user needs. However, this type of a network is hardly efficient or economically viable because of limited resources, high maintenance costs, and round-trip delays. Unmanned aerial vehicles (UAVs) contingent aerial communication is developed as an intriguing model for IoT services because of its comparatively low cost and flexibility of mobility. As a result, for future 6th generation (6G) systems, the research circle is debating a triple-tier integrated satellite–aerial–terrestrial (I-SAT) network [3]. Moreover, cognitive radio (CR), which can alleviate the spectrum insufficiency issue and improve spectrum efficiency by intelligent spectrum access, is considered as a propitious solution for IoT applications. To indicate the benefits of integrating UAV-based aerial base station into a CR-based cellular network, a number of literature are published. For example, the authors of [4] optimize the transmission rate and transmit power control of UAVs in order to maximize the network throughput of ground users.

F. Pervez · L. Zhao (✉) · C. Yang
Department of Electrical, Computer & Biomedical Engineering, Toronto Metropolitan University (formerly Ryerson University), Toronto, ON, Canada
e-mail: farhan.pervez@ryerson.ca; l5zhao@ryerson.ca; cungang@ryerson.ca

H. Nguyen et al. (eds.), *30th Biennial Symposium on Communications 2021*,
Signals and Communication Technology, https://doi.org/10.1007/978-3-031-06947-5_11

In contrast, the authors in [5] adjust the UAV's trajectory and transmit power of both the terrestrial BS and the UAV to maximize the achievable rate of terrestrial users while accounting for interference and the movement of UAV. Moreover, [6] enable interference limit requirement for a primary satellite network when supplying secondary ground users with good availability for cognitive satellite–terrestrial networks. Furthermore, [7] present power allocation strategy to maximize the achievable terrestrial user rate with amplify-and-forward relays.

In this chapter, we formulate and solve a throughput maximization problem for autonomous vehicles served by an aerial–terrestrial network in a CR-enabled I-SAT environment, motivated by the benefits of using cognitive communication in satellite–terrestrial and aerial–terrestrial networks, as well as the vast application prospects of IoT. The paper proposes a solution to a joint power allocation and UAV mobility control problem in which on-ground vehicles are secondary users (SUs) serviced by a Ka-band aerial–terrestrial network, while the satellite receiver (SR) is the primary user (PU). As a result, an I-SAT network must use CR intelligence to ensure that the service criteria for PU and SUs are met.

Within the specified constraints, there are still open challenges for a three-tier cognitive system to efficiently handle IoT applications, such as moving intelligent automobiles. On the one hand, in a dense wide area, the mobility of smart vehicles makes it difficult to achieve expected level of QoS. On the other hand, to enforce the required objective, different resources, such as the BS and UAV transmit powers and the trajectories of UAVs, should be assigned jointly instead of being allocated separately. This chapter considers an intelligent I-SAT network that consists of one on-ground base station and multiple unmanned aerial vehicles (UAVs) serving moving vehicles on the ground. Within the same spectrum, the transmitters (BS/UAVs) communicate data to secondary users (on-ground vehicles) in the vicinity of a primary user (satellite receiver). The interference experienced by the SR due to BS/UAVs transmit power must be less than the predefined limit for using the satellite network's radio bands, which means that the transmission power, as well as the trajectory, should be intelligently designed to protect the primary network's performance. On the one hand, if the BS and UAV allot higher transmission power to communicate with on-ground vehicles, the SR will be subjected to more interference. On the other hand, as a UAV moves far from the primary user or gets closer to a vehicle, it provides higher transmission power to boost network throughput. Based on this, the goal of this chapter is to optimize the vehicle's possible throughput by optimizing BS/UAV transmit power allocations and UAV trajectory with SR interference limitations and UAV mobility constraints at the same time.

The remainder of this chapter is organized as follows. Section 2 discusses the system model and problem formulation for the integrated network. In Sect. 3, we propose an efficient iterative algorithm to solve the formulated problem by applying the subproblem method, whereas the performance of the proposed solution is demonstrated in Sect. 4. Finally, we conclude the paper in Sect. 5.

2 System Model and Problem Statement

As shown in Fig. 1, we propose a UAV-enabled terrestrial network in which a single base station (BS) and a group of UAVs function as transmitters to give network connectivity to K vehicles. A single antenna is used by both the transmitters and the users on the ground. Let k be the vehicle index, $k \in \{1, 2, ..., K\}$, whereas q as the unmanned aerial vehicle index, $q \in \{1, 2, ..., Q\}$. Besides, the UAV-assisted aerial–terrestrial network and the satellite network share the same spectrum, with the satellite connecting to the satellite receiver and the BS/UAVs sending data to the vehicular users. As shown in the figure, UAV q provides service to its connected users through a range of channels. Further, the UAV may cause interference to other unrelated vehicles as well as the satellite receiver. Moreover, in between a time span of C seconds, UAVs can move freely and modify their horizontal path while maintaining an altitude, H_u, that is fixed. Aerial stations are also linked to the terrestrial station via free space optical (FSO) communications, as shown in the model. The network optimization of the considered model, in accordance with the objective function, on the other hand, can be done in the network cloud. To make the issue more simplified, the time span is split into T time slots, $t \in \{1, 2, ..., T\}$, with all slots having a duration of $\delta = C/T$ each. As a result, UAV q has horizontal coordinates at time slot t as $f_q[t]$, $\forall t$. Moreover, due to its great mobility and stationary qualities, a rotary-wing UAV is selected in our work.

The total transmitters are expressed by M, with $m \in \{1, 2, ..., M\}$, that includes Q UAVs and a BS. A user can link to the transmitter that offers the best downlink

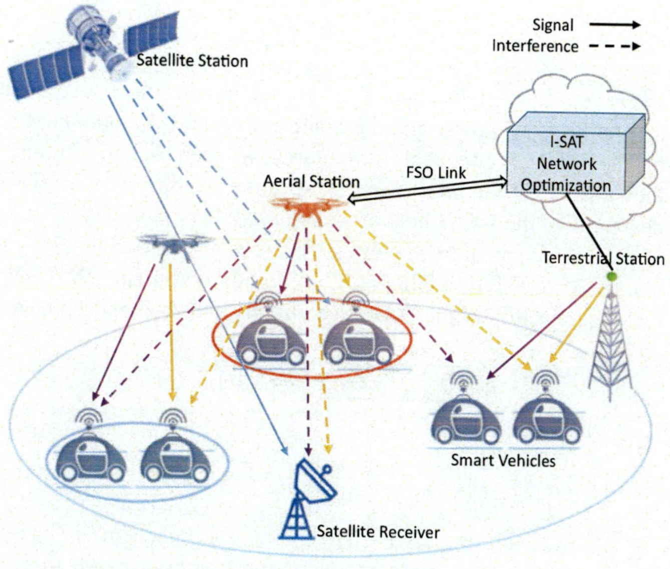

Fig. 1 Schematic of the integrated model under consideration

channel quality. The fast fading is averaged out since the association is taken out over a long time scale in relation to the change in channel condition. Let n be the number of available channels, with $n \in \{1, 2, ..., N\}$. Each transmitter has a maximum of N channels; however, each user is only allotted one channel. As a result, at time t, the possible rate achieved for vehicle k linked to transmitter m via channel n is

$$R_{kmn}[t] = B_m \cdot \log_2 (1 + \gamma_{kmn}[t]), \tag{1}$$

where B_m represents the transmission bandwidth and γ_{kmn} denotes the SINR for vehicle k given as

$$\gamma_{kmn}[t] = \frac{G_m p_{mn}[t] h_{km}[t]}{\sum_{j \neq m} G_j p_{jn}[t] h_{kj}[t] + G(\theta_{ks}) P_{SS} h_{ks} + \sigma_k^2}, \tag{2}$$

where p_{mn} gives the transmission power for m over n, h_{km} denotes the channel gain from the transmitter to vehicle, and G_m represents the transmitter's directional antenna gain. Moreover, h_{ks} denotes the channel gain from the satellite to vehicle k, whereas P_{SS} gives the transmit power of the satellite, which is considered uniform over each time slot. Furthermore, σ_k^2 denotes the power of additive white Gaussian noise at the vehicle, and $G(\theta_{ks})$ gives the satellite's antenna gain at transmission, where θ_{ks} is the angle between the satellite station and the vehicle.

Additionally, within time slot t, the interference from the BS/UAVs to the SR r is represented as

$$\zeta^r[t] = \sum_m \sum_n G_m p_{mn}[t] g_{rm}[t], \tag{3}$$

where g_{rm} denotes the channel gain from the transmitter to the satellite receiver.

The BS/UAV power and the trajectory of the UAV both have a significant effect on the vehicle's attainable throughput, as seen in (1), (2), and (3). In addition, they also affect the interference toward the satellite receiver. Therefore, intelligent designing of the I-SAT network is required.

Our objective is to maximize the sum of achievable data rate of K vehicles connected over a time horizon C by jointly optimizing BS/UAV power allocation and the trajectory of the UAV. Accordingly, the problem can be stated as follows:

$$\max_{\mathbf{p}, \mathbf{f}} \sum_t \sum_k \sum_m \sum_n R_{kmn}[t], \tag{4a}$$

s.t.

$$\zeta^r[t] \leq \zeta_{th}, \quad \forall t, \tag{4b}$$

$$0 \leq p_{mn}[t] \leq p_m^{max}, \quad \forall m, n \tag{4c}$$

$$f_q[1] = f_q[T], \quad \forall q, \tag{4d}$$

$$\|f_q[t+1] - f_q[t]\|^2 \leq V_q^{max}\delta, \quad t = 1, \ldots, T-1, \tag{4e}$$

$$\|f_q[t] - f_i[t]\|^2 \geq \chi_{min}^2, \quad \forall t, i, i \neq q, \tag{4f}$$

where (4b) denotes the interference limit, ζ_{th}, that the satellite receiver can handle, and (4c) defines the range of transmit power for UAVs and BS, with 0 as minimum and p_m^{max} as maximum transmission power. Constraint (4d) states that at the completion of time span C, the UAV gets back to its original location. Furthermore, UAV trajectories are influenced by the maximum speed (4e) and collision avoidance (4f). The V_q^{max} in (4e) specifies the maximum velocity or speed in meters per second (m/s), and the χ_{min} in (4f) denotes the minimum distance in meters in between UAVs in order to avoid collisions. The location of the UAV is assumed to be unchanging inside each interval by selecting a reasonably small time slot, δ.

3 Proposed Solution

The solution to the problem described in Sect. 2 is the topic of this section. The problem formulated can be divided into two main subproblems: transmit power allocation and UAV trajectory optimization. The subproblems are solved in turn to enhance the downlink network throughput. The following subsections investigate each of these subproblems separately. Finally, in the last subsection, a joint optimization algorithm is presented that integrates both subproblems.

3.1 *Transmission Power Optimization*

Given the locations of the BS and UAVs and the fact that maximizing for each interval, t, leads to maximizing the rate across the time span C, the transmit power in (4) can be optimized by resolving the following subproblem for each time window t:

$$\max_{\mathbf{p}} \sum_m \sum_n R_{mn}, \tag{5a}$$

s.t.

$$\zeta^r \leq \zeta_{th}, \tag{5b}$$

$$0 \leq p_{mn} \leq p_m^{max}, \quad \forall m, n. \tag{5c}$$

As the objective function (5a) is strictly concave, the subproblem (5) is therefore strictly quasi-concave for the stated constraints. Moreover, as a result, a local maximum is a global maximum, and because it is rigorously quasi-concave, (5) has only one maximum.

The solution to (5) can be found by splitting the problem into two layers. The first layer assigns power at the transmitter level to obtain power for every transmitter as $\mathbf{s} = \{s_1, ..., s_M\}$. The second, denoted as $\boldsymbol{\varrho} = \{\varrho_1, ..., \varrho_N\}$, offers power distribution in between distinct channels for every transmitter m. We use a simplistic and efficient geometric water-filling (GWF) method [8, 9] to resolve the two sub-layers. In contrast to the conventional water-filling (CWF) method, it does not require solving a non-linear system from the target problem's Karush–Kuhn–Tucker (KKT) conditions in order to determine the water level. Furthermore, it avoids complex derivations found in traditional optimization approaches, such as derivative or gradient procedures. The first layer allocating power, s_m, at the transmitter level for each transmitter m is expressed by

$$\max_{\mathbf{s}} \sum_{m} \log_2\left(1 + \overline{a_m} s_m\right), \tag{6a}$$

s.t.

$$\sum_{m} s_m g_{rm} \leq \zeta_{th}, \tag{6b}$$

$$0 \leq s_m \leq s_m^{max}, \quad \forall m, \tag{6c}$$

where antenna gain is set to one and $\overline{a_m}$ is the average or mean channel gain of N_m on-ground users linked to transmitter m and is expressed as

$$\overline{a_m} = \frac{1}{N_m} \cdot \sum_{k} h_{km}, \quad \forall m. \tag{7}$$

Figure 2a–b shows an explanation of the GWF algorithm. Inside a water tank, we assume four steps/transmitters ($M = 4$) having unit width. We selected M to indicate the number of steps in the analysis here because the number of steps in the GWF approach is same as the transmitters in the network. The step depth of the stairs denoted as $\{1, ..., M\}$ is monotonically rising because the sequence $\overline{a_m}$ is ordered as monotonically decreasing. Rather than attempting to discover ξ, a non-negative real number, the GWF technique seeks to identify the water-level step, which is an integer value ranging from 1 to M. This is the highest step in water, represented by the symbol m^*.

Let $P_A(m_z)$ indicate the volume of the water exceeding step m_z for $z = \{1, \ldots, |E|\}$, where E is a subsequence of the sequence $\{1, \ldots, M\}$, and $|E|$ is the cardinality of set E. By eliminating the water volume beneath step m_z from the interference threshold ζ_{th}, the value of $P_A(m_z)$ is denoted as

$$P_A(m_z) = \left[\zeta_{th} - \sum_{y=1}^{|E|-1} \left(d_{m_z} - d_{m_y}\right)\right]^+, \quad 1 \leq z \leq |E|, \tag{8}$$

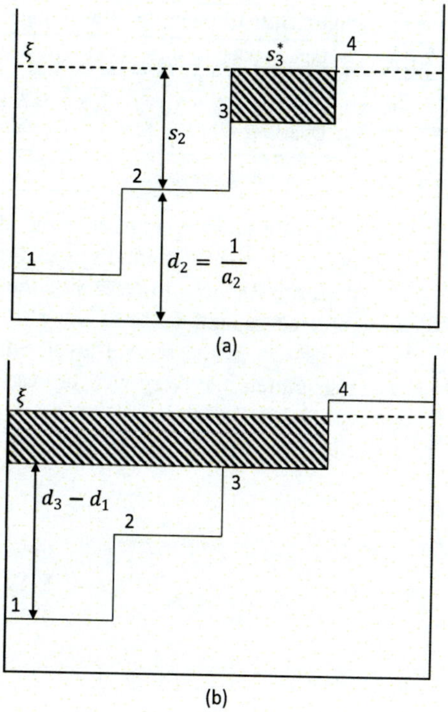

Fig. 2 Sketch for the GWF technique. (**a**) Illustration of water-level step $m^* = 3$, assigned power for the third step s_3^*, and step/stair depth $d_m = 1/\overline{a_m}$. (**b**) Illustration of $P_A(m)$ (the shadowed area showing the water/power above step m) when $m = 3$

where the step depth d_{m_z} is expressed as $\frac{1}{\overline{a_m}}$.

Hence, the explicit solution to (6) is

$$s_m = \begin{cases} \frac{1}{g_{rm}} \cdot [s_{z^*} + (d_{z^*} - d_m)], & 1 \leq m \leq z^* \\ 0, \; z^* < m \leq M \end{cases}, \tag{9}$$

where

$$z^* = max \{ z \mid P_A(m_z) > 0, \;\; 1 \leq z \leq |E| \}, \tag{10}$$

and the power level for the step can be expressed as

$$s_{z^*} = \frac{1}{z^*} \cdot P_A(m_{z^*}). \tag{11}$$

At the transmitter layer, Algorithm 1 shows the GWF-based dynamic power allocation mechanism. The state of this process is determined by the difference between the individualized peak power sequence and the present power distribution sequence. The process control is determined by (9)–(11) and is dependent on the

Algorithm 1 Power distribution at transmitter level

1: Input: vector d_m, s_m^{max} for $m = 1, 2, ..., M$, the set $E = 1, 2, ..., M$, and ζ_{th}.
2: Use (9)–(11) to get s_m.
3: $\Lambda \Rightarrow \{m \mid s_m > s_m^{max}, m \in E\}$. If Λ is null output is $\{s_m\}_{m=1}^{M}$ else $s_m = s_m^{max}$.
4: Update E with $E \setminus \Lambda$, update ζ_{th} with $\zeta_{th} - \sum_{x \in \Lambda} s_x$. Then go back to 2:.

state indicated above. As a result, at the next time stage, a new state occurs, and an optimal power distribution procedure with a state feedback emerges. The technique performs M loops to determine the optimal solution as the finite set E shrinks until the set Λ is null.

Moreover, the second layer, which assigns power, ϱ_n, to each channel n of a transmitter, may very well be resolved individually for each transmitter m by the subproblem given as

$$\max_{\varrho} \sum_{n} \log_2 (1 + a_n \varrho_n), \tag{12a}$$

s.t.

$$\sum_{n} \varrho_n = s_m, \tag{12b}$$

$$\varrho_n > 0, \ \forall n \tag{12c}$$

where

$$a_n = \frac{h_{kn}}{\sum_{j \neq m} \varrho_{jn} h_{kj} + G(\theta_{ks}) P_S s h_{ks} + \sigma_k^2}. \tag{13}$$

Using local partial zero-forcing precoding [10], the first term in the denominator can be omitted, while the step depth is defined as $d_n = 1/a_n$. Therefore, as a result of the analysis in Fig. 2, it is simple to deduce that the explicit answer to (12) is

$$\varrho_n = \begin{cases} \varrho_{z^*} + (d_{z^*} - d_n), & 1 \leq n \leq z^* \\ 0, & z^* < n \leq N \end{cases}, \tag{14}$$

where z^* and the power level for the step, ϱ_{z^*}, can be determined by using (10) and (11), accordingly. The value of $P_A(z)$, on the other hand, can be calculated by subtracting the water volume in step z from the overall transmitter power s_m as

$$P_A(z) = \left[s_m - \sum_{n=1}^{z-1} (d_z - d_n) \right]^+, \quad 1 \leq z \leq N. \tag{15}$$

Algorithm 2 Water-level step search for transmitter "m" channel power allocation

1: Initialize $s_m^X = s_{m^*} = s_m; n = 1$.
2: Compute $s_{m^*} \leq s_{m^*} - (d_{n+1} - d_n)$. Then $n \leq n + 1$, where "$\leq$" represents the assignment operation.
3: If $s_{m^*} > 0$ and $n \leq N$, $s_m^X = s_{m^*}$, and go back to step 2); else, output $z^* = n - 1$ and $\varrho_{z^*} = s_m^X$.

We can see from the above explicit solution that when z^* is obtained, $P_A(z^*)$ is given. To compute ϱ_{z^*}, it is first recorded and then scaled with a constant. As a result, the important aspect of the suggested GWF technique is how to search z^*, and Algorithm 2 explains how to do so.

3.2 UAV Trajectory Control

For the given transmission power of BS/UAVs, the UAV trajectory in problem in (4) may be optimized by solving the subsequent subproblem:

$$\max_{\mathbf{f}} \sum_t \sum_q R_q[t], \tag{16a}$$

s.t.

$$f_q[1] = f_q[T], \quad \forall q, \tag{16b}$$

$$\| f_q[t+1] - f_q[t] \|^2 \leq V_q^{max} \delta, \quad t = 1, \ldots, T-1, \tag{16c}$$

$$\| f_q[t] - f_i[t] \|^2 \geq \chi_{min}^2, \quad \forall t, i, i \neq q, \tag{16d}$$

where $R_q[t] = \sum_k R_{kq}$ indicates that the rate attained by UAV q is contingent on the linked vehicles in the time interval t.

Due to the non-convex constraint in (16d), (16) is neither concave nor quasi-concave. As a result, Lemma 1 is utilized to modify the non-convexity, and an iteration-based successive convex approximation approach is utilized, in which the original function is estimated by a more manageable function at each iteration at a given local point [11].

Lemma 1 *We define a quadratic function as given below:*

$$g(x) = x^2. \tag{17}$$

At iteration η, the following inequality can be expressed for a given local location x^η:

$$x^2 \geq 2x^\eta \times (x - x^\eta) + (x^\eta)^2. \tag{18}$$

Proof From (17), we can see that $g(x)$ is a convex function. The inequality holds in (18) because any convex function can determine its lower bound by first-order Taylor approximation at a local point.

In the ηth iteration, we represent UAV trajectory as $F^{\eta} = \{f_q^{\eta}[t], \forall q, t\}$. Since $\| f_q[t] - f_i[t] \|^2$ is a convex function with regard to $f_q[t]$, the following inequality based on Lemma 1 for constraint (16d) can be given:

$$\| f_q[t] - f_i[t] \|^2 \geq - \| f_q^{\eta}[t] - f_i^{\eta}[t] \|^2 + 2(f_q^{\eta}[t] - f_i^{\eta}[t])^{\perp} \times (f_q[t] - f_i[t]), \quad \forall i \neq q, t, \tag{19}$$

where the matrix transpose is given by $\perp$. For any given local point F^{η} and the lower bound defined in (19), the problem can be estimated by recalling the first-order Taylor approximation:

$$\max_{\mathbf{F}} \sum_t \sum_q R_q^{\eta}[t], \tag{20a}$$

s.t.

$$f_q[1] = f_q[T], \quad \forall q, \tag{20b}$$

$$\| f_q[t+1] - f_q[t] \|^2 \leq V_q^{max} \delta, \quad t = 1, \ldots, T-1, \tag{20c}$$

$$\chi_{min}^2 \leq -\| f_q^{\eta}[t] - f_i^{\eta}[t] \|^2 + 2(f_q^{\eta}[t] - f_i^{\eta}[t])^{\perp} \times (f_q[t] - f_i[t]), \quad \forall t, i, i \neq q, \tag{20d}$$

where constraint (20c) is convex quadratic and constraints (20b) and (20d) are both linear. Therefore, problem (20), being a convex optimization problem, can be efficiently solved by standard convex optimization solvers [12]. The lower bounds adopted in (20d) suggest that any feasible solution for (20) is also feasible for (16); however, the reverse does not hold. Thus, the optimal objective value obtained from (20) in general caters as a lower bound of (16).

3.3 *Alternating Optimization Algorithm*

Based on the discussed solutions for the two subproblems, we propose an iterative approach for the problem in (4) utilizing the block descent method [13], also referred as the alternating optimization method. The optimization parameters in the original problem, $\{\mathbf{p}, \mathbf{f}\}$, are divided into two blocks. Then, by solving problems (5) and (16) consecutively, while keeping the other block of variables unchanged, the transmit power $\mathbf{p}$ and UAV trajectory $\mathbf{f}$ are alternately optimized. In addition, each iteration's solution is used as the input for the following iteration. Algorithm 3 summarizes the framework. The computational complexity of solving (5) can be expressed as

Algorithm 3 Joint optimization using an alternating algorithm

1: Initialize the BS/UAVs power allocation and UAVs initial locations;
2: **while** The increase in downlink throughput exceeds a predetermined tolerance **do**
3: Optimize BS/UAV power allocation with fixed UAV deployment using Algorithm 1 and Algorithm 2;
4: Optimize UAV trajectory with fixed BS/UAV power using standard convex optimization;
5: **end while**;
6: Output UAVs' optimized location, and the BS and UAVs' transmission power as the optimal solution to (4).

$\mathcal{O}(|M| + |M||N|)$. In Section III-B, (16) is estimated as a convex function and so can be resolved by using interior point technique with complexity $\mathcal{O}(H(2|Q|)^3)$, where H is the number of iterations. As a result, $\mathcal{O}(|M|+|M||N|+H(2|Q|)^3)$ can be used to compute the overall computational complexity of Algorithm 3 for every interval.

4 Numerical Results

The simulation results are introduced and investigated in this section to establish the effectiveness of the proposed algorithm. Assume a 1500 m × 1500 m area where a BS at (0,0) and a different number of UAVs provide connectivity for terrestrial vehicles. Moreover, we consider a unit bandwidth, while other network parameters are as follows: $\sigma_k^2 = -169$ dBm/Hz, $H_u = 100$ m, $V_q^{max} = 100$ m/s, $C = 100$ s, $p_m^{max} = 100$ W, $\chi_{min} = 1$ m, and $\zeta_{th} = 400$ W. The time slot length, δ, is chosen to be 1 s, and thus the number of discrete intervals is $T = 100$. The channel models given in [14, 15] are considered for terrestrial and non-terrestrial links, respectively. Multiple scenarios are investigated, such as varying the speed of vehicles, the number of UAVs, and vehicles' movement. For comparison, we consider the benchmark in which UAV positions are fixed and no power optimization is carried out. The comparison is presented in Fig. 3 that shows the cumulative distribution function (CDF) for the average throughput of vehicles connected. We observe that the proposed joint trajectory and power optimization scheme performs much better compared to the benchmark scheme.

Following that, as shown in Fig. 4, we investigate the variation in average throughput by altering the satellite-receiver interference limit, ζ_{th}, with $M = 4$ (1 BS, 3 UAVs). Our suggested approach is compared to a framework in which the power of BS/UAVs is optimized, while their positions are fixed. It is evident that the proposed method outperforms the opponent. The main reason being the optimized UAV trajectory or horizontal position that provides better link quality to the vehicles connected. Moreover, we present the effect of the change of vertical height of UAV on the average throughput in Fig. 5, showing that the 3D trajectory design of UAV including both altitude and horizontal position optimization can be further leveraged.

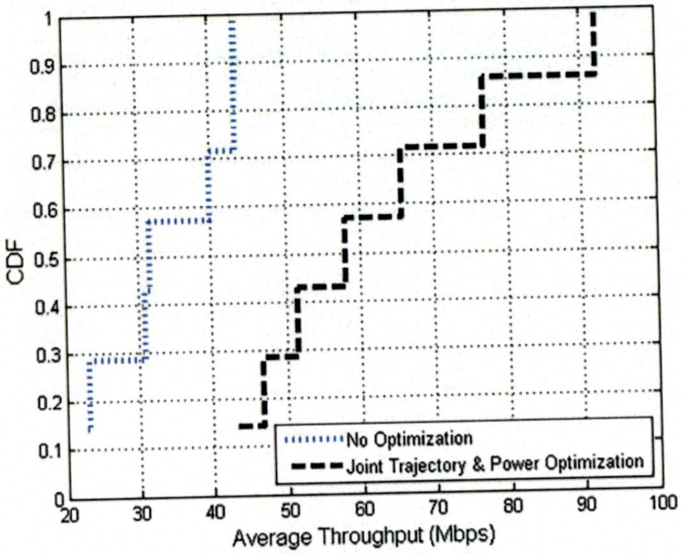

Fig. 3 CDF for the average throughput of vehicles

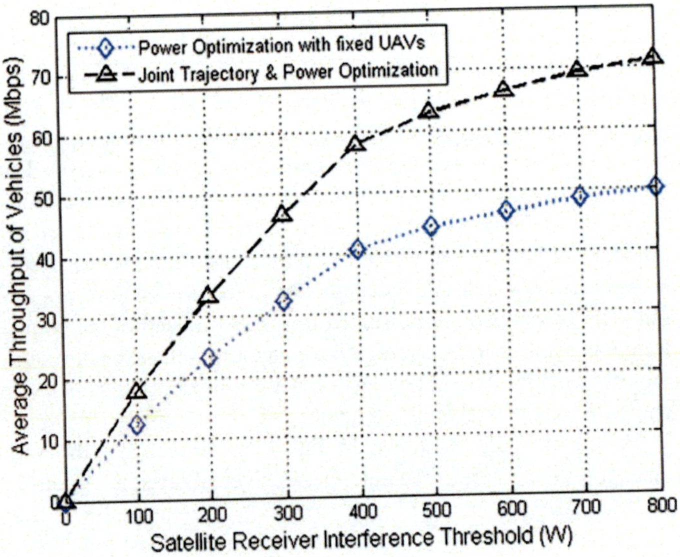

Fig. 4 Average throughput vs. interference threshold for satellite receivers

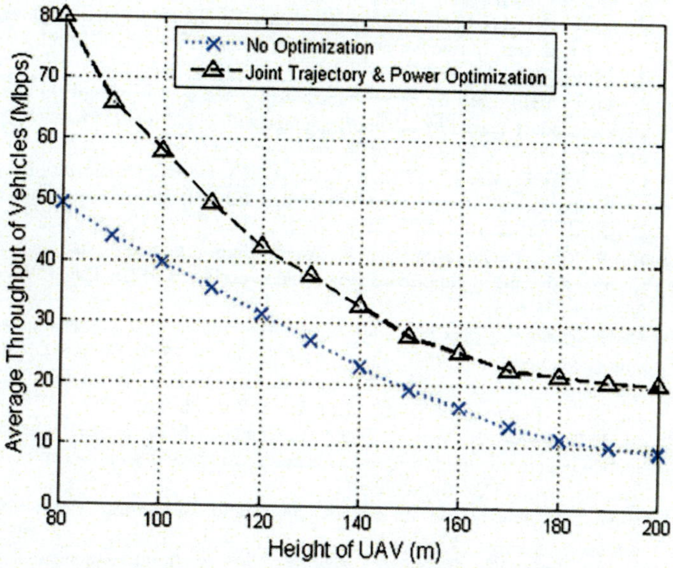

Fig. 5 Average throughput of vehicles with varying UAV height

5 Conclusion

In this chapter, we formulate a throughput maximization problem for a cognitive radio-based integrated satellite–aerial–terrestrial network under a few limitations. The block descent approach is used to break the original non-convex challenge into two subproblems: power optimization and UAV trajectory control. The formulated original problem is addressed iteratively utilizing the solutions to the subproblems, providing a low-complexity joint multi-domain optimization method. Finally, the performance of the proposed technique is investigated, and the viability is confirmed by simulations.

References

1. Liang, Y.-C., Tan, J., Jia, H., Zhang, J., Zhao, L.: Realizing intelligent spectrum management for integrated satellite and terrestrial networks. J. Commun. Inf. Netw. **6**(1), 32–43 (2021)
2. Pervez, F., Yang, C., Zhao, L.: Dynamic resource management to enhance video streaming experience in a C-V2X network. In: IEEE 92nd Vehicular Technology Conference, Victoria, BC, pp. 1–5 (2020)
3. Zhang, S., Zhang, H., Song, L.: Beyond D2D: full dimension UAV-to-everything communications in 6G. IEEE Trans. Vehic. Technol. **69**(6), 6592–6602 (2020)
4. Nguyen, H.T., Tuan, H.D., Duong, T.Q., Poor, H.V., Hwang, W.-J.: Joint D2D assignment, bandwidth and power allocation in cognitive UAV-enabled networks. IEEE Trans. Cognitive Commun. Netw. **6**(3), 1084–1095 (2020)

5. Wang, H., Ren, G., Chen, J., Ding, G., Yang, Y.: Unmanned aerial vehicle-aided communications: joint transmit power and trajectory optimization. IEEE Wirel. Commun. Lett. **7**(4), 522–525 (2018)
6. Sagduyu, Y.E., Shi, Y., MacKenzie, A.B., Hou, Y.T.: Regret minimization for primary/secondary access to satellite resources with cognitive interference. IEEE Trans. Wirel. Commun. **17**(5), 3512–3523 (2018)
7. Li, Z., Xiao, F., Wang, S., Pei, T., Li, J.: Achievable rate maximization for cognitive hybrid satellite-terrestrial networks with AF-relays. IEEE J. Sel. Areas Commun. **36**(2), 304–313 (2018)
8. He, P., Zhao, L., Zhou, S., Niu, Z.: Water-filling: a geometric approach and its application to solve generalized radio resource allocation problems. IEEE Trans. Wirel. Commun. **12**(7), 3637–3647 (2013)
9. He, P., Zhang, S., Zhao, L., Shen, X.: Energy-efficient power allocation with individual and sum power constraints. IEEE Trans. Wirel. Commun. **17**(8), 5353–5366 (2018)
10. Interdonato, G., Karlsson, M., Björnson, E., Larsson, E.G.: Local partial zero-forcing precoding for cell-free massive MIMO. IEEE Trans. Wirel. Comm. **19**(7), 4758–4774 (2020)
11. Li, M.S., Cheng, N., Gao, J., Wang, Y., Zhao, L., Shen, X.: Energy-efficient UAV-assisted mobile edge computing: resource allocation and trajectory optimization. IEEE Trans. Vehic. Technol. **69**(3), 3424–3438 (2020)
12. Boyd, S., Vandenberghe, L.: Convex Optimization. Cambridge University Press, Cambridge (2004)
13. Fu, S., Tang, Y.J., Zhang, N., Zhao, L., Wu, S.: Joint unmanned aerial vehicle (UAV) deployment and power control for internet of things networks. IEEE Trans. Vehic. Technol. **69**(4), 4367–4378 (2020)
14. Lyu, J., Zeng, Y., Zhang, R.: UAV-aided offloading for cellular hotspot. IEEE Trans. Wirel. Commun. **17**(6), 3988–4001 (2018)
15. Guo, K., Lin, M., Zhang, B., Wang, J.-B., Wu, Y., Zhu, W.-P., Cheng, J.: Performance analysis of hybrid satellite-terrestrial cooperative networks with relay selection. IEEE Trans. Vehic. Technol. **69**(8), 9053–9067 (2020)

Wireless Device Authentication Using LSTM-Based Autoencoders

M. A. Haji Bagheri Fard, Jean-Yves Chouinard, Bernard Lebel, and Paul Fortier

1 Introduction

The mobility of devices in wireless networks, for instance, ZigBee devices, often leads to a direct trade-off with a given level of trust in the communication itself. The increased risk of being targeted by an unauthorized wireless rogue device is a major concern in high-rate wireless communication systems. Besides the intrinsic seriousness of physical (PHY) attacks, a vast range of approaches used for attacking devices, from Sybil attacks to wireless traffic sniffing and man-in-the-middle attacks, make authenticating devices a core problem.

Modern attacks present a serious and immediate kind of security threats, since the strength of the underlying algorithms and protocols is rendered largely irrelevant [1]. Focusing on the identity spoofing done through falsifying bit-level credentials of attackers to match the identity of one of the known/authorized members, the most critical point is the discrimination of spoofing and spoofed members. In addition, spoofing physical characteristics from a source may prove a much harder technical challenge than spoofing data. Thus enhancing the security of the communication using features from the physical layer is an essential key, as faking physical features of a signal requires highly specialized equipment that may be challenging to acquire and configure, making it very hard for spoofing devices to mitigate the authorized members' identification. Therefore, for increasing the security of data communication, RF-DNA features at the PHY layer [2] is a powerful

M. A. H. B. Fard (✉) · J.-Y. Chouinard · P. Fortier
Department of Electrical and Computer Engineering, Université Laval, Québec, QC, Canada
e-mail: mohammad-amin.haji-bagheri-fard.1@ulaval.ca; jean-yves.chouinard.1@ulaval.ca; paul.fortier.2@ulaval.ca

B. Lebel
Thales Canada Inc. - TRT, Québec, QC, Canada
e-mail: bernard.lebel@ca.thalesgroup.com

H. Nguyen et al. (eds.), *30th Biennial Symposium on Communications 2021*,
Signals and Communication Technology, https://doi.org/10.1007/978-3-031-06947-5_12

concept for detecting unauthorized devices. These RF-DNA features are unique for a given device [3]. This provides efficient means to detect and reject attackers attempting to interact with authorized devices on a wireless network. The fingerprint extracted using this approach relies on multiple statistical transformations that may be too crude to capture the subtleties from identical devices. Recent advances have demonstrated the efficiency of deep learning in extracting relevant features. On the other hand, recent technology improvements in the computational power of hardware processing units made it possible to use the potential power of deep learning in cybersecurity [4], for ZigBee networks [5], and wireless channel quality prediction [6]. In addition, the ability of autoencoder structures for feature extraction has been recently used as an approach to increase the classification rate in acceptance (rejection) of authorized (rogue) devices [7]. However, the model designed in this chapter is different from [7]. It has a higher accuracy level of classification of the rogue/authorized devices by exploiting the time dependency of transmitted signals.

During the last decades, the theory behind machine learning has made giant leaps with discoveries such as convolutional neural networks (CNN) [8], operating on time-domain complex baseband signals. Fixed length preamble [9–11], [12] and transient behavior of the device [13] are two approaches that leverage deep learning to extract radio frequency signal features for classification purposes. Recent works [4–7] focused on deep learning by extraction of a signal from the steady-state part of the initial signal. In [4], a useful strategy is employed for signal frequency mismatch compensation, useful for increasing the classification accuracy. Besides, while the results from [4] were promising, the training dataset, i.e., the dataset used for training the model toward achieving its tasks, contained data points from devices that were also included in the test set, i.e., the dataset used for evaluating the results of the model. In [7], the authors tried to overcome this issue by keeping new devices for testing. Another point that needs to be addressed is the time dependency of extracted signals in the dataset. The proposed work of this chapter addresses this issue.

This chapter is organized as follows. Section 2 describes the proposed system and methodology used in this chapter. In Sect. 3, experimental results are presented. Section 4 summarizes the advantages and limitations of the proposed method.

2 Proposed LSTM-Based Autoencoder and Classifier

The main idea behind the proposed scenario of this chapter is *one-vs-all*, as used by [7, 14], requiring that the network is able to distinguish between the authorized and unauthorized devices by providing two classes: legitimate devices and (unauthorized) rogue devices. The reason for such an approach is that when we want to differentiate between two authorized devices from each other, and also from unauthorized ones, it is impossible to collect all authorized and rogue/unauthorized devices in the world for model training. Therefore, it is better to make a model for each device to distinguish it from all other devices in the world. Thus, training a model for one authorized device and some unauthorized ones can guarantee its good performance in real cases.

2.1 Dataset Acquisition

Based on the IEEE 802.15.4 protocol, the ZigBee PHY message may be transmitted through each of 11 channels, from 2.4 GHz to 2.48355 GHz. The message preamble includes 8 successive symbols (of 16 μs duration), resulting in whole preamble length of 128 μs. During data transmission, baseband bursts that were generated before transmission are upconverted to mid-band in the modulation section. Then, inside the receiver, the first step is to downconvert the data. Since the clock of the demodulator is slightly different from the modulator upconversion frequency, this frequency difference can lead to a serious phase and frequency distortion of the received baseband burst, which can falsify the classifier and needs to be compensated before use. After preamble extraction, since the quality of the received signals is low, a set of phase and frequency compensation procedures are needed, as described in [7].

2.2 Model Definition

In [7], the efficiency of extracted features is investigated. Should the feature extraction be limited to a known set of statistical information of PHY parameters [9–11, 15], or can a new set of features with maximum inter-class and minimum intra-class distances [16] be more effective for classification purposes? In [7], the presented methodology is based on the automatic feature extraction of autoencoders.

However, another factor that can be taken into account is the time dependency of preambles in the dataset. Time dependency of preambles results in the correlation of adjacent samples in the data point samples that is used as a key factor in the efficiency of extracted features. This is an important factor that is addressed in this chapter through the implementation of LSTM layers in the proposed autoencoder. The model used here is an autoencoder [17] with an LSTM layer structure in combination with two fully connected layers as "Classifier," as shown in Fig. 1.

An autoencoder consists of two parts: an encoder, including an LSTM layer, and a decoder. The decoder part is required by the autoencoder model during training to make sure that the most meaningful features are extracted for the classifier. After training, during testing, the classification procedure just needs the encoder and classifier parts since the weights of the encoder and classifier are already set. At the input, $\mathbf{x}_u$ $(1 \leq u \leq U)$ refers to a data point vector from the dataset with size U. It is worth mentioning that the features that are extracted at the output of the encoder are not necessarily known RF-DNA features. The extracted features are the

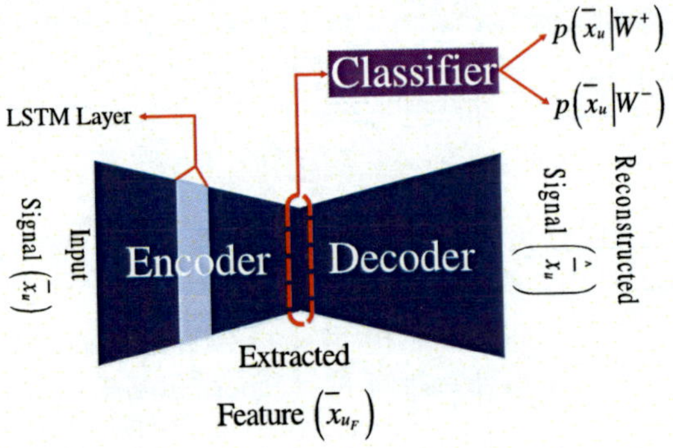

Fig. 1 LSTM-based autoencoder and classifier

best ones detected by the autoencoder as the most effective factor in discrimination of authorized and unauthorized devices with a high accuracy.

After their extraction from the input data, the features are fed to the classifier section as shown in Fig. 1. The common classification structure used in the literature involves two fully connected (dense) layers. However, this structure may overfit the training data, unless the training dataset is considerably large [18]. The architecture introduced in this chapter requires a significant amount of data points (higher than 10 000) for each device in the training, validation, and test datasets to avoid overfitting.

Since the strategy adopted for this chapter is the *one-vs-all* strategy, there are two outputs for the classifier, each presenting the probability of being from a specific device or not, as indicated in Fig. 1. In this structure, W^+ and W^- identify the positive and negative classes, respectively. The positive class refers to a specific device, whereas the negative class refers to all other devices.

2.3 Model Training

During the training, the decoder and classifier outputs should be trained. For this purpose, in the first step, the mean squared error (MSE) between the input and the reconstructed data points, $\mathbf{x}_u$ and $\hat{\mathbf{x}}_u$, at the decoder output, and also the MSE for the training/validation phases ($MSE_{tr/val}$) are expressed as

$$MSE_{(\mathbf{x}_u,\hat{\mathbf{x}}_u)} = \frac{1}{N}\sum_{n=0}^{N-1}\left(x_{u_n} - \hat{x}_{u_n}\right)^2 \tag{1}$$

$$MSE_{tr/val} = \frac{1}{U_{tr/val}}\sum_{u_{tr/val}} MSE_{(\mathbf{x}_u,\hat{\mathbf{x}}_u)}. \tag{2}$$

After feeding the training/validation dataset to the model, it classifies these data points and assigns the positive class label (+1) or the negative class label (−1) to each of these data points. In an ideal case, the extracted labels by the model and the real labels of these data points must be the same, but in the real world, there is a difference between these two sets of labels. The binary cross-entropy $H(y, p)$ between the distribution of the extracted labels $y(\mathbf{x}_u)$ and the distribution of the input data point $p(\mathbf{x}_u)$ from a dataset with size $U_{tr/val}$ is used to assess the accuracy of classifier output during the training process [19]:

$$H = \sum_{\mathbf{x}_u \in \{\mathbf{x}_0, \mathbf{x}_1, \cdots, \mathbf{x}_{U_{tr/val}-1}\}} H(y, p), \tag{3}$$

where

$$H(y, p) = - \sum_{W \in \{W^+, W^-\}} y_W(\mathbf{x}_u) \log\left(p(\mathbf{x}_u)\right). \tag{4}$$

The point is that Eq. (1) used for training of the autoencoder and Eq. (3) at the classifier output evolve simultaneously during training. We can ensure that the accuracy of the decoder output improves in such a way that it provides meaningful features for the classifier at the output of the encoder and makes the label allocation more accurate. This is the reason why we can claim that the feature that is extracted at the output of the encoder part is the most meaningful feature for our proposed classifier structure.

3 Experimental Results

3.1 Signal Acquisition Setup

To assess the performance of the proposed classifier with experimental data, a Zynq FPGA *XC7Z020* was used as the signal receiver. Eight different devices were used: five RZUSBSticks (*RZs*, model *ATAVRRZUSBSTICK*), one XBEE Digi module (*AR*, model *XB24CZ7SITB003*, and two Texas Instruments (*TIs*, model *CC2650*). The data acquisition setup, inside a 0.5 m^3 ahechois chamber, is illustrated in Fig. 2.

After extraction of the preambles from the testing devices, these extracted preambles are compared to a reference preamble for phase compensation [7] in order to improve the quality of the datasets.

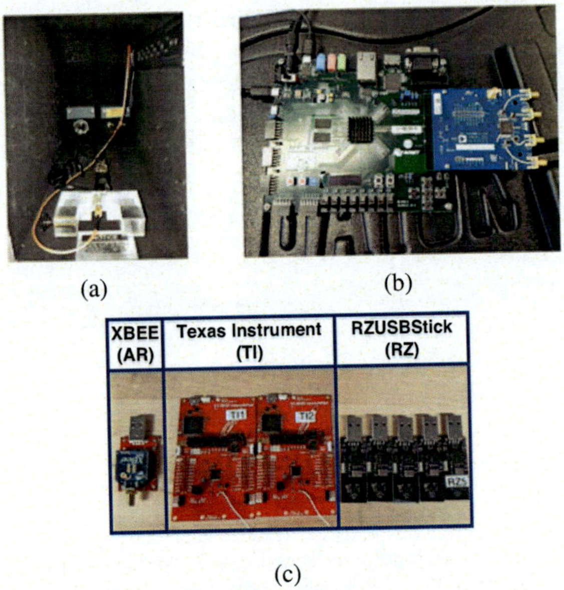

Fig. 2 Data acquisition setup: (**a**) Anechoic chamber, (**b**) Zynq FPGA receiver, (**c**) ZigBee testing devices

3.2 *Proposed LSTM-Based Autoencoder Model*

Table 1 summarizes the implemented classifier model of Fig. 1, including the output shape. $(None, a, b, c)$ refers to the batch size *None* and the size of output signal as (a, b, c). There are three sections in this table, assigned by two blue and one magenta headers. Blue headers are related to the autoencoder structure in Fig. 1, whereas the magenta header in Table 1 corresponds to the classifier in Fig. 1. As can be seen, the model has 1 023 718 trainable parameters. Software and hardware specifications for model training are: GPU NVIDIA Quadro K620, Python 3.6, TensorFlow 1.0.8, Keras 2.2.0, and a batch size of 20.

Autoencoder Structure Rows 0–34 of Table 1 show the autoencoder structure used in this chapter. This structure includes successive convolutional layers (Conv1D or Conv2D), pooling layers (MaxPooling1D), upsampling layers (UpSampling1D), dropout layers, reshape layers, an LSTM layer, and no batch normalization layer.

Classifier Structure Rows 35–40 of Table 1 are allocated to the classification structure. The active function in the last layer of the classification structure is set to a *Sigmoid function* as a nonlinear mapping converter.

For training, an *Adam optimizer* [20] is employed with learning rate $lr = 0.0001$. The number of *epochs* for training is 100. The loss functions used for training the decoder and classification outputs are the mean squared error (MSE) and the binary

Table 1 LSTM-based autoencoder training model

Index	Name	Type	Output shape
0	En.layer_0	InputLayer	(None, 5120, 2, 1)
1	En.layer_1	Conv2D	(None, 4800, 1, 1)
2	En.layer_2	Reshape	(None, 4800, 1)
3	En.layer_3	CuDNNLSTM	(None, 4800, 320)
4	En.layer_4	Dropout	(None, 4800, 320)
5	En.layer_5	Reshape	(None, 4800, 320, 1)
6	En.layer_6	MaxPooling1D	(None, 960, 1, 1)
7	En.layer_7	Dropout	(None, 960, 1, 1)
8	En.layer_8	Conv1D	(None, 800, 1, 32)
9	En.layer_9	Dropout	(None, 800, 1, 32)
10	En.layer_10	Conv1D	(None, 160, 1, 32)
11	En.layer_11	Dropout	(None, 160, 1, 32)
12	En.layer_12	Conv1D	(None, 80, 1, 32)
13	En.layer_13	Dropout	(None, 40, 1, 32)
14	En.layer_14	Conv1D	(None, 40, 1, 32)
15	En.layer_15	Dropout	(None, 40, 1, 32)
16	En.layer_16	Conv1D	(None, 20, 1, 32)
17	En.layer_17	Dropout	(None, 20, 1, 32)
18	En.layer_18	Conv1D	(None, 10, 1, 32)
19	En.layer_19_decoder	Dropout	(None, 10, 1, 32)
20	Dec.layer_0	Conv1DTranspose	(None, 10, 1, 32)
21	Dec.layer_1	Dropout	(None, 20, 1, 32)
22	Dec.layer_2	Conv1DTranspose	(None, 40, 1, 32)
23	Dec.layer_3	Conv1DTranspose	(None, 80, 1, 32)
24	Dec.layer_4	Dropout	(None, 80, 1, 32)
25	Dec.layer_5	Conv1DTranspose	(None, 160, 1, 32)
26	Dec.layer_6	Dropout	(None, 160, 1, 32)
27	Dec.layer_7	UpSampling1D	(None, 800, 1, 32)
28	Dec.layer_8	Dropout	(None, 800, 1, 32)
29	Dec.layer_9	Conv1DTranspose	(None, 960, 1, 32)
30	Dec.layer_10	Dropout	(None, 960, 1, 32)
31	Dec.layer_11	UpSampling1D	(None, 4800, 1, 32)
32	Dec.layer_12	Dropout	(None, 4800, 1, 32)
33	Dec.layer_13	Conv1DTranspose	(None, 5120, 2, 1)
34	Dec.layer_14_decoder	Dropout	(None, 5120, 2, 1)
35	Cl.layer_0	Flatten	(None, 320)
36	Cl.layer_1	Dropout	(None, 320)
37	Cl.layer_2	Dense	(None, 320)
38	Cl.layer_3	Dropout	(None, 320)
39	Cl.layer_4	Dense	(None, 266666)
40	Cl.layer_5_classifier	Dropout	(None, 2)
Total Parameters		1,023,718	
Trainable Parameters		1,023,718	

cross-entropy, respectively. Also, the dimensionality of the feature vector (row 19) is (10, 1, 32).

3.3 Model Validation and Testing

Table 2 shows the different scenarios for training, validation, and testing. For testing, group *A* consists of devices that have been seen by the model during training. Group *B* includes new family members of devices used in the training, and group *C* corresponds to unseen members that do not have family members in the training.

The total number of data points for each device in the dataset is higher than 11,000. 60% of training devices' data points are related to training, 10% of validation devices' data points correspond to validation, and the remaining 30% of the data points are allocated to testing. Using Table 2, four devices are used for training, five devices for validation, and eight devices for testing.

Figures 3 and 4 show the confusion matrices (CM) and the receiver operating characteristics (ROC) obtained from the devices. Before testing, the model is trained for each of the 8 scenarios described in Table 2. As shown, the results are comparable based on the device group (A, B, or C).

Table 2 Different scenarios for label allocation

	Positive Device	**Training**		**Validation**	
1	RZ_1	RZ_1	RZ_2, RZ_3, TI_2	RZ_1	RZ_2, RZ_3, TI_1, TI_2
2	RZ_2	RZ_2	RZ_1, RZ_3, TI_2	RZ_2	RZ_1, RZ_3, TI_1, TI_2
3	RZ_3	RZ_3	RZ_1, RZ_2, TI_2	RZ_3	RZ_1, RZ_2, TI_1, TI_2
4	RZ_4	RZ_4	RZ_1, RZ_3, TI_2	RZ_4	RZ_1, RZ_3, TI_1, TI_2
5	RZ_5	RZ_5	RZ_1, RZ_3, TI_2	RZ_5	RZ_1, RZ_3, TI_1, TI_2
6	TI_1	TI_1	RZ_1, RZ_2, RZ_3	TI_1	RZ_1, RZ_2, RZ_3, TI_2
7	TI_2	TI_2	RZ_1, RZ_2, RZ_3	TI_2	RZ_1, RZ_2, RZ_3, TI_1
8	AR_1	AR_1	RZ_1, RZ_2, TI_2	AR_1	RZ_1, RZ_2, TI_1, TI_2
	Positive Device	**Testing**			
			A	**B**	**C**
1	RZ_1	RZ_1	RZ_2, RZ_3, TI_2	RZ_4, RZ_5, TI_1	AR_1
2	RZ_2	RZ_2	RZ_1, RZ_3, TI_2	RZ_4, RZ_5, TI_1	AR_1
3	RZ_3	RZ_3	RZ_1, RZ_2, TI_2	RZ_4, RZ_5, TI_1	AR_1
4	RZ_4	RZ_4	RZ_1, RZ_3, TI_2	RZ_2, RZ_5, TI_1	AR_1
5	RZ_5	RZ_5	RZ_1, RZ_3, TI_2	RZ_2, RZ_4, TI_1	AR_1
6	TI_1	TI_1	RZ_1, RZ_2, RZ_3	RZ_4, RZ_5, TI_2	AR_1
7	TI_2	TI_2	RZ_1, RZ_2, RZ_3	RZ_4, RZ_5, TI_1	AR_1
8	AR_1	AR_1	RZ_1, RZ_2, TI_2	RZ_3, RZ_4, RZ_5, TI_1	-

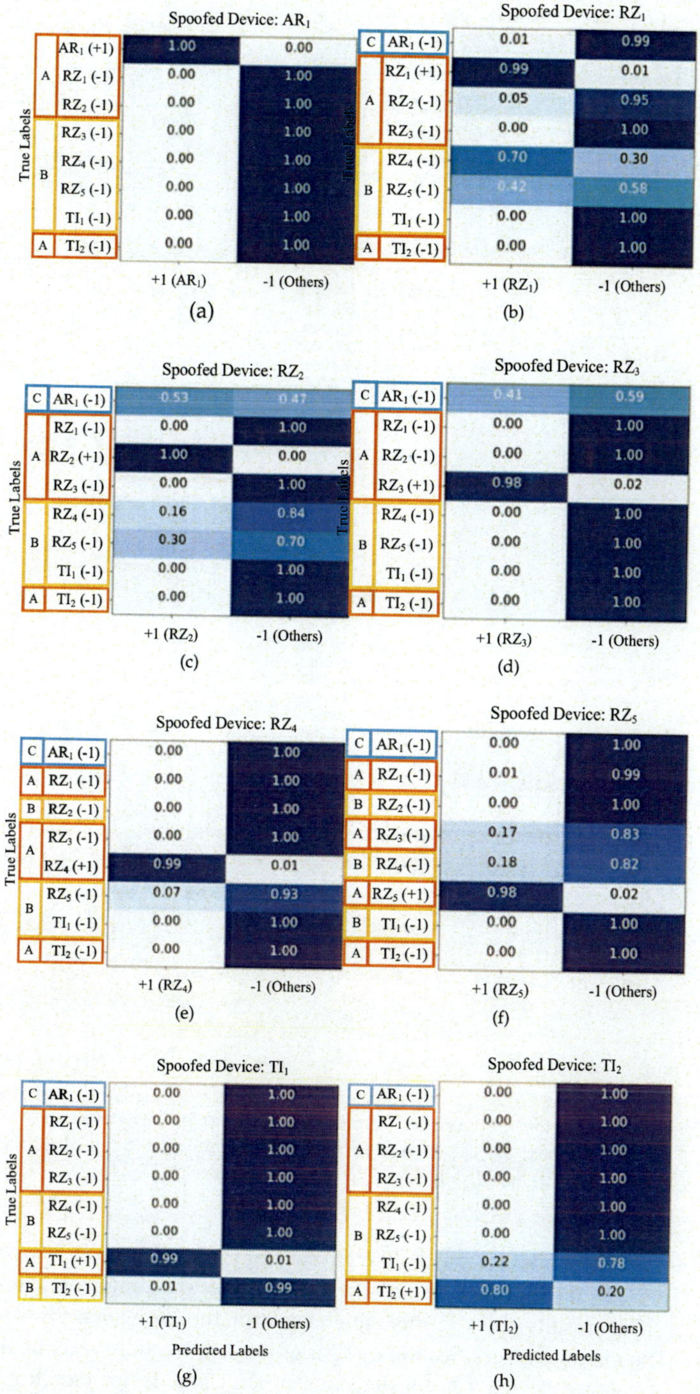

Fig. 3 Confusion matrices for the eight scenarios of Table 2

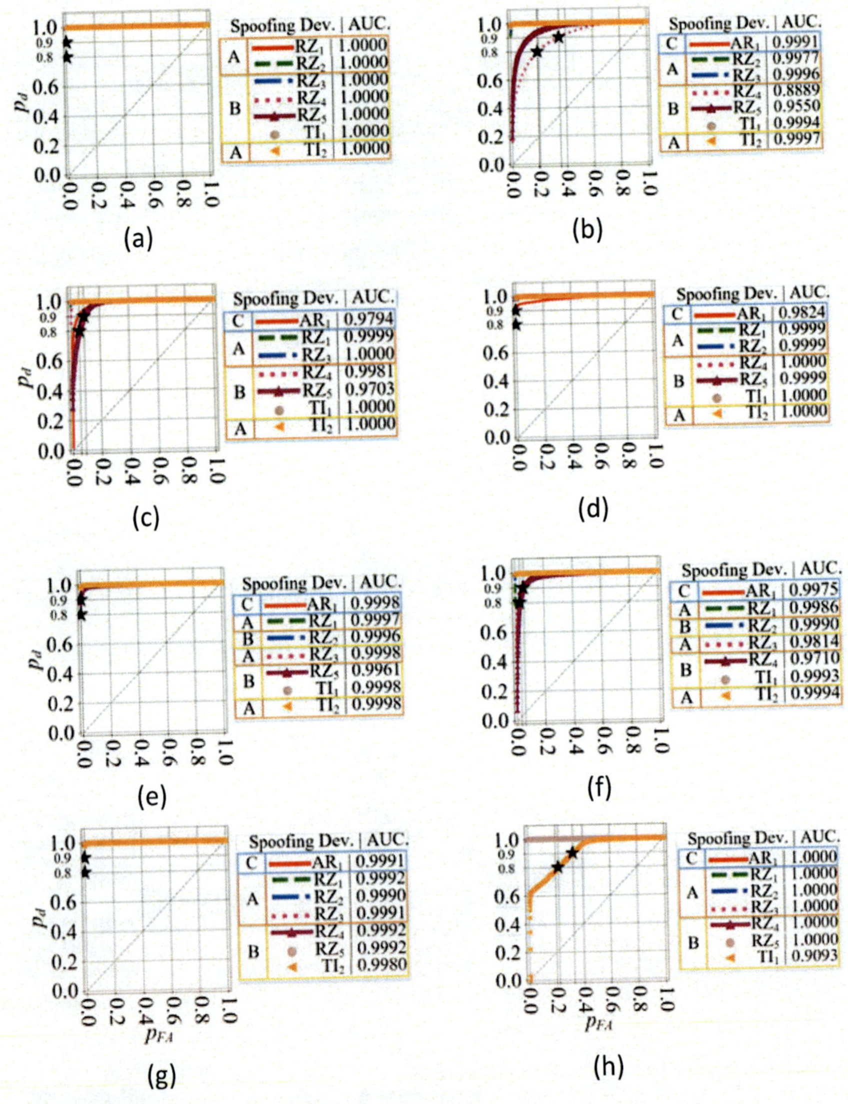

Fig. 4 ROC plots for the eight scenarios of Table 2. (**a**) Spoofed device: AR_1. (**b**) Spoofed device: RZ_1. (**c**) Spoofed device: RZ_2. (**d**) Spoofed device: RZ_3. (**e**) Spoofed device: RZ_4. (**f**) Spoofed device: RZ_5. (**g**) Spoofed device: TI_1. (**h**) Spoofed device: TI_2

Confusion matrices for all eight scenarios of Table 2 are shown in Fig. 3. Table 3 presents the summary of classification rates for the confusion matrices of Fig. 3.

The results given in this table correspond to the *worst cases for classification*. The minimum classification rate for group A from all models belongs to the model of spoofed device TI_2 for detection of the same device (see Fig. 3h) and is equal

Table 3 Confusion matrices summary

	Positive Device	Classification Rate A	Classification Rate B	Classification Rate C		Positive Device	Classification Rate A	Classification Rate B	Classification Rate C
1	RZ_1	≥ 0.95	≥ 0.30	0.99	5	RZ_5	≥ 0.83	≥ 0.82	1.00
2	RZ_2	1.00	≥ 0.70	0.47	6	TI_1	≥ 0.99	0.99	1.00
3	RZ_3	≥ 0.98	1.00	0.59	7	TI_2	≥ 0.80	≥ 0.78	1.00
4	RZ_4	≥ 0.99	≥ 0.93	1.00	8	AR_1	1.00	1.00	-

Table 4 ROC plots summary

	Positive Device	p_d A	p_d B	p_d C
1	RZ_1	≥ 0.9 *for* $p_{FA} \in [0.01, 1]$	≥ 0.9 *for* $p_{FA} \in [0.353, 1]$	≥ 0.95 *for* $p_{FA} \in [0.01, 1]$
2	RZ_2	~ 1.0 *for* $p_{FA} \in [0, 1]$	≥ 0.9 *for* $p_{FA} \in [0.09, 1]$	≥ 0.9 *for* $p_{FA} \in [0.057, 1]$
3	RZ_3	≥ 0.95 *for* $p_{FA} \in [0.01, 1]$	≥ 0.95 *for* $p_{FA} \in [0.01, 1]$	≥ 0.9 *for* $p_{FA} \in [0.01, 1]$
4	RZ_4	≥ 0.95 *for* $p_{FA} \in [0.01, 1]$	≥ 0.9 *for* $p_{FA} \in [0.01, 1]$	≥ 0.95 *for* $p_{FA} \in [0.01, 1]$
5	RZ_5	≥ 0.95 *for* $p_{FA} \in [0.01, 1]$	≥ 0.9 *for* $p_{FA} \in [0.059, 1]$	≥ 0.95 *for* $p_{FA} \in [0.01, 1]$
6	TI_1	≥ 0.95 *for* $p_{FA} \in [0.01, 1]$	≥ 0.95 *for* $p_{FA} \in [0.01, 1]$	≥ 0.95 *for* $p_{FA} \in [0.01, 1]$
7	TI_2	~ 1.0 *for* $p_{FA} \in [0, 1]$	≥ 0.9 *for* $p_{FA} \in [0.328, 1]$	~ 1.0 *for* $p_{FA} \in [0, 1]$
8	AR_1	~ 1.0 *for* $p_{FA} \in [0, 1]$	~ 1.0 *for* $p_{FA} \in [0, 1]$	-

to 0.80. On the other hand, the minimum classification rates for groups B, and C correspond, respectively, to the rogue device RZ_4 (Fig. 3b) (obtained from the model of device RZ_1), that is, 0.30, and the rogue device AR_1 (Fig. 3c) (given by the model of spoofed device RZ_2), that is 0.47.

Figure 4 presents the ROC plots for groups A, B, and C (Table 2) for each of the 8 tested scenarios. Table 4 provides the p_d (correct classification rate or detection probability of authorized device) and its corresponding p_{FA} (misclassification rate or false alarm probability of rogue device) of groups A, B, and C. Again, the results in this table correspond to the worst cases for classification. The best classification is related to the spoofed device AR_1 with a $p_d = 1.0$ for all the rogue devices with a $p_{FA} \in [0, 1]$ (Fig. 4a). The worst classification rate is for spoofed device RZ_1, with the highest $p_{FA} = 0.353$ for $p_d = 0.9$ (Fig. 4b), for rogue device RZ_4.

3.4 *Performance Comparison*

A performance comparison of the curren work and [4, 7] is as follows (Table 5). Since in [4] no device is kept out of the training loop, for the testing, the comparison is done with the model for group A of Table 2. Referring to Fig. 4 for group A of devices, the *minimum area under the curve* of the ROC plot, AUC_{min}, is allocated to rogue device RZ_3 in Fig. 4f and is equal to 0.9814. Moreover, the maximum AUC,

Table 5 Comparison of ROC plot results for [4, 7], and the current work

	Group A		Group B	
	AUC_{min}	AUC_{max}	AUC_{min}	AUC_{max}
[4]	0.9653	0.9971	-	-
[7]	0.9947	1.0000	0.3926	1.0000
Current work	0.9814	1.0000	0.8889	1.0000
	Group C			
	AUC_{min}	AUC_{max}		
[4]	-	-		
[7]	0.9430	1.0000		
Current work	0.9794	1.0000		

$AUC_{max} = 1.0000$, is allocated to nine cases of rogue devices in Fig. 4a, c, d, h. The AUC_{min} and AUC_{max} in [4] are 0.9653 and 0.9971, respectively.

Since in [7] the model has been tested by the unseen devices, the comparison is done in three groups (A, B, and C) of the RAW dataset in [7] and our current work. Although $AUC_{min} = 0.9947$ in [7] is slightly higher than $AUC_{min} = 0.9814$ (RZ_3 in Fig. 4f) in this chapter for group A, $AUC_{min} = 0.8889$ for group B of our current work (RZ_4 in Fig. 4b) has a remarkable improvement compared to $AUC_{min} = 0.3926$ of group B in [7]. Then, the worst performance of the model in this chapter is considerably better than [7]. In addition, for group C, $AUC_{min} = 0.9794$ of our current work (AR_1 in Fig. 4c) is higher than $AUC_{min} = 0.9430$ in [7]. Finally, AUC_{max} in both works for groups A, B, and C is 1.0000.

4 Conclusion

In this chapter, an LSTM-based autoencoder for feature extraction and rogue device detection was proposed. The results led to AUC values of the ROC plots for all spoofing devices, higher than 0.8889. The false alarm probability, p_{FA}, for a 90% detection probability, p_d, is lower than 36%. Also, the trained model is tested only on unseen devices. However, despite the advantages, the degrading accuracy for devices from the same manufacturer, large training datasets, and a unique model for each device are challenges that have to be overcome.

Acknowledgments The authors wish to thank L. Bélanger, M. Lévesque, G. Godbout, and S. Brin-Marquis and S. Carrier for his helpful suggestions on the data analysis. The research project is financially supported by NSERC, MITACS, and Thales TRT, Canada.

References

1. Cobb, W.E., Laspe, E.D., Baldwin, R.O., Temple, M.A., Kim, Y.C.: Intrinsic physical-layer authentication of integrated circuits. IEEE Trans. Inf. Forensics Secur. **7**(1), 14–24 (2011)
2. Ramsey, B.W.: Improved wireless security through physical layer protocol manipulation and radio frequency fingerprinting, Ph.D. Dissertation, Air Force Institute of Technology Wright-Patterson AFB OH Graduate School of Engineering and Management (2014)
3. Reising, D.R., Temple, M.A.: WiMAX mobile subscriber verification using gabor-based RF-DNA fingerprints. In: 2012 IEEE International Conference on Communications (ICC), pp. 1005–1010. IEEE, Piscataway (2012)
4. Merchant, K., Revay, S., Stantchev, G., Nousain, B.: Deep learning for RF device fingerprinting in cognitive communication networks. IEEE J. Selec. Top. Signal Process. **12**(1), 160–167 (2018)
5. Deniz, E., Samet, R.: A new model for secure joining to ZigBee 3.0 networks in the internet of things. In: 2018 International Congress on Big Data, Deep Learning and Fighting Cyber Terrorism (IBIGDELFT), vol. , no. , 2018, pp. 102–106
6. Herath, J.D., Seetharam, A., Ramesh, A.: A deep learning model for wireless channel quality prediction. In: ICC 2019 - 2019 IEEE International Conference on Communications (ICC), vol. , no. , 2019, pp. 1–6
7. Haji Bagheri Fard, M.A., Chouinard, J.-Y., Lebel, B.: Rogue device discrimination in ZigBee networks using wavelet transform and autoencoders. Ann. Telecommun. **76**, 1–16 (2020)
8. O'Shea, T.J., Corgan, J., Clancy, T.C.: Convolutional radio modulation recognition networks. In: International Conference on Engineering Applications of Neural Networks, pp. 213–226. Springer, Berlin (2016)
9. Reising, D.R., Temple, M.A., Jackson, J.A.: Authorized and rogue device discrimination using dimensionally reduced RF-DNA fingerprints. IEEE Trans. Inf. Forens. Security **10**(6), 1180–1192 (2015)
10. Patel, H.: Non-parametric feature generation for RF-fingerprinting on ZigBee devices. In: 2015 IEEE Symposium on Computational Intelligence for Security and Defense Applications (CISDA), pp. 1–5. IEEE, Piscatawa (2015)
11. Bihl, T.J., Bauer, K.W., Temple, M.A.: Feature selection for RF fingerprinting with multiple discriminant analysis and using ZigBee device emissions. IEEE Trans. Inf. Forens. Security **11**(8), 1862–1874 (2016)
12. Yuan, Y., Huang, Z., Wang, F., Wang, X.: Radio specific emitter identification based on nonlinear characteristics of signal. In: 2015 IEEE International Black Sea Conference on Communications and Networking (BlackSeaCom), pp. 77–81. IEEE, Piscataway (2015)
13. Ureten, O., Serinken, N.: Wireless security through RF fingerprinting. Canadian J. Electr. Comput. Eng. **32**(1), 27–33 (2007)
14. Lebel, B., Bélanger, L.-N., Haji Bagheri Fard, M.A., Chouinard, J.-Y.: RF fingerprinting for 802.15.4 devices: Combining convolutional neural networks and RF-DNA. In: CYBER 2017 : The Second International Conference on Cyber-Technologies and Cyber-Systems. ThinkMind, Barcelona, pp. 70–74 (2017)
15. Dubendorfer, C., Ramsey, B.W., Temple, M.A.: ZigBee device verification for securing industrial control and building automation systems. In: International Conference on Critical Infrastructure Protection, pp. 47–62. Springer, Berlin (2013)
16. Boehmke, B., Greenwell, B.M.: Hands-On Machine Learning with R. CRC Press, Boca Raton (2019)
17. Meng, Q., Catchpoole, D., Skillicom, D., Kennedy, P.J.: Relational autoencoder for feature extraction. In: 2017 International Joint Conference on Neural Networks (IJCNN), pp. 364–371. IEEE, Piscataway (2017)
18. Cao, X.: A practical theory for designing very deep convolutional neural networks. Technical Report (2015)
19. Goodfellow, I., Bengio, Y., Courville, A.: Deep Learning, vol. 1. MIT Press Cambridge, Cambridge (2016)
20. Kingma, D.P., Ba, J.: Adam: A method for stochastic optimization (2014). Preprint arXiv:1412.6980

Energy-Efficient Resource Allocation in D2D-Assisted Cache-Enabled Fog Networks

Rabeea Basir, Mudassar Ali, Saad Qaisar, Muhammad Naeem, and Alagan Anpalagan

1 Introduction

One of the major critical challenges of our future intelligent communication networks is to find their energy consumption metric. Almost 3% of the world's energy is used by communication networks according to [1], and this is increasing day by day with the advent of future technologies. The potential problems of latency, location awareness, and reliability, for future Internet-of-Things (IoT) applications, can be catered using fog computing over traditional cloud computing [2, 3].

R. Basir (✉)
School of Electrical Engineering and Computer Science, National University of Science and Technology, Islamabad, Pakistan
e-mail: rbasir.dphd17seecs@seecs.edu.pk

M. Ali
School of Electrical Engineering and Computer Science, National University of Science and Technology, Islamabad, Pakistan

Department of Telecommunication Engineering, University of Engineering and Technology, Taxila, Pakistan

S. Qaisar
School of Electrical Engineering and Computer Science, National University of Science and Technology, Islamabad, Pakistan

Department of Electrical and Electronic Engineering, University of Jeddah, Jeddah, Saudi Arabia
e-mail: saad.qaisar@seecs.edu.pk

M. Naeem
Department of Electrical Engineering, COMSATS University Islamabad, Wah Cantt, Pakistan

A. Anpalagan
WINCORE Laboratory, Department of Electrical and Computer Engineering, Ryerson University, Toronto, ON, Canada
e-mail: alagan@ee.ryerson.ca

H. Nguyen et al. (eds.), *30th Biennial Symposium on Communications 2021*,
Signals and Communication Technology, https://doi.org/10.1007/978-3-031-06947-5_13

To provide enhanced quality of service to IoT devices in terms of low latency, high data rate, and more energy efficiency, fog computing can be used. It is a virtualized platform, in which intensive services are offloaded in the nearby vicinity of end users. This storage, computation, and processing in closely distant servers help in saving the backhaul bandwidth and reduce the latency and energy consumption [4, 5]. Due to limited resources and energy conversion constraints, there is a need for an optimized fog architecture by leveraging the cache availability and device-to-device (D2D) communication technology. D2D communication can be used to reduce energy consumption and enhance throughput [6]. D2D communication along with cache-enabled fog nodes in architecture gives more energy-efficient solutions by reducing the latency, as some of the user nodes may get requested contents from the nearby D2D pairing device rather than experiencing other delays (transmission/processing at fog/or maybe propagation delay via the cloud).

Keeping in view the comparative analysis of some previous work mentioned in Table 1, we believe that this is the first research contribution that shows how to maximize the energy efficiency under power, rate, and cache constraints for D2D-assisted fog network. Introducing D2D mode with cache-enabled fog nodes will result in more energy-efficient node association by reducing the latency:

- In this chapter, we have used non-linear optimization by outer approximation algorithm (Bonmin) for the formulated constrained problem. We formulated a mixed-integer non-linear programming problem that maximizes the energy efficiency, which jointly considers the power allocation, mode selection, and cache availability at fog nodes.
- To verify ε-optimal solution, extensive simulations have been carried out.

This chapter has the following parts, Sect. 2 has a description of our proposed system architecture, which involves the problem formulation. Section 3 explains the proposed algorithm used for solving the problem, followed by simulation results in Sect. 4. Finally, we conclude our work in Sect. 5.

2 System Model and Problem Formulation

Consider a fog network shown in Fig. 1, which consists of a cloud server and a set $f \in \mathcal{F} = \{1, 2, \ldots, F\}$ of fog nodes each serving a subset of IoT nodes. There are $k \in \mathcal{K} = \{1, 2, \ldots, K\}$ IoT nodes that are served either by a fog node or they are in D2D mode. Let x_m^k be the binary indicator to show IoT node k is connected either in fog mode or in D2D mode, i.e., $m \in \mathcal{M} = \{f, d\}$, where f = fog and d = D2D. For mode selection, we can write binary indicator as $x_m^k = [0, 1]\ \forall k \in \mathcal{K}, m \in \mathcal{M}$. Each fog node $f \in \mathcal{F}$ is equipped with a data storage of capacity S_f that contains $c \in \mathcal{C}$ data available in the system cache. IoT node k requests a file Λ^c that has a request arrival rate with Poisson distribution. This distribution has a mean of $\lambda_{k,c}$ where higher mean arrival rate of a file shows that it is more popular in the system. If a node k requests a file c from fog and the file is available at fog's cache, it is directly

Table 1 Comparative analysis of previous work (U.A = User Association, P.A = Power Allocation, C.A = Cache Availability, M.S = Mode selection)

Previous contribution	Key point	Factors				
		U.A	P.A	C.A	M.S	D2D
[7]	Content Placement Genetic Algorithm is proposed for minimizing the energy consumption under cache limit size threshold for a fog network assisted by D2D and social networking-layered architecture	✗	✓	✓	✗	✓
[8]	Lagrange dual function and a heuristic approach are proposed for maximizing the harvested energy in fog radio access networks (F-RAN). Constraints were power and throughput thresholds	✓	✓	✗	✗	✗
[9]	Stackelberg game and many-to-many matching game are proposed as a solution to the formulated concave mixed-integer non-linear programming (MINLP) problem. Objective is to maximize the utility under service and delay constraints	✓	✗	✗	✗	✗
[10]	Three parallel algorithms followed by Lyapunov drift give the solution for MINLP optimal resource allocation problem under threshold of capacity and time	✓	✗	✗	✗	✗
[11]	A non-convex maximizing sum rate problem is formulated under power and rate constraints. Hybrid semi-distributed resource allocation algorithm was proposed to solve the problem for fog radio access network and cloud radio access networks. System was studied under channel state information metric	✓	✓	✗	✗	✗
[12]	An algorithm named as energy-efficient computation offloading and resource allocation (ECORA) scheme is proposed as a solution to mixed-integer minimizing optimization problem. The objective is to minimize the system cost under power, delay, and computational capability	✓	✓	✗	✗	✗
[13]	Iterative algorithms are proposed for non-linear integer maximizing throughput problem under delay thresholds	✓	✗	✗	✓	✗
[14]	To support low latency tactile Internet applications, an optimal tradeoff between two metrics is studied. A dual decomposition optimization framework is developed to observe tradeoff point between latency and fog node's power usage. Case study is implemented on self-driving bus system in Dublin. Tradeoff depends upon the workload distribution among fog nodes	✓	✗	✗	✗	✗

(continued)

Table 1 (continued)

Previous contribution	Key point	Factors				
		U.A	P.A	C.A	M.S	D2D
[15]	Fog radio access system model with millimeter-wave (mmWave) communication is proposed. Under power, interference, caching, and subchannel scheduling constraints, the system's energy is maximized. The alternative direction method of multipliers (ADMM) is proposed in comparison with the iterative many-to-one matching algorithm, which is of low complexity	✓	✓	✓	✗	✗
This work	A mixed-integer non-linear maximization problem is formulated, which maximizes the energy efficiency of the network under power, cache, and rate constraints. An outer approximation algorithm is used for solving the problem	✓	✓	✓	✓	✓

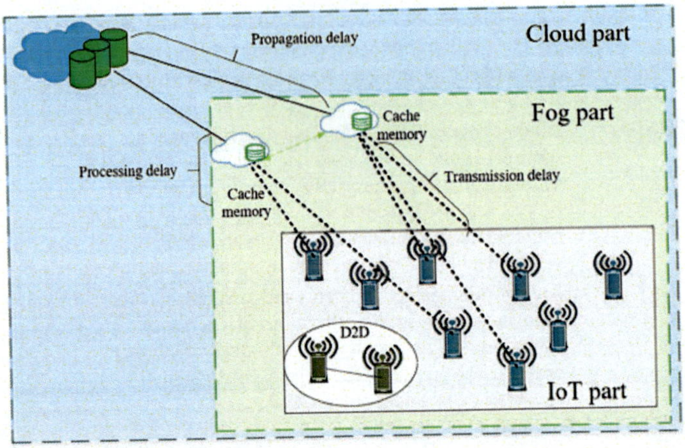

Fig. 1 D2D-assisted fog network scenario

delivered to the node; otherwise, the request will be handled by the cloud itself. This cloud request will increase latency in communication and result in increase in cost and energy consumption. We assume a certain file content c with a variable size of Λ^c for node k. The possible rate of the kth node in connection with fog node in terms of latency is calculated as

$$r_f^k = \frac{\Lambda^c}{l_f^k}. \tag{1}$$

Latency in (1) can be calculated in terms of transmission, processing, and propagation delay as

$$l_f^k = \frac{\Lambda^c}{\log_2\left(1 + \frac{p_f^k h_k}{N_o}\right)} + \left(a_f^k\right)\left(l^P + l^D\right), \tag{2}$$

where $a_f^k = [0, 1]$ is a binary indicator representing the presence of requested file in the cache memory of fog node. It is equal to 0, when the file is directly passed to node from fog node. Therefore, it nullifies the second factor of (2), which calculates the delay incurred by cloud in terms of processing and propagation delays.

We denote p_f^k and p_d^k, the kth node's transmitted power in downlink transmission and the kth node's power in D2D mode. The channel gain between the fog and IoT nodes for downlink and k^{th} node D2D pair is denoted as h_k and g_k, respectively. Let the antenna gain be denoted as G_o and $\zeta\, 10^{\zeta/2}$ be lognormal shadowing, where ζ is zero-mean Gaussian random variable with standard deviation σ [16]. The channel gain h_k with d_o antenna far-field reference distance and d distance between the fog node and IoT node is modeled as [17]

$$h_k = \bar{h}_k \zeta G_o \left(\frac{d_o}{d}\right)^{\alpha}, \tag{3}$$

where the path-loss exponent and the Rayleigh random variable are denoted by α and $\bar{h}_k$, respectively. The rate for the kth pair in D2D communication is r_d^k and is denoted as

$$r_d^k = \log_2\left(1 + \frac{p_d^k g_k}{N_o}\right). \tag{4}$$

The total data rate, according to the proposed mode selection policy, is calculated as

$$r_m^k = x_m^k r_f^k + \left(1 - x_m^k\right) r_d^k. \tag{5}$$

Introduction of binary indicator x_m^k represents the mode selection. Its value of 1 represents that one of the fog nodes is serving the IoT node, and rate is calculated using (1) and (2). Otherwise, ($x_m^k = 0$), node is in D2D mode, and rate is calculated using (4). A summary of symbol notation is given in Table 2.

Energy efficiency is the ratio of the total data rate (bits/sec) to the total power consumed (Watts). In this chapter, we formulate an optimization problem that maximizes energy efficiency based on mode selection, cache availability, and power allocation in D2D-assisted fog networks. Mathematically, joint mode selection and energy-efficiency maximization problem with objective function (6) subject to constraints (7)–(14) can be written as

Table 2 Notations

Notation	Description
$\mathscr{F}, \mathscr{K}$	Sets of fog f and IoT k nodes, respectively
$\mathscr{M}$	Set of modes (fog and D2D), $m \in \mathscr{M} = \{f, d\}$
x_m^k	Binary indicator for connection mode of an IoT node k in mode m (fog or D2D)
a_f^k	Binary indicator for the presence of requested files in cache of fog node f requested by an IoT node k
p_f^k, p_d^k	Power allocated to IoT node k when served by any fog node f or in D2D mode, respectively
h_k, g_k	Channel gain when an IoT node k is connected to any fog node f or in D2D mode, respectively
$\bar{h}_k$	Rayleigh random variable when an IoT node k is in connection
ζ	Zero-mean Gaussian variable
G_o	Antenna gain of a fog node f
d_o	Antenna far-field reference distance
d	Distance of receiving IoT node k from any fog node f
r_m^k	Data rate of an IoT node k in mode m (fog or D2D)
S_f	Maximum storage capacity of a fog node f
l_P	Processing time at fog node f
l_D	Propagation delay due to backhaul links between fog node f and cloud
l_f^k	Latency experienced by an IoT node k in connection with fog node f
P_f, P_d	Maximum transmit power of a fog node f and in D2D mode, respectively
P_c	Circuit power

$$P1: \quad \max \ \mathscr{U}_{(x,p)} = \frac{\sum_{m\in\mathscr{M}} \sum_{k\in\mathscr{K}} r_m^k}{P_c + \sum_{k\in\mathscr{K}}(p_d^k + \sum_{f\in\mathscr{F}} p_f^k)} \tag{6}$$

subject to:

$$C1: \quad \sum_{m\in\mathscr{M}} x_m^k \leq 1, \qquad \forall k \in \mathscr{K} \tag{7}$$

$$C2: \quad \sum_{k\in\mathscr{K}} p_f^k \leq P_f, \qquad \forall f \in \mathscr{F} \tag{8}$$

$$C3: \quad \sum_{k\in\mathscr{K}} p_d^k \leq P_d, \tag{9}$$

$$C4: \quad p_d^k \leq x_m^k P_d, \qquad \forall k \in \mathcal{K} \tag{10}$$

$$C5: \quad p_f^k \leq x_m^k P_f, \qquad \forall k \in \mathcal{K}, f \in \mathcal{F} \tag{11}$$

$$C6: \quad p_m^k \geq 0, \qquad \forall k \in \mathcal{K}, f \in \mathcal{F} \tag{12}$$

$$C7: \quad r_m^k \geq x_m^k \mathcal{R}_{min}^k \qquad \forall k \in \mathcal{K} \tag{13}$$

$$C8: \quad \sum_{k \in \mathcal{K}} a_f^k \Lambda^c \leq S_f; \qquad \forall f \in \mathcal{F}. \tag{14}$$

The objective function $P1$ subject to (C1–C8) constraints in (6) is to maximize energy efficiency with mode selection. C1 ensures that an IoT node is connected in D2D mode or any one of the fog nodes at a given time. C2 and C3 are the maximum power constraints for fog nodes and D2D. C4 and C5 are the individual power constraints, which ensure that if a node is not in connection, its power will be zero. Minimum power and minimum rate are ensured in constraints C6 and C7, respectively. C8 represents the storage capacity constraint of fog node f.

2.1 Alternate Problem Formulation

The formulated problem in (6) is a concave fractional programming problem (CFP), in which changes in denominator variable p_m^k result in possible variations in numerator variable r_m^k. These two variables are real-valued functions defined on R^n subset. We can see in (6) that in the numerator we have to maximize the capacity and in the denominator we have to minimize the overall power consumption. To solve this concave function to convex function ratio, we can apply Charnes–Cooper transformation (CCT) [18]. We get a concave programming problem after applying this theorem, where $p = (y/t)$. The equivalent problem can be written as

$$P2: \quad \max \quad \mathcal{U}_{(y,t)} = t \sum_{m \in \mathcal{M}} \sum_{k \in \mathcal{K}} x_m^k \log \left(1 + \frac{y_m^k h_k}{t N_o} \right) \tag{15}$$

subject to following constraints:

$$C1: \quad \sum_{m \in \mathcal{M}} x_m^k \leq 1, \qquad \forall k \in \mathcal{K} \tag{16}$$

$$C2: \quad \sum_{m \in \mathcal{M}} \sum_{k \in \mathcal{K}} y_m^k - P_m t \leq 0 \tag{17}$$

$$C3: \quad y_m^k - x_m^k P_m t \leq 0, \qquad \forall k \in \mathcal{K}, m \in \mathcal{M} \tag{18}$$

$$C4: \quad y_m^k \geq 0, \qquad \forall k \in \mathcal{K}, m \in \mathcal{M} \tag{19}$$

$$C5: \sum_{m\in\mathcal{M}} x_m^k \log\left(1+\frac{y_m^k h_k}{tN_o}\right) \geq \sum_{m\in\mathcal{M}} x_m R^k, \qquad \forall k\in\mathcal{K}, m\in\mathcal{M} \tag{20}$$

$$C6: \quad P_c t + \sum_{m\in\mathcal{M}}\sum_{k\in\mathcal{K}} y_m^k = 1. \tag{21}$$

Equation (15) along with constraints (16)–(21) represents a mixed-integer non-linear problem, which is NP-hard and very complex to solve in polynomial time. C2 and C3 constraints of the original problem (8) and (9) are collectively represented by a single constraint C2 (17) of the altered formulated problem. Similarly, C4 and C5 (10) and (11) are represented by C3 (18).

We have a binary variable x_m as well as the continuous variable y_m, which makes it difficult to find an algorithm for an optimal solution. Search space for the above equation using exhaustive search increases exponentially as the number of IoT and fog nodes increases. One solution is to apply convex optimization using the exhaustive search on the binary variable. Due to the complexity of exhaustive search on 2^k optimization problems, in this chapter, OAA is proposed, which gives a near-optimal converged solution. The detailed implementation of the outer approximation algorithm (OAA) for the formulated energy-efficiency maximization problem is explained in the next section.

3 Proposed Technique

Figure 2 shows the OAA algorithm. To explain the algorithm, consider $\mathcal{E}$ to be the objective function and ϕ_{c1-c6} to be the set of constraints (16)–(21), $Y = \{y_f, y_d\}$ and $X = x \cup Y$. It can be proved that Y is compact, nonempty, and a convex set. The $\mathcal{E}$ and ϕ_{c1-c6} sets are convex in Y. For fixed values of Y, both $\mathcal{E}$ and ϕ_{c1-c6} sets are differentiable. We can get non-linear programming problem, which can be solved easily.

For finding an optimal solution, the alternate problem is divided into primal and master problems. Fixing X variables gives a primal problem, and a sequence of upper bounds is found after solving this primal subproblem. The OAA will provide a converged solution based on lower and upper bound sequences, in a finite number of iterations. At the ith iteration of the algorithm, integer variable X will have the value of X^i, and the primal problem can be described as

$$\min_{Y} \; -\mathcal{E}(X^i, Y)$$

$$\text{subject to}: \; \phi_{c1-c6}(X^i, Y) \leq 0. \tag{22}$$

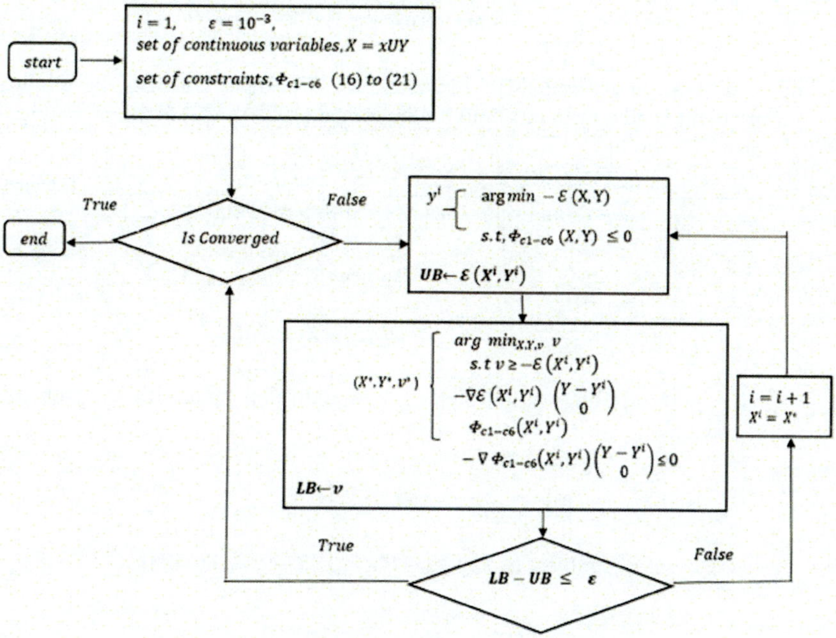

Fig. 2 Outer approximation algorithm (OAA) explanation

In the solution of the above problem, Y^i can be found by making $\mathscr{E}$ and ϕ_{c1-c6} linear, which is used for deriving master subproblem [19]. The algorithm finds the threshold of the upper bound that will not go for the further rise and similarly the threshold of the lower bound that will not go for further decrease. After each iteration, lower and upper bounds get closer to each other. When the difference between the upper and lower bounds is less than ε, the algorithm stops. The problem can be rewritten as

$$\min_X \min_Y \; -\mathscr{E}(X^i, Y)$$
$$\text{subject to : } \; \phi_{c1-c6}(X^i, Y) \leq 0. \tag{23}$$

Equation (23) can be rewritten as

$$\min_X \; -\nu(X), \tag{24}$$

where

$$\nu(X) = \min_Y \; -\mathscr{E}(X^i, Y)$$

$$\text{subject to}: \ \phi_{c1-c6}(X^i, Y) \leq 0. \tag{25}$$

Projection of alternate problem is given in (24) on $\mathscr{X}$ space. The solution to this problem as all constraints hold for the primal problem (22) for all $\mathscr{X}^k$ values is given as

$$\min_{X} \min_{Y} \ -\mathscr{E}(X^i, Y^i) - \nabla\mathscr{E}(X^i, Y^i)\binom{Y-Y^i}{X-X^i}$$
$$\text{subject to}: \ \phi_{c1-c6}(X^i, Y^i)$$
$$-\phi_{c1-c6}(X^i, Y^i)\binom{Y-Y^i}{X-X^i} \leq 0. \tag{26}$$

A new variable ν is introduced, and equivalent minimization problem can be written as

$$\min_{X,Y,\nu} \ -\nu$$
$$\text{subject to}: \ \nu \geq -\mathscr{E}(X^k, Y^k) - \nabla\mathscr{E}(X^i, Y^i)\binom{Y-Y^i}{X-X^i}$$
$$\phi_{c1-c6}(X^i, Y^i) - \phi_{c1-c6}(X^i, Y^i)\binom{Y-Y^i}{X-X^i} \leq 0. \tag{27}$$

The above problem (27) is a master subproblem, which gives lower bounds. This problem is equivalent to problem (15) if Y is compact, nonempty, and a convex set. This problem is a mixed-integer linear programming (MILP) problem, which can be solved using the OAA algorithm.

4 Discussions on Simulation Results

Extensive simulations have been done to study the performance of the proposed algorithm. To solve the formulated MINLP problem, non-linear optimization with OAA (Bonmin) software is used [20]. The problem is studied in terms of mode selection, network throughput, and energy efficiency. Used simulation parameters [21, 22] are summarized in Table 3. IoT nodes are distributed in a uniform manner in the network.

Figure 3 depicts the node distribution between each mode (fog and D2D). The number of nodes selected in the fog mode is more than that of the nodes served in D2D mode. In D2D mode, there is very little chance that an IoT node gets the required files. In fog mode, IoT nodes will surely receive the requested file from cache memory, if not then the request is completed in the cloud part. Figure 4 shows the increasing trend of network throughput. As the number of nodes increases, there will be more data production that increases the throughput. Also, two modes

Table 3 System parameters

Parameter	Value
P_f	12 W
P_d	3 W
d_o	5 m
Processing delay at fog node, l_P	0.1 ms
Fog node's storage capacity, S_f	40 MB
Min. data rate requirement, $\mathscr{R}^k_{min}$	200 kbps
G_o	50
ζ	10 dB
Min. IoT nodes	4
Max. IoT nodes	20
Number of fog nodes	4
Path-loss exponent	2

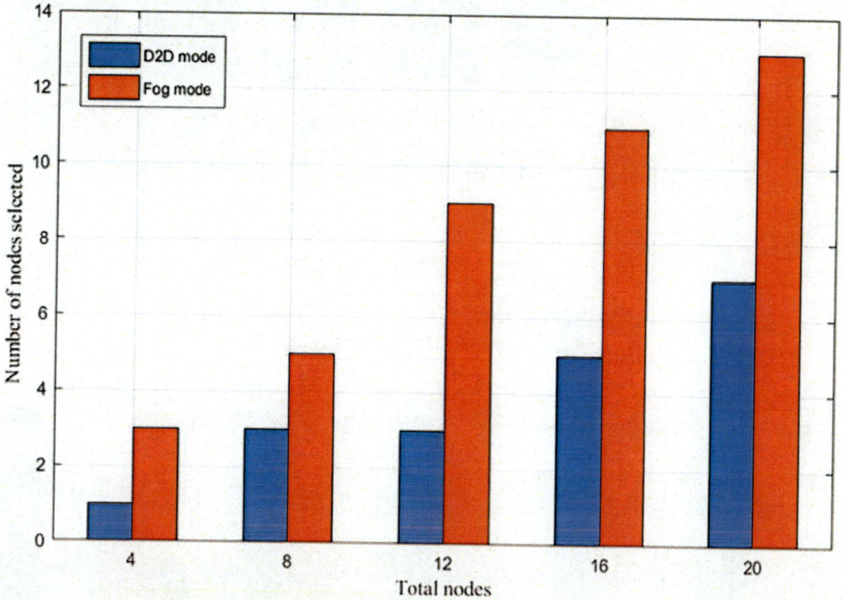

Fig. 3 Node distribution in D2D-assisted fog network

of connection result in the increment of data production. System throughput also depends on the channel state information.

In Fig. 5, we plotted energy efficiency against the number of nodes. The general increasing trend of energy efficiency with an increase in the number of nodes in the system can be seen. Simulations are done with and without D2D mode to see how the performance of the network behaves. The presence of D2D mode helps in more energy-efficient performance, as nodes will get desired files from the nearby node

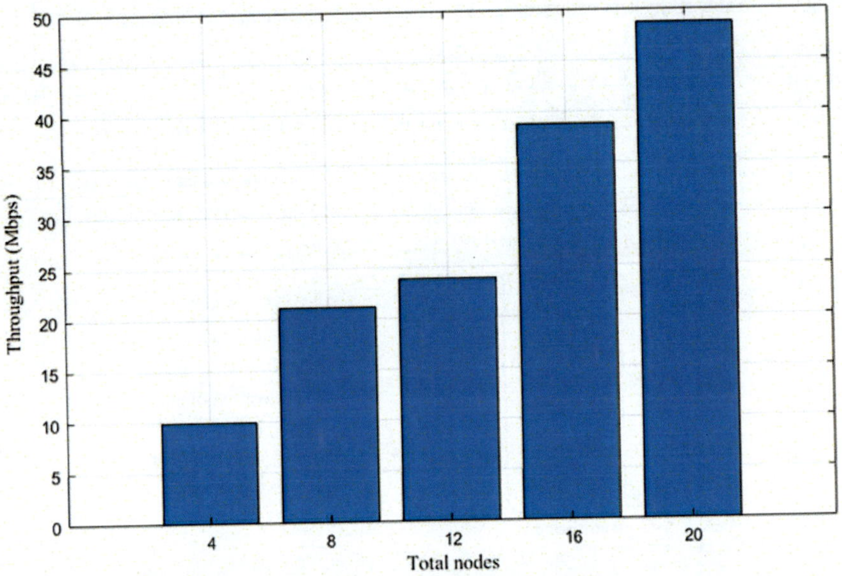

Fig. 4 Throughput of D2D-assisted fog network vs. the number of nodes

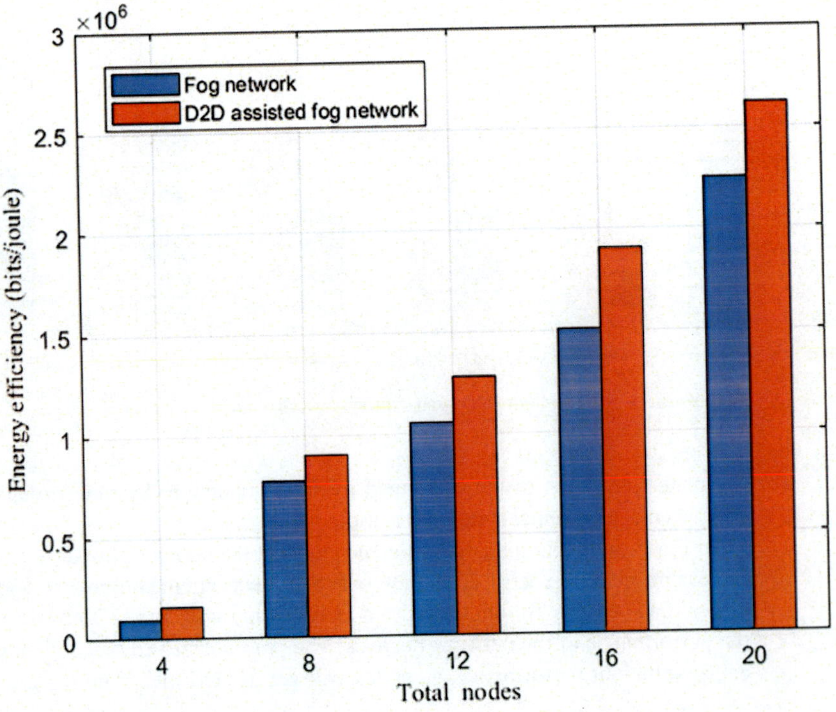

Fig. 5 Energy efficiency vs. the number of nodes

rather than faraway fog nodes. If the requested file is delivered via fog mode, it will cost more energy in terms of experienced latency and computation at resources.

5 Conclusion

In this chapter, we discussed optimal resource management for D2D-assisted fog networks to maximize energy efficiency. The network energy (bits/sec/Hz/watt) is maximized after solving the problem of mode selection and power allocation. The resource allocation in terms of joint mode selection, node association, and power allocation is formulated as an MINLP problem. Using Charnes–Cooper transformation (CCT), the formulated concave fractional programming problem is converted into the concave optimization problem. The solution is found using OAA that gives $\varepsilon = 10^{-3}$—optimal solution after linear convergence. The results show that the availability of cache at the fog node and optional D2D mode for connection increases the energy efficiency of the system.

References

1. Chu, X., Lopez-Perez, D., Yang, Y., Gunnarsson, F.: Heterogeneous Cellular Networks: Theory, Simulation and Deployment. Cambridge University Press, Cambridge (2013)
2. Yi, S., Li, C., Li, Q.: A survey of fog computing: concepts, applications and issues. In: Proceedings of the 2015 Workshop on Mobile Big Data, pp. 37–42 (2015)
3. Basir, R., Qaisar, S., Ali, M., Aldwairi, M., Ashraf, M.I., Mahmood, A., Gidlund, M.: Fog computing enabling industrial internet of things: State-of-the-art and research challenges. Sensors **19**(21), 4807 (2019)
4. Bonomi, F., Milito, R., Zhu, J., Addepalli, S.: Fog computing and its role in the internet of things. In: Proceedings of the First Edition of the MCC Workshop on Mobile Cloud Computing, pp. 13–16. ACM, New York (2012)
5. Akpakwu, G.A., Silva, B.J., Hancke, G.P, Abu-Mahfouz, A.M.: A survey on 5g networks for the internet of things: communication technologies and challenges. IEEE Access **6**, 3619–3647 (2017)
6. Pekka, J., Chia-Hao, Y., Cassio, R., Carl, W., Klaus, H., Olav, T., Visa, K.: Device-to-device communication underlaying cellular communications systems. In: Int'l J. Communications, Network and System Sciences, vol. 2009 (2009)
7. Wang, S., Huang, X., Liu, Y., Yu, R.: Cachinmobile: An energy-efficient users caching scheme for fog computing. In: 2016 IEEE/CIC International Conference on Communications in China (ICCC), pp. 1–6. IEEE, Piscataway (2016)
8. Liu, C., Li, Y., Mao, Z., Sha, T.: On the design of ofdm-based simultaneous wireless information and power transfer in fog-radio access networks. In: 2018 IEEE/CIC International Conference on Communications in China (ICCC Workshops), pp. 272–276. IEEE, Piscataway (2018)
9. Zhang, H., Xiao, Y., Bu, S., Niyato, D., Yu, F.R., Han, Z.: Computing resource allocation in three-tier iot fog networks: a joint optimization approach combining stackelberg game and matching. IEEE Int. Things J. **4**(5), 1204–1215 (2017)

10. Li, L., Guan, Q., Jin, L., and Guo, M., "Resource allocation and task offloading for heterogeneous real-time tasks with uncertain duration time in a fog queueing system," *IEEE Access*, vol. 7, pp. 9912–9925, 2019.
11. Pontois, N., Kaneko, M., T. H. L. Dinh, and Boukhatem, L., "User pre-scheduling and beam-forming with imperfect csi in 5g fog radio access networks," *arXiv preprint arXiv:1802.01104*, 2018.
12. Li, Q., Zhao, J., Gong, Y., Zhang, Q.: Energy-efficient computation offloading and resource allocation in fog computing for internet of everything. China Commun. **16**(3), 32–41 (2019)
13. Li, Z., Sichitiu, M.L., Qiu, X.: Fog radio access network: a new wireless backhaul architecture for small cell networks. IEEE Access **7**, 14150–14161 (2018)
14. Xiao, Y., Krunz, M.: Distributed optimization for energy-efficient fog computing in the tactile internet. IEEE J. Selec. Areas Commun. **36**(11), 2390–2400 (2018)
15. Zhang, H., Zhu, L., Long, K., Li, X.: Energy efficient resource allocation in millimeter-wave-based fog radio access networks. In: 2018 2nd URSI Atlantic Radio Science Meeting (AT-RASC), pp. 1–4. IEEE, Piscataway (2018)
16. Hanif, M.F., Smith, P.J.: On the statistics of cognitive radio capacity in shadowing and fast fading environments. IEEE Trans. Wirel. Commun. **9**(2), 844–852 (2010)
17. Goldsmith, A.: Wireless Communications. Cambridge University Press, Cambridge (2005)
18. Charnes, A., Cooper, W.W.: Programming with linear fractional functionals. Naval Res. Log. Quart. **9**(3–4), 181–186 (1962)
19. Floudas, C.A., Pardalos, P.M.: Encyclopedia of Optimization. Springer Science & Business Media, Berlin (2008)
20. Bonami, P.: Basic Open-Source Nonlinear Mixed Integer Program- ming . http://www.coin-or.org/Bonmin/. Accessed 1 Dec 2019
21. Basir, R., Qaisar, S.B., Ali, M., Naeem, M., Joshi, K.C., Rodriguez, J.: Latency-aware resource allocation in green fog networks for industrial IoT applications. In: 2020 IEEE International Conference on Communications Workshops (ICC Workshops), pp. 1–6. IEEE, Piscataway (2020)
22. Basir, R., Qaisar, S., Ali, M., Pervaiz, H., Naeem, M., Imran, M.A.: Resource allocation and throughput maximization for iot real-time applications. In: 2020 IEEE 91st Vehicular Technology Conference (VTC2020-Spring), pp. 1–5. IEEE, Piscataway (2020)

Wireless Edge Caching with Short-Term Content Popularity Prediction for Cost Minimization

Thinh Duy Tran and Long Bao Le

1 Introduction

By appropriately caching popular contents at the network edge repositories, a mobile edge caching system can mitigate the network congestion, reduce the operation costs, and effectively deal with the skyrocketing growth of mobile video traffic [1]. Such a system is typically designed by leveraging the content popularity, which represents the number of times individual contents are requested or viewed by user equipment (UE). The content popularity, which must be predicted, is thus crucial to perform cache replacements over time.[1] Most existing works on mobile edge caching, however, consider the scenarios with *time-invariant* or slowly varying content popularity [2–4], which may not hold true in environments with rapid changes of view/request patterns [5]. In practice, high-frequency content replacements can be required to cope with *time-varying* content popularity. For example, certain content platform such as YouTube sees a large number of new contents per day with highly dynamic view-per-day patterns, as shown in Fig. 1. In addition, one should perform content replacements at the cache repository based on both current and future predicted content popularity information. Such action helps the content caching system achieve lower operation cost by avoiding unnecessary content replacements.

There are several recent works that have studied the content popularity prediction problem [5–7]. Particularly, the linear regression and deep learning techniques have been employed to predict the future content popularity based on information/data

[1] We use the terms content and file interchangeably in this chapter.

T. D. Tran (✉) · L. B. Le
INRS-EMT, University of Quebec, Montréal, QC, Canada
e-mail: thinh.tran@inrs.ca; long.le@inrs.ca

H. Nguyen et al. (eds.), *30th Biennial Symposium on Communications 2021*,
Signals and Communication Technology, https://doi.org/10.1007/978-3-031-06947-5_14

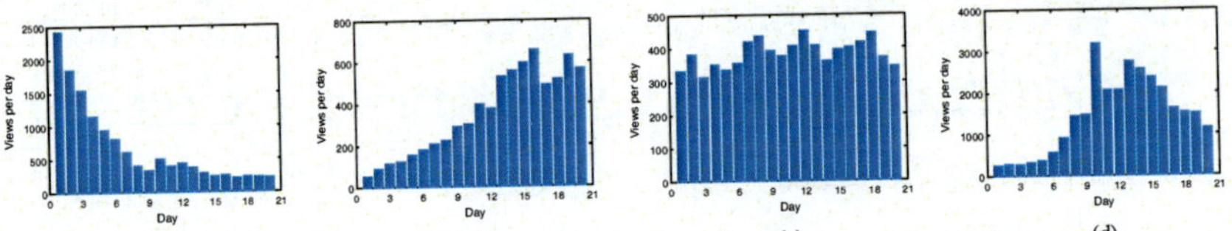

(a) (b) (c) (d)

Fig. 1 Trends of the number of views per day of various YouTube videos. (**a**) Decreasing trend. (**b**) Increasing trend. (**c**) Flat trend. (**d**) Concave-like trend

such as the view patterns of different contents and/or certain features of the contents. Some recent papers have proposed various content caching strategies considering the *time-varying* content popularity [8–10]. Variants of deep neural networks [8, 9] and heuristic online learning [10] techniques are employed to predict the time-varying trends of contents in these works. They, however, only address the backhaul load reduction [8, 9] in their caching designs while not considering content replacement/update cost at storage repositories. In fact, this cache update cost could represent a significant contribution to the total operation cost of systems with limited resource and low-rate bandwidth such as sensor networks [11], or systems with high update overheads such as software-defined networks (SDN) [12].

In this chapter, we propose a rolling-based cache update framework that aims to minimize the total backhaul and content update/replacement cost at the network edge. Our design leverages the content popularity information that is predicted over a window of multiple future time slots.[2] The proposed framework tends to cache contents popular over a long period; thus, it prevents unnecessary cache updates while still effectively reducing the congestion of the backhaul links. Toward this end, the content replacement problem is formulated as a cost minimization optimization problem, which is is then transformed into an integer linear program (ILP). By smartly exploiting several characteristics of the underlying problem, we propose a polynomial-time algorithm, namely Window-Based Content Caching (WBCC) algorithm, to solve it. Numerical results confirm the efficacy of our proposed framework compared to a baseline algorithm in terms of the achievable long-term cost.

[2] There are various models for predicting content popularity in the literature [5, 6], which can be adopted to our caching design. Hence, proposing a novel content popularity prediction model is beyond the scope of this chapter.

2 System Model and Problem Formulation

2.1 System Model

We consider a mobile edge caching system operating in the time slotted fashion. Here, each time slot interval can be several hours or days depending on the dynamics of content popularity. End users are interested in accessing contents/files of the same size in a set denoted as $\mathcal{F} = \{c_f\}, f = 1, 2, \ldots, F$. We assume that the storage repository at the edge network can cache up to S_{max} files of the content set $\mathcal{F}$.

The content replacement is performed in the rolling manner in our design where the content update decision at the end of each time slot is made based on the content popularity information predicted over a window of P future time slots. Such predicted information is used in the cost minimization problem over the prediction window. After solving this problem, the content replacement solution at the current time slot is implemented. Then, the optimization/prediction window is shifted one slot forward, and the cost minimization is performed again over the new (shifted) prediction window.

For a particular prediction window, we denote the current time slot as t_{N} and the time slot prior to t_{N} as t_{O}, i.e., the content-updating time slot of the previous window. At t_{N}, we introduce a binary variable x_f^{N} to capture the caching decision for file $c_f \in \mathcal{F}$, where $x_f^{\text{N}} = 1$ if file c_f is cached in the edge server (ES), and $x_f^{\text{N}} = 0$, otherwise. Similarly, the binary variable x_f^{O} denotes the caching decision for file c_f at t_{O}. $\boldsymbol{x}_{\text{N}}$ and $\boldsymbol{x}_{\text{O}}$ denote the caching decision vectors for all files at t_{N} and t_{O}, respectively. We denote by $\mathcal{C}_{\text{O}} = \{c_f \in \mathcal{F} \mid x_f^{\text{O}} = 1\}$ and $\mathcal{C}_{\text{N}} = \{c_f \in \mathcal{F} \mid x_f^{\text{N}} = 1\}$ the sets of files being cached at the ES at t_{O} and t_{N}, respectively.

In this chapter, content popularity of a particular content is defined as the number of requests/views of this content in the underlying time slot. We also denote by t_i the i-th time slot of the considered window counting from t_{N} where t_0 is also t_{N}. The content popularity of file c_f at t_i is denoted as $v_{f,i}$. Because we do not know the true number of requests of c_f in future time slots of the considered window, $v_{f,i}$ can be obtained from a certain content popularity prediction strategy.[3] For ease to follow, key notations used in this chapter are summarized in Table 1. The considered rolling-based content replacement framework is illustrated in Fig. 2.

2.2 Problem Formulation

We consider the content replacement optimization problem for one particular window that aims to minimize the total cost over the window. The total cost

[3] The Szabo–Huberman (SH) prediction strategy [6] is used in the simulation even though our proposed cache replacement design can be used jointly with any other prediction strategy.

Table 1 Summary of key notations

Notation	Description
$\mathcal{F} = \{c_f\}, f = 1, 2, \ldots, F$	The set of all contents interested by UEs
c_f	Content with index f in the content set $\mathcal{F}$
S_{max}	Maximum number of contents can be stored at the network edge storage repository
t_{N}	The current time slot
t_{O}	The time slot prior to the current time slot t_{N}
P	Number of future time slots counted from t_{N}
t_i	The i-th time slot counting from t_{N}, where t_0 is also t_{N}
x_f^{N}	Caching decision for content c_f at time slot t_{N}
$\boldsymbol{x}_{\mathsf{N}}$	Vector of caching decision at time slot t_{N}
x_f^{O}	Caching decision for content c_f at time slot t_{O}
$\boldsymbol{x}_{\mathsf{O}}$	Vector of caching decision at time slot t_{O}
$\mathcal{C}_{\mathsf{N}} = \{c_f \in \mathcal{F} \mid x_f^{\mathsf{N}} = 1\}$	The set of contents cached at ES at t_{N}
$\mathcal{C}_{\mathsf{O}} = \{c_f \in \mathcal{F} \mid x_f^{\mathsf{O}} = 1\}$	The set of contents not cached at ES at t_{O}
$v_{f,i}$	The predicted content popularity of file c_f at time slot t_i
$\hat{v}_{f,i}$	The actual content popularity of file c_f at time slot t_i
α	The cost per file for transmitting a file over the backhaul link
β	The cost for replacing one content at the ES

comprises two components, namely the cache update cost and backhaul cost, which are described in more detail in the following.

2.2.1 Cache Update Cost

This cost arises from replacing contents currently cached in the ES with different contents at t_{N}. This cost component can account for different factors such as storage price at the edge and energy consumption due to the cache update task [11, 12]. Mathematically, this cost component can be calculated as follows:

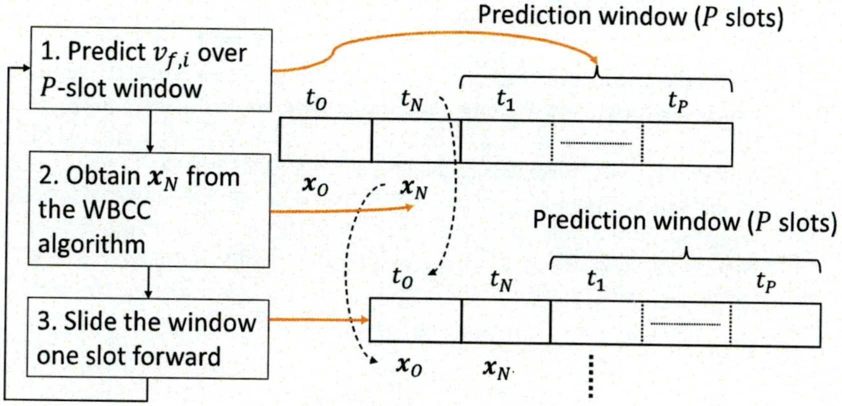

Fig. 2 Operations of the rolling-based content replacement framework

$$R(\boldsymbol{x}_{\mathrm{N}}, \boldsymbol{x}_{\mathrm{O}}) = \beta\,[S_{\max} - |\mathcal{C}_{\mathrm{O}} \cap \mathcal{C}_{\mathrm{N}}|]\,, \tag{1}$$

where β ($/file) represents the cost for replacing one content in the ES. Note that $\boldsymbol{x}_{\mathrm{O}}$ is known so it is an input parameter for the optimization performed at t_{N}. In (1), $|\mathcal{C}_{\mathrm{O}} \cap \mathcal{C}_{\mathrm{N}}|$ represents the size of the set of cached contents at t_{O} not replaced at t_{N}. Therefore, $S_{\max} - |\mathcal{C}_{\mathrm{O}} \cap \mathcal{C}_{\mathrm{N}}|$ represents the number of contents cached at t_{O} to be replaced at t_{N}.[4]

2.2.2 Backhaul Cost

If file c_f is not cached in the ES (i.e., $x_f^{\mathrm{N}} = 0$), it must be transmitted over the core network (CN) and the backhaul link before being delivered to a requesting UE over the wireless access network. In contrast, a file c_f already cached in the ES can be delivered immediately to the requesting UE without using the backhaul link. Therefore, if there are $v_{f,i}$ requests to file f at time slot t_i of the considered window, the total backhaul cost can be expressed as $\alpha(1 - x_f^{\mathrm{N}})v_{f,i}$, where α is the cost per file for transmitting it over the backhaul link. Therefore, if we assume that the same set of files is cached in the ES in each time slot of the considered window, the total backhaul cost is given by

$$B(\boldsymbol{x}_{\mathrm{N}}) = \sum_{i=0}^{P} \sum_{f=1}^{F} \alpha(1 - x_f^{\mathrm{N}})v_{f,i}. \tag{2}$$

---

[4] Note that when the caching system starts for the first time, the cache is assumed to be empty, i.e., $x_f^{\mathrm{O}} = 0$ for all $f \in \mathcal{F}$.

2.2.3 Problem Formulation

If more contents are replaced at the beginning of a particular window, content replacement cost will become larger as can be verified from (2). Intuitively, only contents that are sufficiently popular in several time slots of the window should replace contents currently cached in the ES that tend to become less popular in the future time slots. Therefore, to perform cost-minimization-based content replacement, we assume that the content popularity over P future time slots can be accurately estimated. The cost minimization problem can be stated as

$$\min_{\boldsymbol{x}_\mathsf{N}} \; J(\boldsymbol{x}_\mathsf{N}) = B(\boldsymbol{x}_\mathsf{N}) + R(\boldsymbol{x}_\mathsf{N}, \boldsymbol{x}_\mathsf{O}) \tag{3a}$$

$$\text{s.t.} \; \sum_{f \in \mathcal{F}} x_f^\mathsf{N} \leq S_\mathsf{max} \tag{3b}$$

$$x_f^\mathsf{N} \in \{0, 1\}, \; f = 1, \ldots, F, \tag{3c}$$

where (3b) captures the caching capacity constraint and (3c) indicates the binary nature of the optimization variables. Upon solving this problem, the achieved cache replacement solution is realized for time slot t_0 of the considered window. Then, when the system enters the next time slot, we re-solve (3) for the next window, which is achieved by shifting the previously considered window one time slot forward. Note that the newly predicted content popularity information for the new window must be used for the optimization.

3 Proposed Algorithm

In the following, we first transform problem (3) to a more explicit form based on which we can develop a polynomial-time algorithm to solve it.

3.1 Problem Transformation

We now re-express the cache update cost function (1) to facilitate problem transformation (3). Specifically, the term $|\mathcal{C}_\mathsf{O} \cap \mathcal{C}_\mathsf{N}|$ in (1) can be re-written as follows:

$$|\mathcal{C}_\mathsf{O} \cap \mathcal{C}_\mathsf{N}| = \sum_{f=1}^{F} x_f^\mathsf{O} \times x_f^\mathsf{N}, \tag{4}$$

where x_f^O and x_f^N are Boolean (binary) variables. Using this result, problem (3) can be re-stated as follows:

$$\min_{x_\mathsf{N}} J(x_\mathsf{N}) = \sum_{i=0}^{P}\sum_{f=1}^{F} \alpha(1 - x_f^\mathsf{N})v_{f,i} + \beta\left(S_\mathsf{max} - \sum_{f=1}^{F} x_f^\mathsf{O} x_f^\mathsf{N}\right)$$

$$\text{s.t. constraints (3b) and (3c).} \tag{5}$$

Problem (5) is an integer linear programming (ILP) problem, which can be solved by a commercial solver. However, solving a large-scale problem (5) takes extremely long time with a high dimension x_N for a large file set and a large value of P. To overcome this challenge, we exploit some characteristics of problem (5) that are stated in Proposition 1.

Proposition 1 *To minimize the total cost, the following must hold: (i) The storage capacity of the ES should be fully utilized for caching contents at any time slot. (ii) Any content $c_f \in \mathcal{C}_\mathsf{O}$ cached in the ES at t_O should only be replaced by an un-cached content $c_{f'}$ at time slot t_N if this replacement leads to a lower total cost, i.e., a positive cost reduction.*

Proof In problem (5), we see that: (i) If more files are cached (i.e., large $\sum_{f=1}^{F} x_f^\mathsf{N}$), the larger number of subtrahend terms $x_f^\mathsf{N} v_{f,i}$ will affect $J(x)$, thus resulting in a smaller $J(x)$. (ii) Consider a pair of $c_f \in \mathcal{C}_\mathsf{O}$ and $c_{f'}$, where $f \neq f'$, the cost incurred due to replacing c_f with $c_{f'}$ at t_N is $\beta + \sum_{i=0}^{P} \alpha v_{f,i}$, where $i = 0$ denotes the time index of t_N. Meanwhile, the backhaul cost when keeping content c_f in the ES at time slot t_N is $\sum_{i=0}^{P} \alpha v_{f'}$. We can define a *cost reduction* achieved by replacing c_f by $c_{f'}$ as follows:

$$r(c_f, c_{f'}) \triangleq \beta + \sum_{i=0}^{P} \alpha(v_{f,i} - v_{f',i}). \tag{6}$$

If $r(c_f, c_{f'}) > 0$, then c_f should be replaced by $c_{f'}$ at t_N to reduce the cost. Otherwise, we should keep c_f at t_N in the ES. Thus, we have proved statements (i) and (ii) of Proposition 1. ∎

3.2 Proposed Algorithm

Based on the results in Proposition 1, we propose the Window-Based Content Caching (WBCC) algorithm presented in Algorithm 1, which is based on the predicted content popularity over P future time slots. Specifically, we first define the set $\mathcal{P}_\mathsf{N}$ with S_max contents having highest values V_f calculated in line 2 (total views over $P+1$ slots). Algorithm 1 first calculates the cost reduction factors $r(c_a, c_b)$ for all potential content pairs (c_a, c_b) (line 5), where c_a and c_b belong to sets $\mathcal{G}$ and $\mathcal{H}$

Algorithm 1 Window-based content caching algorithm (WBCC)

1: Predict $v_{f,i}$ for the next P time slots counting from t_{N}.
2: Calculate $V_f := \sum_{i=0}^{P} v_{f,i}$, $\forall f$ and sort all contents f in descending order $(V_f)_{f \in \mathcal{F}}$.
3: Put S_{max} highest-V_f contents to the set $\mathcal{P}_{\mathsf{N}}$.
4: Compute $\mathcal{G} := \mathcal{C}_{\mathsf{O}} \setminus \mathcal{P}_{\mathsf{N}}$, and $\mathcal{H} := \mathcal{P}_{\mathsf{N}} \setminus \mathcal{C}_{\mathsf{O}}$.
5: Use (6) to compute $(r(c_a, c_b))_{c_a \in \mathcal{G}, c_b \in \mathcal{H}}$.
6: **while** $\max\limits_{c_a \in \mathcal{G}, c_b \in \mathcal{H}} r(c_a, c_b) > 0$ AND $\mathcal{G} \neq \emptyset$ **do**
7:     Find $(c_{a^*}, c_{b^*}) := \text{argmax}_{c_a \in \mathcal{G}, c_b \in \mathcal{H}}\, r(c_a, c_b)$.
8:     Replace $c_{b^*} \in \mathcal{P}_{\mathsf{N}}$ with $c_{a^*} \in \mathcal{C}_{\mathsf{O}}$.
9:     Update $\mathcal{G} := \mathcal{G} \setminus \{a^*\}$ and $\mathcal{H} := \mathcal{H} \setminus \{b^*\}$.
10: **end while**
11: Return the caching set $\mathcal{C}_{\mathsf{N}} := \mathcal{P}_{\mathsf{N}}$ at time slot t_{N}.

(defined in line 4), respectively. Then, we iteratively perform content replacements for content pairs achieving the largest and positive cost reduction in each iteration.

For better understanding of the proposed algorithm, let us consider an example. We assume $\mathcal{F} = \{c_1, c_2, c_3, c_4, c_5, c_6\}$, $S_{\mathsf{max}} = 3$, and $\mathcal{C}_{\mathsf{O}} = \{c_1, c_2, c_3\}$. In line 3, suppose Algorithm 1 returns $\mathcal{P}_{\mathsf{N}} = \{c_1, c_4, c_5\}$, which is the best caching set *without* considering the cache replacement cost. Next, line 4 in Algorithm 1 returns $\mathcal{G} = \{c_2, c_3\}$ and $\mathcal{H} = \{c_4, c_5\}$. Considering the content replacement cost, the algorithm computes the cost reduction for each pair $r(c_2, c_4)$, $r(c_2, c_5)$, $r(c_3, c_4)$, and $r(c_3, c_5)$ using (6). With the obtained cost reduction for these content pairs, we can iteratively find the best pair of contents for replacement if it results in a positive cost reduction value over iterations of the "while loop."

Proposition 2 *Algorithm 1 converges given the predicted content popularity in the future P time slots.*

Proof In the "while loop," we can find the best pair of contents (c_{a*}, c_{b*}) for replacement (lines 7 and 8) in each iteration if it results in a highest positive cost reduction value, thus further reducing $J(\boldsymbol{x}_{\mathsf{N}})$. After each replacement, the underlying pair of contents will be excluded from their sets $\mathcal{G}$ and $\mathcal{H}$. Thus, the stopping condition of the "while loop" ($\max\limits_{c_a \in \mathcal{G}, c_b \in \mathcal{H}} r(c_a, c_b) \leq 0$ or $\mathcal{G} = \emptyset$) will be met, which means the proposed algorithm converges. ∎

Algorithm 1 has polynomial-time complexity. Specifically, the sorting procedure in line 3 has complexity of $\mathcal{O}(F \log(F))$ if we use the merge sort algorithm. The calculation of $r(c_a, c_b)$ in line 5 has the worst-case complexity of $\mathcal{O}(S_{\mathsf{max}}^2)$. The "while loop" has the worst-case complexity of $\mathcal{O}(S_{\mathsf{max}} \log(S_{\mathsf{max}}))$. Therefore, the worst-case complexity of Algorithm 1 in each time slot is $\mathcal{O}(F \log(F) + S_{\mathsf{max}} \log(S_{\mathsf{max}}) + 2S_{\mathsf{max}}^2)) \approx \mathcal{O}(F \log(F) + S_{\mathsf{max}}^2)$.

4 Numerical Results

We evaluate our proposed framework by comparing the WBCC algorithm with the per-slot most popular caching (SMPC) algorithm. Specifically, the SMPC algorithm will cache the most popular contents at each time slot, taking as input the content popularity information at this time slot only. The performance metric is *cost gain*, which is defined as $cost\_gain \triangleq J(x_{\mathsf{SMPC}}) - J(x_{\mathsf{WBCC}})$. We average the cost gain over a period of 10 time slots. Unless stated otherwise, the cost factors α and β are 0.01 and 1000, respectively.

Figure 3 shows the cost gain of the WBCC algorithm over the SMPC algorithm when P varies. Moreover, the true/perfect and the predicted future content popularity information are used in our algorithm, which are indicated as WBCC perfect and WBCC predicted in Fig. 3, respectively, in order to evaluate the impact of content popularity prediction accuracy on the proposed framework's performance. The former information is obtained from a dataset $\mathcal{D}$ with 2121 randomly chosen YouTube videos in 2008 [13], while the latter information is obtained from the SH model [6].[5] In [6], the accumulated number of views $v(f, t_r, t_t)$ of file f at the future time slot t_t is predicted based on the known accumulated number of views $\hat{v}(f, t_r)$ at time slot t_r as $v(f, t_r, t_t) = \gamma_{t_r,t_t}\hat{v}(f, t_r)$.[6] Here, γ_{t_r,t_t} represents the linear correlation factor between the content popularity at t_r and t_t, which is estimated by using the known popularity levels of other files in a dataset $\mathcal{D}'$ as

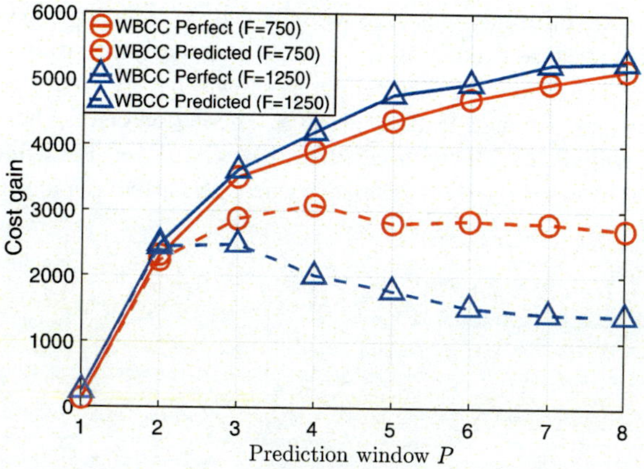

Fig. 3 The cost gain of the proposed WBCC algorithm versus the prediction window P

[5] We emphasize again that any other prediction strategy can be used with our framework.

[6] To apply the SH model in our framework, we set $t_r = t_{\mathsf{N}}$ and $t_t = t_{\mathsf{N}} + t_i$, where t_i is the i-th time slot of the prediction window.

$$\gamma_{t_r,t_t} \triangleq \frac{\sum_{f\in\mathcal{D}'} \hat{v}(f,t_r)/\hat{v}(f,t_t)}{\sum_{f\in\mathcal{D}'} \left(\hat{v}(f,t_r)/\hat{v}(f,t_t)\right)^2}. \tag{7}$$

The training set $\mathcal{D}'$ contains 1900 videos randomly selected from $\mathcal{D}$, and the remaining 200 videos are used for testing the prediction accuracy. We use the performance metric called mean relative squared error (mRSE) defined as

$$mRSE = \frac{1}{|\mathcal{D}|}\sum_{f\in\mathcal{D}} \left(\frac{v(f,t_r,t_t)}{\hat{v}(f,t_r,t_t)} - 1\right)^2, \tag{8}$$

where $v(f,t_r,t_t)$ and $\hat{v}(f,t_r,t_t)$, respectively, denote the predicted and actual content popularity.

The results in Fig. 3 are averaged over different values of S_{max} ranging from 50 to 150 videos. The cost gain increases when P becomes larger before becoming saturated with perfect information of content popularity (WBCC perfect curves). If we can predict accurately the content popularity in future time slots, the knowledge of content popularity enables the WBCC algorithm to cache longer-term popular contents, thus preventing unnecessary frequent content replacements and lowering the operation cost. However, the accuracy level of the content popularity predictor greatly affects the performance of the WBCC algorithm. Any practical prediction strategy cannot perfectly predict the content popularity over the prediction window. Therefore, a tradeoff between P and the cost gain exists when the predicted content popularity information is used with the WBCC algorithm (WBCC predicted curves). This is because predicting the content popularity in farther future time slots (i.e., large t_t) tends to be less accurate, as shown in Fig. 4. Further, the larger size of the content set $\mathcal{F}$ could lead to the larger prediction error. Therefore, the highest cost gain with $F = 750$ is larger than that with $F = 1250$ under the SH prediction strategy even though the reverse is true with perfect popularity information.

Fig. 4 The mRSE of the SH model versus t_t, given $t_r = 1$

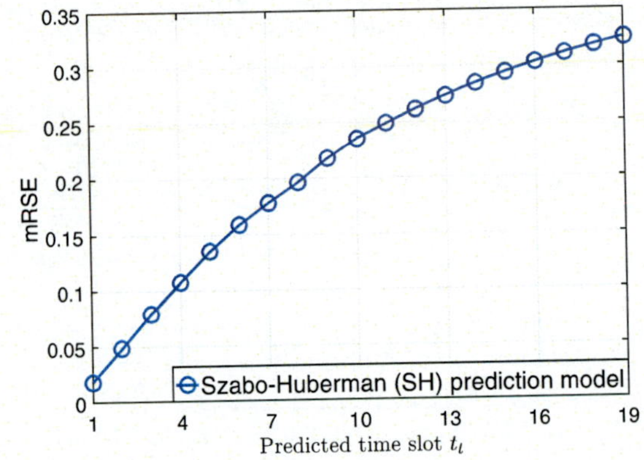

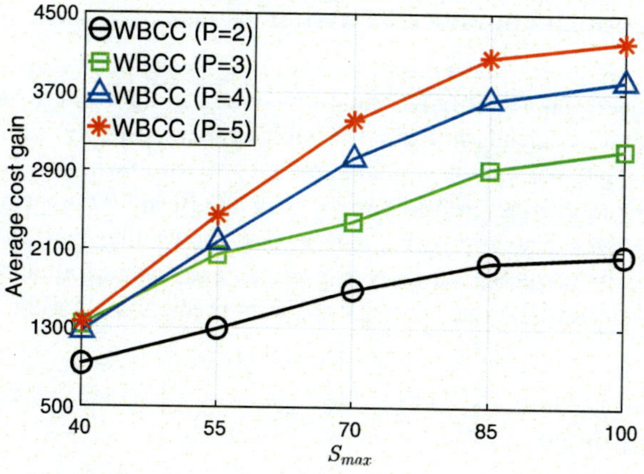

Fig. 5 The average cost gain versus $S_{\max}$

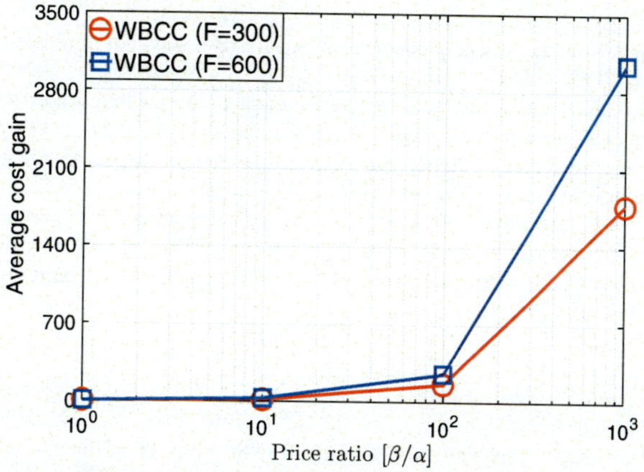

Fig. 6 The average cost gain versus the price ratio β/α

We show the cost gain due to the WBCC algorithm with perfect content popularity information versus $S_{\max}$ in Fig. 5, where the results are obtained by averaging over different values of content set size $F = \{300, 600, 750, 1000, 1250\}$. The cost gain increases with $S_{\max}$ because the system can cache more popular contents, thus reducing the backhaul and content replacement costs with larger $S_{\max}$. This figure also confirms the benefit of choosing the larger prediction window in optimizing the content replacement decisions. Finally, Fig. 6 shows the impact of the price ratio (β/α) in log-scale on the cost gain where the results are obtained by averaging over different values of $S_{\max}$ ranging from 50 to 150. We can see that the larger price ratio results in the higher cost gain for the WBCC algorithm (with perfect content popularity information), which reflects the importance of considering the content replacement cost.

5 Conclusion and Discussion

We have proposed a polynomial-time caching algorithm that aims at minimizing the total backhaul and content replacement costs by exploiting the predicted content popularity over a predetermined prediction window. Numerical results have confirmed that the proposed framework significantly outperforms the conventional MPC algorithm. Moreover, it has been shown that the accuracy of the content popularity prediction strongly impacts the caching performance. Therefore, we plan to develop an accurate content popularity prediction algorithm in our future work.

References

1. Cisco, Cisco Visual Networking Index: Forecast and Trends, 2017–2022 White Paper, 2018. Available: https://www.cisco.com/c/en/us/solutions/collateral/service-provider/visual-networking-index-vni/white-paper-c11-741490.html
2. Yang, Z. et al.: Cache placement in two-tier HetNets with limited storage capacity: cache or buffer? IEEE Trans. Commun. **66**(11), 5415–5429 (2018)
3. Vu, T.X., Chatzinotas, S., Ottersten, B.: Edge-caching wireless networks: performance analysis and optimization. IEEE Trans. Wirel. Commun. **17**(4), 2827–2839 (2018)
4. Tran, T.D., Le, L.B.: Joint resource allocation and content caching in virtualized content-centric wireless networks. IEEE Access **6**, 11329–11341 (2018)
5. Pinto, H., Almeida, J.M., Gonçalves, M.A.: Using early view patterns to predict the popularity of YouTube videos. In: Proceeding of ACM WSDM, pp. 365–374, Rome, Italy (2013)
6. Szabor, G., Huberman, B.A.: Predicting the popularity of online content. Commun. ACM **53**(8), 80–88 (2010)
7. Zhang, Y., Tan, X., Li, W.: PPC: popularity prediction caching in ICN. IEEE Commun. Lett. **22**(1), 5–8 (2018)
8. Liu, W., Zhang, J., Liang, Z., Peng, L., Cai, J.: Content popularity prediction and caching for ICN: a deep learning approach with SDN. IEEE Access **6**, 5075–5089 (2018)
9. Doan, K.N., Van Nguyen, T., Quek, T.Q.S., Shin, H.: Content-aware proactive caching for backhaul offloading in cellular network. IEEE Trans. Wirel. Commun. **17**(5), 3128–3140 (2018)
10. Muller, S., Atan, O., van der Schaar, M., Klein, A.: Context-aware proactive content caching with service differentiation in wireless networks. IEEE Trans. Wirel. Commun. **16**(2), 1024–1036 (2017)
11. Ding, Z., Fan, X., Yu, J., Bi, J.: Update cost-aware cache replacement for wildcard rules in software-defined networking. In: proceedings of the IEEE Symposium on Computers and Communications (ISCC), Natal, pp. 457–463 (2018)
12. Ha, M., Kim, D.: On-demand cache placement protocol for content delivery sensor networks. In: Proceedings of the International Conference on Computing, Networking and Communications (ICNC), Santa Clara, CA, pp. 207–216 (2017)
13. Cheng, X., Dale, C., Liu, J.: Statistics and Social Network of YouTube Videos Dataset. Available: http://netsg.cs.sfu.ca/youtubedata/

Semi-supervised Learning with Self-training Classifier for Cache Placement in Mobile Edge Networks

Lubna B. Mohammed, Alagan Anpalagan, Ahmed S. Khwaja, and Muhammad Jaseemuddin

1 Introduction

With the broad use of various mobile applications and services, mobile traffic is expanding at an exponential rate. The mobile core network and backhaul links are under high traffic, and solutions are needed to ease the situation. A promising solution is mobile edge networks that offer mobile edges with cache storage. By placing content closer to users at user terminals (UTs) and small base stations (SBSs), the transmission load on both the backhaul links and the network will be reduced dramatically. Many cache placement strategies have been developed for edge caching in wireless networks. The majority of present research on cache content placement algorithms aims to increase cache hit rates by placing contents depending on changes in content popularity. This is because a few popular files are responsible for a huge number of content requests.

Cache placement algorithms based on prior user requests were designed using supervised learning methods. Singular-value decomposition (SVD) was used to estimate content preferences in [1]. In [2], non-negative matrix factorization machine learning was used to estimate content popularity using users-to-content association. In [3], a deep neural network was utilized to offer multi-layer learning, which gives a nonlinear model for the association between users and content requests. In [4], the conditional variational autoencoder is employed to produce a model-free mobility prediction technique based on unsupervised learning. They investigated users' motions in SBSs across a dense small-cell network, from their past trajectory to proactively caching contents. The authors of [5] used unsupervised

L. B. Mohammed · A. Anpalagan (✉) · A. S. Khwaja · M. Jaseemuddin
Dept. of Electrical, Computer and Biomedical, Engineering, Ryerson University, Toronto, ON, Canada
e-mail: lubnabadri@ryerson.ca; alagan@ryerson.ca; akhwaja@ryerson.ca; jaseem@ryerson.ca

H. Nguyen et al. (eds.), *30th Biennial Symposium on Communications 2021*,
Signals and Communication Technology, https://doi.org/10.1007/978-3-031-06947-5_15

learning and feature extraction to classify the popularity of content. Then, using deep reinforcement learning, the classification results were used to place the content.

Labeled data (input characteristics and the associated target output) is applied to train the model in supervised classification problems. In practical applications, however, it is difficult, costly, and time-consuming to collect labeled samples. When a user requests content while moving around a cell, unlabeled data is available and simple to access with each user request. Overfitting of the training dataset may also be a problem with supervised learning. This happens when the supervised classifier tries to conform itself to the statistical noise in the training sample. Because wireless networks have so many distinct and changing input properties, these sorts of mistakes are almost always unavoidable and unexpected. The generated classifier will not accurately represent the underlying association between input characteristics and output attributes. The benefit of applying semi-supervised learning to increase classifier performance and reduce overfitting was shown in [6].

In [7], the authors suggest that a semi-supervised self-training classifier model for disease classification will be built by combining a small number of labeled datasets with a large number of unlabeled datasets. The efficacy of their experimental results in resolving the disease classification problem was shown.

In this research, we present a semi-supervised self-trained learning approach for designing a cache placement binary classifier in mobile edge networks. Using a small number of labeled data, we first train a supervised classifier using the logistic regression (LR) learning approach. We employ the label propagation approach described in [8] to self-train the classifier and produce pseudo-labels by predicting class labels for all unlabeled data using the trained classifier. Then, to the labeled training set, the pseudo-labels with the greatest likelihood of being correct are added as pseudo-labels. Finally, we used the results to assess the prediction accuracy after training the classifier to predict cache contents for the labeled test dataset. Our results show that for cache placement, the semi-supervised self-trained (SSST) learning approach outperforms other methods. When there is not enough data to train the model, the suggested approach has the capacity to self-train and enhance classification accuracy.

The following is how the rest of the chapter is organized: The system model and assumptions are described in Sect. 2. In Sect. 3, we use a binary classification technique to formulate a multi-objective cache placement problem. Semi-supervised self-training for binary classifiers is discussed in Sect. 4. The simulation results and discussion are presented in Sect. 5, followed by the conclusion in Sect. 6.

2 System Model

Each macrocell consists of one MBS connected to a central network gateway through a high-speed interface, N SBSs connected to the MBS via backhaul lines, and M UTs connected to neighboring devices and SBSs via D2D communication.

As shown in Fig. 1, we investigate a MEN with several SBSs and UTs. Each macrocell is composed of one MBS connected to a central network gateway through a high-speed interface, N SBSs connected to the MBS via backhaul lines, and M UTs connected to neighboring devices and SBSs via D2D communication. s=$s1, s2, cdots, sN$ and u=$u1, u2, cdots, uM$ denote the sets of SBSs and UTs, respectively. Each MBS, SBS, and UT has cache storage units with varying storage capabilities. To determine cache storage capabilities inside one macrocell for SBSs and UTs caches, two sets of $c_s = c_{s1}, c_{s2}, cdots, c_{sN}$ and $c_u = c_{u1}, c_{u2}, cdots, c_{uN}$ may be used. The MBS is expected to have adequate cache capacity to hold the F files described by f=$\{f_1, f_2, \ldots, f_F\}$.

As demonstrated in Fig. 1, there are four different kinds of cache content delivery systems. We explore the following techniques out of these: local cache, SBS cache, and MBS cache.

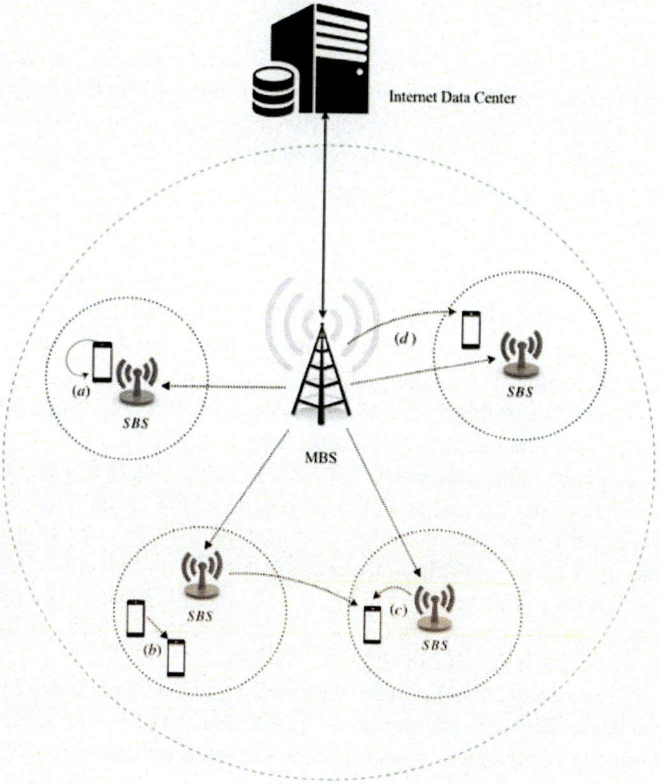

Fig. 1 An illustration of the network model showing the four cache content delivery approaches: (**a**) local cache, (**b**) D2D cache, (**c**) SBS cache, and (**d**) MBS cache

The content library is made up of F files that are cached in the MBS. Each file is indicated by fz for $z = 1, \ldots, F$. Based on their popularity distribution, the files are requested from the main library. The set Λ denotes the popularity of the F files, where $\Lambda = \{\Lambda 1, \Lambda 2, \ldots, \Lambda F\}$. The Zipf popularity distribution [9] can be used to describe the set Λ. Users' connectivity in D2D communication and the small-cell network (communication between mobile UT and SBS) are assumed to be a peer-to-peer connectivity model [10]. The mobility of the user is characterized by spatial and temporal characteristics. The spatial aspects of user mobility patterns give physical location information, whereas the temporal features provide time-related information [11]. A pairwise contact process that follows an independent Poisson process can be used to describe user mobility.

3 Multi-objective Cache Placement

The goal of the cache placement policy is to optimize the cache hit rate while minimizing content delivery time. The weighted-sum cache placement method offers information about the properties of the input data. For developing effective content delivery to UTs, a cache placement strategy that may offer insight into which input properties might increase the hit rate is essential. For M UTs and N SBSs, $i, k = 1, 2, \ldots, M$ and $j = 1, 2, \ldots, N$, the contact durations between all UTs and other UTs and SBSs within a single cell are represented by $(\mathcal{T}^{UT,SBS})_{M\times(M+N)}$. A matrix $(\mathcal{D}^{UT,SBS})_{M\times(M+N)}$ represents the communication ranges at time t when both UT $u_i$and UT $u_k$are within the D2D communication range d^{D2D}, and UT u_i and SBS s_j are within the SBS communication range d^{SBS}. The probability of a UT to contact other UTs or SBSs $P_m^{UT,SBS}$ is a matrix of size $(M \times (M + N))$, which represents the probability of UT u_i to contact with UT u_k, and for the probability of UT u_i to contact with SBS s_j, respectively.

Assume a matrix $A = [A^{UT} : A^{SBS}]_{(M+N)\times F}$ as the cache strategy matrix that needs to be determined by the cache placement algorithm. The contents of matrix A are either one or zero (cached or not cached). Let the matrix $(A^{UT})_{M\times F}$ represent the cache strategy of file f_z that is cached in UTs, where $a_{k,z} \in A^{UT}$ represents file f_z that can be served by UT u_k, $(k = 1, \ldots, M)$. Then, the matrix $(A^{SBS})_{N\times F}$ represents the cache strategy of the files cached in SBSs, where $a_{j,z} \in A^{SBS}$ represents file f_z that can be served by SBS s_j $\forall j \in \{1, \ldots, N\}$.

The latency efficient cache placement algorithm's objective function maximizes the probability of UT u_i downloading the required files from adjacent UTs and SBSs within the contact durations [12] and is expressed as follows:

$$\underset{<A, w_x>}{\text{maximize}} \sum_{x=1}^{4} g^*(A, w_x) \tag{1a}$$

subject to

$$\sum_{z=1}^{F} a_{i,z}.fl_z \leq C_{u_i}, \forall i \in \{1, \ldots, M\} \quad (1b)$$

$$\sum_{z=1}^{F} a_{j,z}.fl_z \leq C_{s_j}, \forall j \in \{1, \ldots, N\} \quad (1c)$$

$$a_{k,z}^{UT} \in [0, 1], \forall k \in \{1, \ldots, M\} \quad (1d)$$

$$a_{j,z}^{SBS} \in [0, 1], \forall j \in \{1, \ldots, N\} \quad (1e)$$

$$\sum_{x=1}^{4} w_x = 1 \quad (1f)$$

$$w_1, w_2, w_3, \text{ and } w_4 > 0, \quad (1g)$$

and the weighted sum of the normalized objective values is

$$\begin{aligned} g^*(A, w_x) = {} & w_1. \sum_{i=1}^{M+N} A(i, :) \times \Lambda_z^T \ \forall z \in \{1, \ldots, F\}, \\ & + w_2. \sum_{i=1}^{M} \sum_{z=1}^{F} \mathcal{T}^{*UT,SBS}(i, :) \times A(:, z) \\ & - w_3. \sum_{i=1}^{M} \sum_{z=1}^{F} \mathcal{D}^{*UT,SBS}(i, :) \times A(:, z) \\ & + w_4. \sum_{i=1}^{M} \sum_{z=1}^{F} P_m^{UT,SBS}(i, :) \times A(:, z), \end{aligned} \quad (2)$$

where $w_1, w_2, w_3,$ and w_4 are the weights of each objective function. The file popularity probability, normalized contact duration, normalized communication range, and contact probability are represented by Λ_z^T, $\mathcal{T}^{*UT,SBS}$, $\mathcal{D}^{*UT,SBS}$, and $P_m^{UT,SBS}$, respectively.

4 Classification-Based Influence Weights

The cache placement decision is a binary decision (place or not to place the contents) that results in one of two cases: either the content is found ($Hit = 1$) or the content is not found ($Hit = 0$) in UT and SBS caches when the user requests the content.

We will use the binary classification conditional probabilities presented in [13]. Consider that the cache hit is positive and the probabilities can be mapped between the ranges 0 and +1. Then, the cache miss is negative, and the probabilities can be mapped between the ranges −1 and 0. The range of mapped probabilities for any input attribute is between −1 and +1. Since each weight represents its influence on both probabilities, we can compute the influence weight for an input attribute for both cache states as follows:

$$w_i = Pr(Hit \mid I_i) + Pr(Miss \mid I_i). \tag{3}$$

Assume that there are $\mathcal{V}$ user requests for a given period of time. If each request produces a hit or a miss, the hit rate (h) may be calculated by dividing the total number of hits by the total number of requests.

The aim of classification is to determine which class a set of input attributes belongs to. The classifier's goal is to define a decision surface $k(I, w) = 0$ such that, given a set of input attributes I, the classifier can predict which class the input attributes belong to. Then, given the value of I, its class label y is predicted according to the following:

$$y\prime = k(I, w), \tag{4}$$

where $k(I, w)$ is used to decide on which side of the decision surface $k(I, w) = 0$ lies the input attributes I [14]. The aim is to create a hyperplane that separates these two classes for a given training set with $\mathcal{V}$ samples that result in either $y = -1$ in a miss or $y = 1$ in a hit.

$$\begin{aligned} k(I, w) &= f(I, w) \\ &= f(I, w_0, w_1, w_2, w_3, w_4) = 0, \end{aligned} \tag{5}$$

where w_0 is the bias and f is the classifier function, which can be either linear or nonlinear. In the following section, an SSST learning approach is provided that can be used to estimate the values of influence weights and design the cache placement model.

5 Semi-supervised Self-training (SSST) for Binary Classifier

There are two primary stages to semi-supervised self-training. To develop the binary classifier, a typical supervised classifier is trained on a small collection of labeled data in the first stage. In this chapter, we use a limited number of labeled training datasets to train a logistic regression classifier and then predict on the test dataset. The second stage is to employ a self-training technique, which is a semi-supervised learning method that is iterative. The prediction probabilities are used in self-training to identify unlabeled data. The labels with the highest probability prediction values are added to the dataset as new labels and used in the following iteration.

Consider the n input features $(I_1, I_2, \ldots, I_n)$ and the corresponding class y for the $\mathcal{V}$ number of observations in the training dataset, which is commonly referred to as the observations. The classifier output y might be 1, indicating that the observation belongs to the class cache, or 0, indicating that the observation does not belong to that class. Assume the dataset contains m samples $m = m_1 + m_2$, with m_1 labeled samples and m_2 unlabeled samples.

5.1 *Logistic Regression (LR) for Supervised Classification*

A dataset of input/output pairs is required for a supervised classifier. The observations (input attributes), also known as the vector of features, and the corresponding class y are represented by each row in the dataset. The classification function aims to train a model with the input dataset and then estimate the output $y\prime$ for new input features. When given new input features, the classifier uses past datasets to predict the output. The classifier's output is a decision to place the content (cache) or not place the content (not cache).

The binary logistic regression (LR) algorithm's objective is to train a classifier to make a binary decision on a new input observation's class. The LR model predicts the probability $P_r(y = 1 \mid I)$ that this observation belongs to the classcache or $P_r(y = 0 \mid I)$ and that it does not belong to the class cache. The LR acquires knowledge from the training set by updating the weights $w_1, w_2, \ldots, w_4$, as well as the bias w_0. The weights denote the significance of each input feature in solving the classification problem. The LR learns by updating the weights and bias parameters from a training set. Each weight, w_i, corresponds to a specific input attribute, I_i. The weight encapsulates the input attribute's significance in the classification decision. The bias terms are added to the weighted inputs.

Let us consider LR model for a two-class (y_1, y_2) classification task with the posterior probabilities modeled as [15]

$$
\begin{aligned}
P_r(y_1 \mid I) =& \frac{1}{1 + \exp(-Iw)} \\
P_r(y_2 \mid I) =& 1 - P(y_1 \mid I).
\end{aligned} \tag{6}
$$

After complete training of the weights and bias, the classifier multiplies each input I_j by its updated weight w_j, sums up the weighted features, and adds the updated bias term w_0.

Algorithm 1 Semi-supervised self-training classifier for cache placement

1: **Input:** Labeled dataset m_1 with input features I_j and one output label y and m_2 unlabeled dataset with input features I_j.
2: **Output:** Classifier model $k(I, w)$ given in (4).
3: **Initialize** bias w_0=0, weights $w_j = 0$, and learning rate $\alpha = 0.3$.
4: **Set** training dataset $M = m_1$.
5: **while** not converge OR iterations $\leq$ Maximum limit **do**
6: **Train** the LR classifier using labeled dataset M.
7: **Generate** pseudo labels with classifier predictions.
8: **Add** the pseudo labels with 99% probability of prediction to the M dataset.
9: **end while**

5.2 *Self-training Technique*

The trained classifier is used to predict the class label for each sample of unlabeled data. To determine which data points I have similar labels, a fully connected graph is constructed with nodes representing all the data points, both labeled and unlabeled. Any two nodes in the network are connected by a weight d_{ij}, with the weight increasing as the nodes get closer in the Euclidean distance. The parameter σ controls the weights in such a way that

$$t_{ij} = \exp\left(-\frac{d_{ij}^2}{\sigma^2}\right) = \exp\left(-\frac{\sum_{d=1}^{D}(I_i^d - I_j^d)}{\sigma^2}\right). \tag{7}$$

The objective is to estimate the output variable y' using both labeled and unlabeled data points. The model predicts labels for all unlabeled data and their associated probabilities. Then choose the labels with the highest probability (in our work, we consider 99% as the greater probability). The adopted labels are referred to as "pseudo-labels," and they are added to the dataset for subsequent iterations. The updated dataset is used to retrain the classifier model. Iterations continue until no more predictions with a probability greater than 99% remain, or until no unlabeled data remains [8]. Algorithm 1 shows the processes involved in developing a semi-supervised self-training classifier for the cache placement technique.

6 Experimental Results

In this section, we conduct simulations to assess the performance of the classifier-based SSST for cache placement.

Table 1 Simulation parameters

Parameter	Value
Number of files in MBS	100
SBSs cache size	(10–100)% of MBS library
File size	(1–8) MB
SBS data rate	(2–16) Mbps
SBS communication range	(50–200) m
User speed	(10–30) m/sec
Number of user requests	10,000

6.1 Simulation Setup

Our simulation was performed on an area with a high density of SBSs, as indicated in [16]. We assume that the macrocell contains one MBS, 15 SBSs, and three distinct paths along which people pass, each with 11 locations. The results are computed using 10,000 random user locations and 10,000 random user requests within a single wireless network cell. Each request includes the file ID for the requested file. The requests for files are distributed according to the Zipf popularity distribution. A series of experiments is conducted to determine the effect of various system parameters on the performance of various cache placement algorithms. During each experiment, one parameter is modified, while the others remain constant. Table 1 illustrates the parameters. We assume that the user mobility parameters are similar to the work described in [10]. It is characterized as an independent Poisson process; the pairwise contact time between mobile UT and SBS follows an exponential distribution defined by the parameter $\lambda_{i,j}^{SBS}$, with $\Gamma(10, 1/100)$ representing the contact rate between mobile UT and SBS.

The dataset is divided into 70% data for training and 30% data for testing. Python was used to run the simulation on a Jupyter notebook running on the Anaconda environment [17].

6.2 Cache Placement Algorithms

To evaluate the proposed SSST classifier-based cache placement algorithm and to study the relationship between various input attributes, we compare the impact of various variables on six alternative caching strategies. The following techniques are used for cache placement: semi-supervised self-training (SSST), support vector machine (SVM) with SGD learning, logistic regression (LR) with SGD learning, and artificial neural network with SGD learning.

6.3 *Performance Evaluation*

In this chapter, we provide only the results of caching in SBS caches. The cache hit rate for several trials is used to evaluate and compare the four techniques. User mobility is evaluated when users request files while moving between locations inside the wireless network. Users can download files directly from nearby SBSs without requiring a backhaul connection. This results in a cache hit. The cache hit rate is calculated by dividing the total number of user requests by the number of cache hits.

6.3.1 Impact of Cache Size

To investigate the effect of SBS cache size on cache hit rate, we considered SBS cache sizes ranging from 10 to 100% of the total file library size. Figure 2 illustrates the effect of adjusting the size of the SBS cache on the SBS hit rate. Increases in cache size result in an increase in the cache hit rate for all cache placement techniques. This is because increasing the size of the cache allows for the caching of more files in SBS caches, which increases the probability of obtaining the requested content.

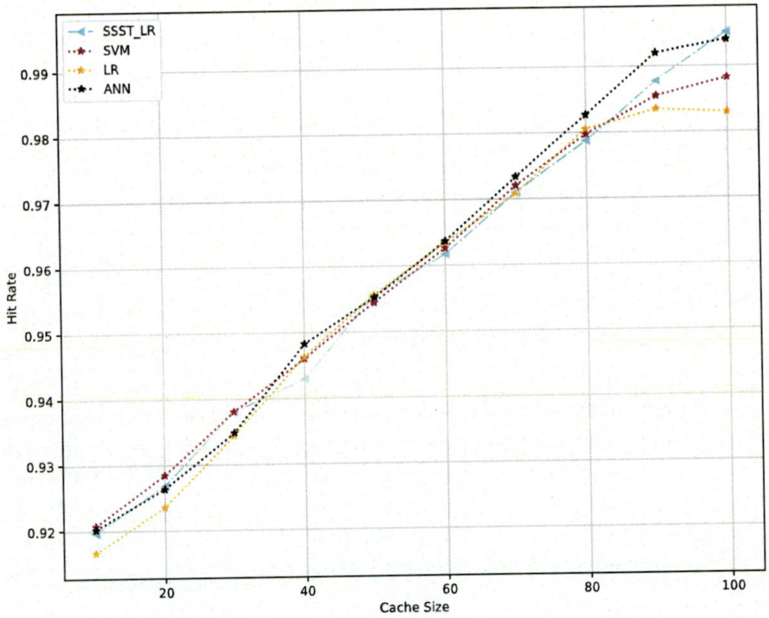

Fig. 2 Impact of cache size on the hit rate

Certain values for the variables user speed, SBS communication range, SBS data rate, and file size are assumed to be 10 m/sec, 100 m, 8 mbps, and 1 mb, respectively. Regardless of the cache size, the trained models in the four learning techniques are capable of extracting information from the dataset and making accurate predictions about the cache content placement.

The performance of the learning algorithms SSST, ANN, SVM, and LR is very similar across a range of cache sizes, from 10 to 80%. At cache sizes of 90% and greater, ANN and SSST algorithms behave similarly and exceed SVM and LR algorithms.

6.3.2 Impact of User Speed

Increasing user speed has an effect on all algorithms. Keeping in mind that in our scenario, the speed is increased just to the extent where the UT can still download content from SBS, while the user is on the go. Figure 3 illustrates the effect of SBS cache size on hit rate at various user speeds. The figure illustrates the performance of four cache placement techniques at 10 and 25 m/sec: SSST, ANN, SVM, and LR. We noticed that when the user moves at a slow speed (10 m/sec), the four approaches produce comparable outcomes. When user speed is increased to 25 m/sec, the SSST outperforms the LR, SVM, and ANN in terms of hit rate.

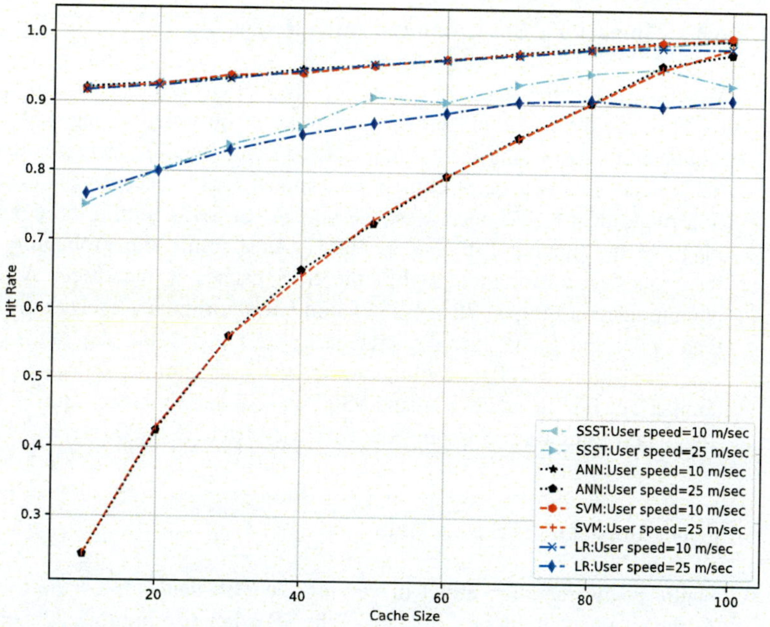

Fig. 3 Impact of cache size on the hit rate with different values of user speed

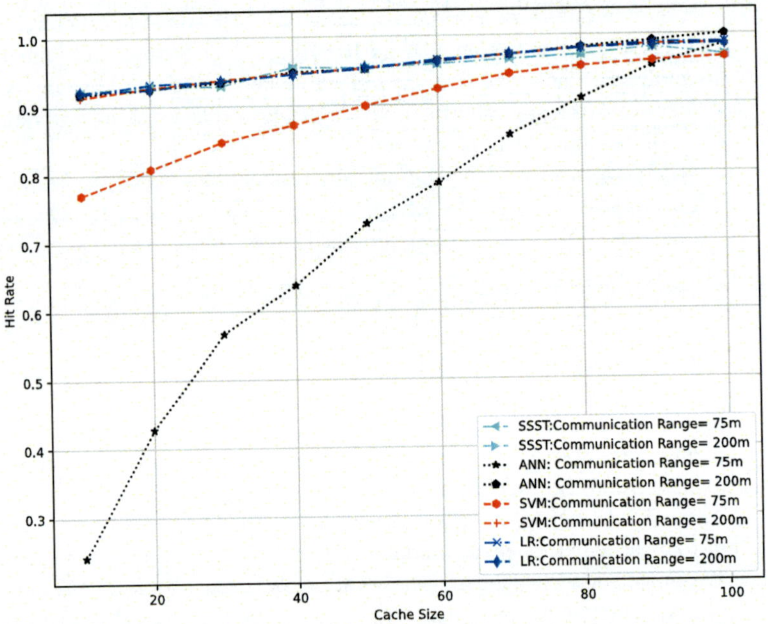

Fig. 4 Impact of cache size on the hit rate with different values of SBSs' communication range

6.3.3 Impact of SBSs Communication Range

We investigate the effect of increasing the SBS communication range from 75 to 200 m on the cache placement performance illustrated in Fig. 4. As the results demonstrate, increasing the communication range of SBSs increases the hit rate of SBS caches in all cache placement algorithms. This is because as the communication range of the SBS increases, the number of SBSs within the communication range of the moving UT also increases. As a result, the probability of finding the requested information within the SBS caches is increased. At lower SBS communication ranges (75 m), SSST and LR algorithms achieve a greater hit rate than ANN and SVM caching strategies. After increasing the range of the SBS communication to 200 m, the hit rate for the four learning strategies is relatively similar. The hit rate achieved using SSST and LR learning techniques was relatively high when communication ranges were varied from 75 to 200 m.

6.3.4 Impact of SBS Data Rate

Figure 5 illustrates the effect of varying the SBS data rate on the cache hit rate. As the results demonstrate, increasing the SBS data rate improves the hit rate for all cache placement approaches, as SBSs can transfer more data during contact with the

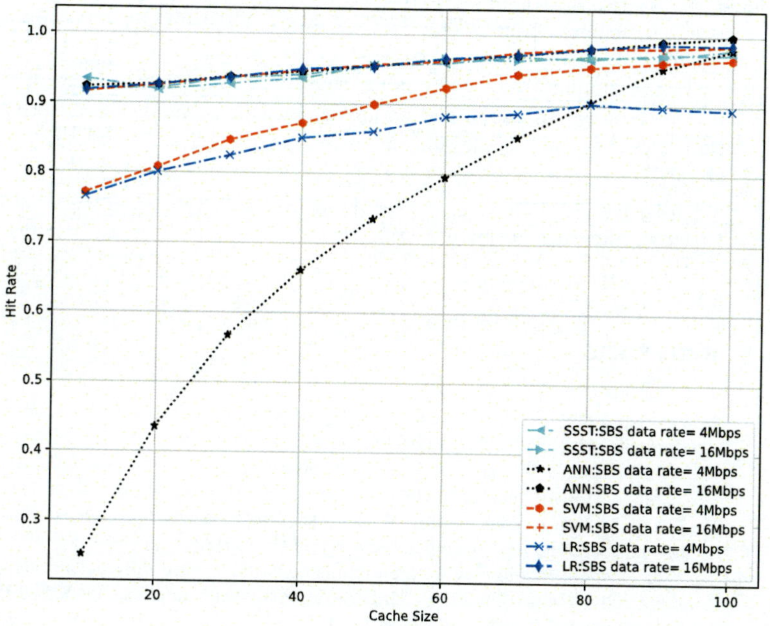

Fig. 5 Impact of cache size on the hit rate with different values of SBSs data rate

UT. By increasing the SBS data rate, the user can complete the download within the available time while moving from one location to another. When using SBS data at a rate of 4 Mbps, the SSST approach outperforms the SVM, LR, and ANN caching algorithms in terms of hit rate. While the performance of the learning approaches SSST, ANN, SVM, and LR is very similar at greater data rates (16 Mbps), the performance of the learning techniques SSST, ANN, SVM, and LR is very similar at lower data rates. As illustrated in the figure, the SSST approach has no effect on the resulting hit rate values when the SBS data rates are changed.

7 Conclusion

In this chapter, we present a semi-supervised self-training classification model for the cache placement problem. We evaluated the proposed SSST method using datasets generated using various machine learning techniques. We compared the performances of various machine learning models using the same datasets.

It has been observed that when system parameters are changed while using the SSST cache placement algorithm, the SBS cache hit rate increases with increased SBS cache size, decreases with increased user speed, and remains unchanged when SBS data rate and communication ranges are changed.

As a result, we conclude that the SSST algorithm is more robust and can significantly enhance the performance of the cache placement problem.

The accumulation of continuous datasets is suggested to better understand the influence of classification and parameter modifications on the adaptivity of SSST for cache placement in mobile edge networks.

Acknowledgement This work is funded in part by a research grant from the National Science and Engineering Research Council of Canada.

References

1. Zeydan, E., et al.: Big data caching for networking: moving from cloud to edge. IEEE Commun. Mag. **54**(9), 36–42 (2016)
2. Ahn, J., Jeon, S.H., Park, H.-S.: A novel proactive caching strategy with community-aware learning in CoMP-enabled small-cell networks. IEEE Commun. Lett. **22**(9), 1918–1921 (2018)
3. Zhang, S., Yao, L., Sun, A., Tay, Y.: Deep learning based recommender system: a survey and new perspectives. ACM Comput. Surv. **1**(1), 1–35 (2018)
4. Yao, Z., et al.: Mobility-aware pre-caching based on unsupervised deep generative model for small cell networks. In: 2019 13th IEEE International Conference on Signal Processing and Communication Systems (ICSPCS), pp. 1–7 (2019)
5. Zhou, Y., et al.: Content placement with unknown popularity in fog radio access networks. In: 2019 IEEE International Conference on Industrial Internet (ICII), pp. 361–367 (2019)
6. Zhu, X., Goldberg, A.B.: Introduction to semi-supervised learning. Synth. Lect. Artif. Intell. Mach. Learn. **3**(1), 1–130 (2009)
7. Chai, H., Liang, Y., Wang, S., Shen, H.W.: A novel logistic regression model combining semi-supervised learning and active learning for disease classification. Sci. Rep. **8**(1), 1–10 (2018)
8. Zhu, X., Ghahramani, Z.: Learning from labeled and unlabeled data with label propagation, CMU Tech report, CMU-CALD-02-107 (2002)
9. Tan, Y., Yuan, Y., Yang, T., Xu, Y., Hu, B.: Femtocaching in wireless video networks: Distributed framework based on exact potential game. In: 2016 IEEE/CIC International Conference on Communications in China (ICCC), pp. 1–6. IEEE, Piscataway (2016)
10. Chen, M., Hao, Y., Hu, L., Huang, K., Lau, K.V.: Green and mobility-aware caching in 5G networks. IEEE Trans. Wirel. Commun. **16**(12), 8347–61 (2017)
11. Wang, R., Peng, X., Zhang, J., Letaief, K.B.: Mobility-aware caching for content-centric wireless networks: modeling and methodology. IEEE Commun. Mag. **54**(8), 77–83 (2016)
12. Mohammed, L.B., Anpalagan, A., Khwaja, A.S., Jaseemuddin, M.: Mobility-aware latency-efficient cache placement in mobile edge networks. In: 22nd IEEE International Conference on High Performance Computing and Communications; HPCC 2020, Fiji (2020)
13. Quinn, A., Stranieri, A., Yearwood, J.: Classification for accuracy and insight: A weighted sum approach. In: Proceedings of the Sixth Australasian Conference on Data Mining and Analytics, vol. 70, pp. 203–208 (2007)
14. Theodoridis, S.: Machine Learning : A Bayesian and Optimization Perspective, Elsevier Science and Technology. Academic, New York (2020)
15. Kolosova, T., Berestizhevsky, S.: Supervised Machine Learning: Optimization Framework and Applications with SAS and R. Chapman and Hall/CRC, London (2020)
16. McNett, M., Voelker, G.M.: Access and mobility of wireless PDA users. ACM Sigmobile Mobile Comput. Commun. Rev. **9**(2), 40–55 (2005)
17. Pedregosa, F., et al.: Scikit-learn: Machine learning in Python. J. Mach. Learn. Res. **12**, 2825–30 (2011)

On Enhancing Low Bit-Rate Performance of an Image Codec Using Deep Learning-Based Nonlinear Processing

Susmita Saha and Pradeepa Yahampath

1 Introduction

Learning-based image compression has a long history [1]. However, the recent success of convolutional neural networks (CNNs) in many image processing problems has reignited the interest. The recent work on CNN-related image coding methods can be classified into two broad categories: those that use only CNNs for *end-to-end* image compression [2–4] and those that use CNNs to improve a conventional image compression algorithm [5–7]. In end-to-end compression, a single deep CNN is used to learn the best mapping between the original and a coded image. Although learning a single mapping from data seems to be the best strategy, end-to-end learning of image compression possesses many challenges. First, it requires one to carefully choose a proper CNN architecture which can fully capture the underlying, potentially complex, mapping of an image vector quantizer in a manner that preserves perceptual quality of the output. Second, and more critically, training a single CNN to perform near-optimal coding at multiple bit rates can be difficult. Being able to produce good compression performance over a wide range of bit rates is an essential feature of an image compression algorithm. Modern algorithms such as JPEG/JPEG2000 achieve this by incorporating elaborate rate control algorithms, and building rate control into end-to-end CNN architectures remains an open problem.

On the other hand, incorporating CNNs into the conventional transform coding framework provides an opportunity to combine the strengths of CNNs in solving various image processing tasks with highly optimized coding capabilities of algorithms such as JPEG. One such approach is using CNNs for pre-processing

S. Saha · P. Yahampath (✉)
Department of Electrical and Computer Engineering, University of Manitoba, Winnipeg, MB, Canada
e-mail: sahas1@myumanitoba.ca; Pradeepa.Yahampath@umanitoba.ca

H. Nguyen et al. (eds.), *30th Biennial Symposium on Communications 2021*,
Signals and Communication Technology, https://doi.org/10.1007/978-3-031-06947-5_16

of images prior to coding to reduce redundancies and post-processing the decoded image to reduce any artifacts introduced by pre-processing and coding. The basic idea itself is not new and predates CNN-related methods. One such approach, sub-sampling of an image prior to coding (using, for example, a JPEG encoder) to reduce the spatial redundancies and subsequent restoration of the decoded images to the original size, has been investigated by several authors, and some improvements in coding gains have been reported, see, for example, [8–10]. A CNN-based implementation of this idea has also been recently considered in [5, 6].

The main goal of the work presented in this chapter has been to investigate the efficacy of using state-of-the-art CNN-based image sub-sampling, super-resolution, and denoising techniques to enhance the performance of standard image codec at low bit rates. A similar study is presented in [6], where a pair of CNNs for sub-sampling the input of a standard JPEG codec and restoring the decoded image, respectively, has been jointly designed using an iterative training procedure. Experimental results presented in [6] show that fully CNN-based pre- and post-processing of codec input and output are able to produce images with notable quality improvements over those produced by restoring the directly JPEG coded (without sub-sampling and up-sampling) images using many previously reported coding artifact removal methods [11–17].

In our study, we adopted the same sub-sampling network as used in [6]. However, with avoiding complicated iterative training as in [6] in mind, we considered a novel CNN architecture for restoring the coded images which is derived from recently reported state-of-the-art image super-resolution and denoising CNN architectures [18, 19]. Despite the lower training complexity, extensive experimental results show that objective quality improvements achieved by our method is comparable to that in [6]. Furthermore, it is our observation that much of the advantage of CNN-based processing in relation to improving the image codec performance comes from coding artifact removal, rather than more accurate image sub-sampling and super-resolution. Another advantage of one-pass training as we do is that the overall system is robust against the QF setting and hence the bit rate variations of the image codec. We present experimental results to demonstrate the robustness of our system against bit rate variations.

In [7], a CNN-based approach to specifically enhance the JPEG-2000 codec (ours is codec independent) by using a bit-stream re-compression method and a coded image enhancement method has been presented. This method however does not use re-sampling for rate reduction. The chapter only presents a limited set of results and does not provide sufficient details to enable a direct comparison.

2 Description of the Method

Even though an image can be compressed to a lower bit rate by first obtaining a more compact representation via sub-sampling, the difficulty with this approach is that enlarging the image after decoding will lead to amplification of errors due

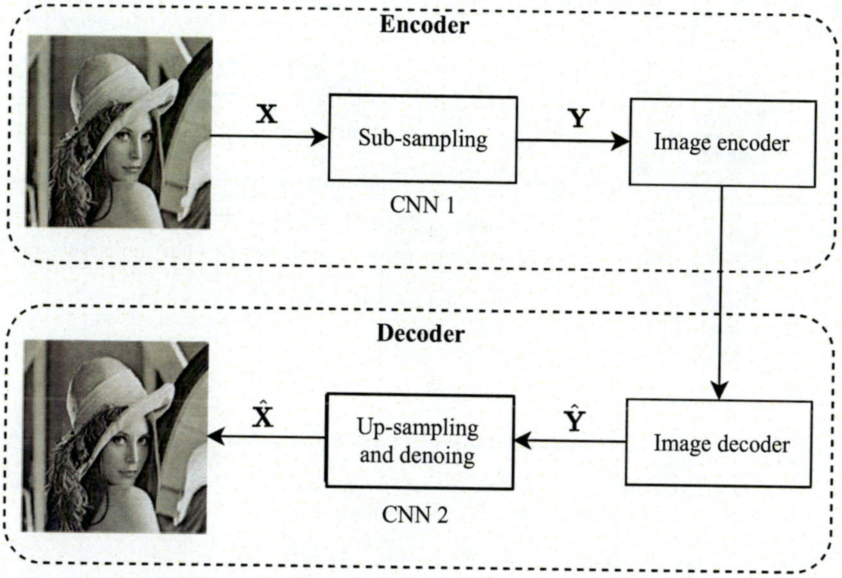

Fig. 1 Block diagram representation of the investigated method

to both imperfect sub-sampling and quantization. The success of this approach therefore heavily depends on the optimality of sub-sampling and restoration procedures applied. Recent research has shown that CNNs can substantially outperform conventional model-based methods in nonlinear image processing tasks such as sub-sampling, super-resolution, and denoising. Motivated by these observations, we have investigated the image coding system shown in Fig. 1.

In the above scheme, we have to ideally use a sub-sampling method such that the error between the original image and an image interpolated from the sub-sampled version is minimized. A CNN can be directly trained to achieve this operation. In this work, we have used the ComCNN network described in [6, Sec. III-A] to realize CNN-1 in Fig. 1. Below, we describe and justify the architecture that we have used for CNN-2.

The purpose of CNN-2 is to perform image super-resolution while minimizing the aforementioned error amplification. While it seems prudent to apply image denoising prior to super-resolution, there is a difficulty with this approach—it is not possible to directly compare the original (noise-free) image and the image codec output due to the size mismatch. In order to get around this problem, we jointly train a cascade of two networks: the first one performing super-resolution and the second one denoising. To this end, we have considered an architecture that combines two state-of-the-art CNN architectures for image super-resolution and denoising, as shown in Fig. 2. This architecture consists of 18 weight layers, where the first 8 layers (perform super-resolution) follow the FSRCNN architecture in [18] and the last 10 layers (perform denoising) follow the DnCNN architecture in [19]. The

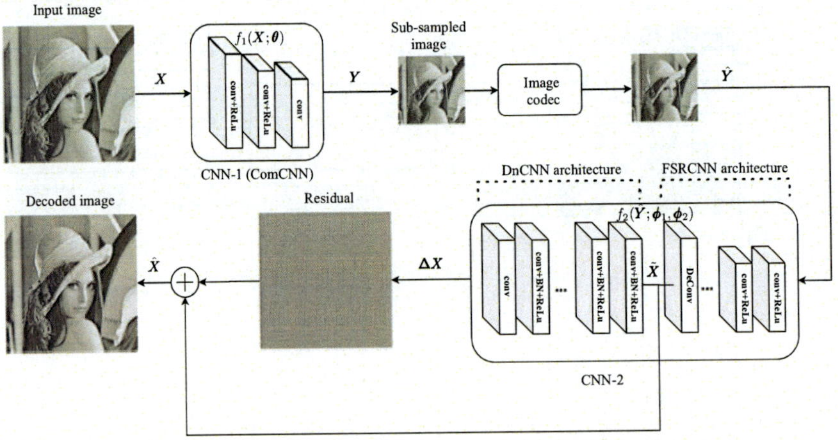

Fig. 2 The investigated system comprises a sub-sampling CNN (CNN-1) and a restoration CNN (CNN-2), where CNN-2 is trained to predict the residual

figure also emphasizes the use of a *residual learning* applied to CNN-2. That is, CNN-2 is trained to learn the residual $\boldsymbol{\Delta X} = \boldsymbol{X} - \tilde{\boldsymbol{X}}$, which is the difference between the up-sampled version $\tilde{\boldsymbol{X}}$ of $\hat{\boldsymbol{Y}}$ produced by the deconvolution layer in the FSRCNN part of CNN-2 and the original image. The importance of residual learning of deep networks to ensure better performance and faster convergence of the training algorithm is well known [20].

2.1 Training Procedures

Given a training set of images $\boldsymbol{x}_i$, $i = 1, \ldots, N$, we first train the CNN-1, independently of CNN-2. Subsequently, CNN-2 is trained for a fixed CNN-1.

1. *CNN-1*: let $\boldsymbol{\theta}$ be the set of weights associated with CNN-1 whose architecture is shown in [6, Fig. 3]. Also, let $f_1(\boldsymbol{X}; \boldsymbol{\theta})$ be the nonlinear function realized by CNN-1. Note that the network output $\boldsymbol{Y} = f_1(\boldsymbol{X}, \boldsymbol{\theta})$ is a sub-sampled version of the input $\boldsymbol{X}$. Therefore, in order to compute the error between the network input and the output as required for training, we used bicubic interpolation to up-sample $\boldsymbol{Y}$ to the size of $\boldsymbol{X}$. Let the interpolated version of $\boldsymbol{Y}$ be given by $g(\boldsymbol{Y})$. We estimated the optimal weights by minimizing the mean square error for the training set

$$L_1(\boldsymbol{\theta}) = \frac{1}{N} \sum_{i=1}^{N} \|\boldsymbol{x}_i - g\left(f_1(\boldsymbol{x}_i; \boldsymbol{\theta})\right)\|^2 .$$

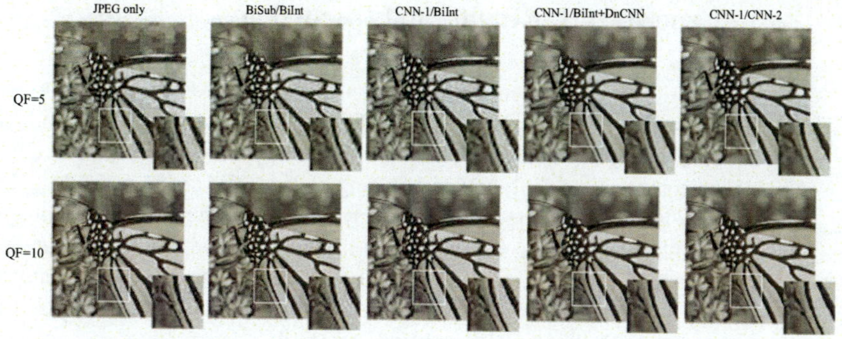

Fig. 3 A visual comparison of images reproduced by coding and restoration methods shown in Tables 1 and 2

2. *CNN-2*: given the trained CNN-1 with optimal parameters $\boldsymbol{\theta}_1^*$, we generate subsampled versions $\boldsymbol{y}_i = f_1(\boldsymbol{x}_i, \boldsymbol{\theta}_1^*)$, $i = 1, \ldots, N$ of the original training set. Let $\hat{\boldsymbol{y}}_i$ be the coded version of the $\boldsymbol{y}_i$. We used the residual learning method for CNN-2, as indicated in Fig. 2. Let the output of the deconvolution layer of CNN-2 for the input $\hat{\boldsymbol{y}}_i$ be $\tilde{\boldsymbol{x}}_i = h(\hat{\boldsymbol{y}}_i; \boldsymbol{\phi}_1)$, where $\boldsymbol{\phi}_1$ is the set of weights associated with the first 8 layers of CNN-2. Then, the residual to be predicted by CNN-2 is

$$\boldsymbol{\Delta x}_i = \boldsymbol{x}_i - \tilde{\boldsymbol{x}}_i.$$

Let $f_2(\hat{\boldsymbol{Y}}; \boldsymbol{\phi}_1, \boldsymbol{\phi}_2)$ be the nonlinear function realized by CNN-2, where $\boldsymbol{\phi}_2$ is the set of weights associated with the last 10 layers. Then, CNN-2 is trained to minimize the mean square error

$$\begin{aligned} L_2(\boldsymbol{\phi}_1, \boldsymbol{\phi}_2) &= \frac{1}{N} \sum_{i=1}^{N} \left\| \boldsymbol{\Delta x}_i - f_2(\hat{\boldsymbol{y}}_i; \boldsymbol{\phi}_1, \boldsymbol{\phi}_2) \right\|^2 \\ &= \frac{1}{N} \sum_{i=1}^{N} \left\| \boldsymbol{x}_i - h(\hat{\boldsymbol{y}}_i; \boldsymbol{\phi}_1) - f_2(\hat{\boldsymbol{y}}_i; \boldsymbol{\phi}_1, \boldsymbol{\phi}_2) \right\|^2. \end{aligned}$$

It should be noted that the trained CNN-1 is not affected by the bit rate of the image codec. However, since the CNN-2 is trained for the image codec output, its performance can degrade if it is used to restore codec outputs whose quality deviates substantially from the inputs seen during the training phase. However, as we will demonstrate in the next section, the proposed coding system remains robust over a range of bit rates.

3 Experiments, Results, and Discussion

All experiments described in this section use a standard JPEG codec to compress the sub-sampled images. In a JPEG codec, the output bit rate is determined by the quality factor (QF) setting which can be in the 0–100 range (lower QF implies higher compression). We consider coding at low QF values (5, 10, and 15), as it is at low bit rates that the coding artifacts will be significant. All experiments have been implemented using Python, Keras library [21], and the TensorFlow backend. Training has been performed on an ASUS ROG STRIX G531G computer running Windows 10 and equipped with an Intel Core i7-9750H 2.60GHz CPU, 16 GB of RAM, and an NVIDIA GeForce GTX 1650 GPU.

Both CNN-1 and CNN-2 have been trained using the stochastic gradient descent (SGD) algorithm with an Adam optimizer [22]. The optimizer is initialized with the default parameters $\beta_1 = 0.9$, $\beta_2 = 0.999$, and $\epsilon = 10^{-8}$. The initial learning rate has been set to 0.001. The filter weights have been initialized using the He initializer [23]. In each case, 50 training epochs have been used. Due to memory constraints of the computer, the batch size has been set to 64.

For the training of our system, we have used 400 images of size 180×180. Instead of training a network using whole images, each image has been divided into multiple patches of size 40×40 taken at a stride of 20. Furthermore, each patch has been augmented 8 times by random horizontal and vertical flips as well as rotations by random angles to obtain a total of 204,800 image patches for training. For testing, we have used 8 well-known images (not included in the training set), which are commonly used in the literature to benchmark the performance of compression algorithms.

First, it would be interesting to compare the proposed coding scheme incorporating CNN-based pre- and -post processing with a suitable conventional technique for image sub-sampling and up-sampling. One of the most popular techniques of image resolution conversion is bicubic re-sampling. Tables 1 and 2 compare the commonly used image quality metrics peak signal-to-noise ratio (PSNR) and structural similarity index measure (SSIM) of the proposed methods with systems which use (a) bicubic sub-sampling (BiSub) and bicubic interpolation (BiInt), (b) CNN-1 for sub-sampling and BiInt to restore coded images, and (c) CNN-1 for sub-sampling, BiInt for up-sampling, and DnCNN for denoising. The results in Tables 1 and 2 show that the fully CNN-based approach does have a definite advantage over using simpler bicubic re-sampling. However, the relatively minor differences between the CNN-1/BiInt+DnCNN and CNN-1/CNN-2 suggest that most of the advantage of using CNN-based learning here may be in denoising (artifact removal). The visual inspection of example images reproduced by each method shown in Fig. 3 seems to support this observation.

We also compare, in Tables 3 and 4, the objective performance metrics of our fully CNN-based method with an extensive set of experimental results reported in [6]. These results include performance achievable with JPEG coded images (using the same QP) being restored by several well-known deblocking methods

Table 1 Comparison of PSNR (dB) and SSIM achieved with CNN-based processing and bicubic re-sampling for JPEG QF = 5. The bold face numbers signify the best observed performance

Test image	JPEG only		BiSub/ BiInt		CNN-1/ BiInt		CNN-1/ BiInt+DnCNN		CNN-1/ CNN-2	
	PSNR	SSIM	PSNR	SSIM	PSNR	SSIM	PSNR	SSIM	PSNR	SSIM
Cameraman	24.45	0.7262	24.73	0.7569	24.89	0.7602	25.76	0.8030	**26.21**	**0.8111**
House	27.77	0.7731	29.61	0.8003	29.70	0.8012	30.81	0.8359	**31.11**	**0.8388**
Peppers	27.26	0.7118	29.20	0.7977	29.34	0.7964	30.35	0.8329	**30.64**	**0.8358**
Butterfly	23.80	0.7412	24.39	0.7872	24.46	0.7871	26.86	0.8647	**27.07**	**0.8673**
Lena	27.32	0.7324	29.84	0.8226	29.88	0.8236	30.89	0.8487	**31.08**	**0.8510**
Ship	25.56	0.6582	26.49	0.7163	26.58	0.7193	27.41	0.7472	**27.56**	**0.7492**
Starfish	23.99	0.7083	24.71	0.7513	24.75	0.7540	25.83	0.7984	**26.04**	**0.8002**
Parrot	24.51	0.7124	25.00	0.7926	25.15	0.7957	25.58	0.8276	**25.58**	**0.8309**
Average	25.59	0.7205	26.75	0.7781	26.84	0.7797	27.94	0.8198	**28.16**	**0.8231**

Table 2 Comparison of PSNR (dB) and SSIM achieved with CNN-based processing and bicubic re-sampling for JPEG QF = 10. The bold face numbers signify the best observed performance

Test image	JPEG only		BiSub/ BiInt		CNN-1/ BiInt		CNN-1/ BiInt+DnCNN		CNN-1/ CNN-2	
	PSNR	SSIM	PSNR	SSIM	PSNR	SSIM	PSNR	SSIM	PSNR	SSIM
Cameraman	26.47	0.7951	25.44	0.7948	25.73	0.7982	27.23	0.8412	**27.33**	**0.8429**
House	30.56	0.8184	30.69	0.8259	30.92	0.8301	32.63	0.8609	**32.75**	**0.8623**
Peppers	30.15	0.7916	30.22	0.8282	30.39	0.8276	31.61	**0.8570**	**31.63**	0.8567
Butterfly	26.67	0.8417	26.57	0.8551	26.72	0.8566	29.22	0.9173	**29.32**	**0.9187**
Lena	30.41	0.8214	31.35	0.8586	31.44	0.8608	32.37	0.8792	**32.51**	**0.8802**
Ship	28.13	0.7673	28.23	0.7891	28.45	0.7942	29.44	**0.8167**	**29.53**	**0.8167**
Starfish	26.72	0.8162	26.57	0.8292	26.73	0.8314	28.34	0.8701	**28.55**	**0.8724**
Parrot	26.84	0.8136	26.01	0.8329	26.28	0.8378	26.99	0.8634	**27.09**	**0.8652**
Average	28.24	0.8082	28.14	0.8267	28.33	0.8296	29.73	0.8633	**29.84**	**0.8644**

Table 3 Comparison of PSNR (dB) and SSIM of the proposed method with other methods reported in the literature at JPEG QF = 5. The bold face numbers signify the best observed performance

Method	Cameraman		House		Peppers		Lena		Average	
	PSNR	SSIM	PSNR	SSIM	PSNR	SSIM	PSNR	SSIM	PSNR	SSIM
JPEG	24.45	0.7262	27.77	0.7731	27.26	0.7118	27.32	0.7324	26.70	0.7258
Zhang [11]	25.39	0.7672	29.24	0.8141	29.07	0.7895	29.00	0.8035	28.18	0.7936
Ren [12]	25.46	0.7666	29.66	0.8197	29.07	0.7876	29.07	0.8010	28.32	0.7937
Dabov [13] (BM3D)	25.27	0.7607	29.21	0.8082	28.52	0.7639	28.63	0.7837	27.91	0.7791
Chang [14] (DicTV)	24.54	0.6658	28.45	0.7963	27.95	0.7456	28.07	0.7744	27.50	0.7455
Gu [15] (WNNM)	25.49	0.7674	29.62	0.8178	28.99	0.7827	28.95	0.7947	28.26	0.7906
Zhang [16]	25.61	0.7666	30.12	0.8285	29.61	0.8031	29.51	0.8169	27.63	0.8308
Dong [17] (ARCNN)	25.27	0.7674	29.68	0.8209	29.02	0.7961	29.31	0.8142	28.32	7996
Jiang [6]	**26.53**	**0.8167**	**31.45**	**0.8456**	**30.84**	0.8328	**31.14**	0.8486	**29.99**	**0.8359**
Proposed	26.21	0.8111	31.11	0.8388	30.64	**0.8358**	31.08	**0.8510**	29.76	0.8342

Table 4 Comparison of PSNR (dB) and SSIM of the proposed method with other methods reported in the literature at JPEG QF = 10. The bold face numbers signify the best observed performance

Method	Cameraman		House		Peppers		Lena		Average	
	PSNR	SSIM	PSNR	SSIM	PSNR	SSIM	PSNR	SSIM	PSNR	SSIM
JPEG	26.47	0.7951	30.56	0.8184	30.15	0.7916	30.41	0.8214	29.38	0.8066
Zhang [11]	27.45	0.8329	32.11	0.8513	31.68	0.8317	31.92	0.8597	30.79	0.8439
Ren [12]	27.43	0.8259	32.41	0.8526	31.63	0.8300	31.92	0.8571	30.85	0.8414
Dabov [13] (BM3D)	27.25	0.8240	32.07	0.8492	31.42	0.8250	31.77	0.8549	30.63	0.8383
Chang [14] (DicTV)	26.92	0.8046	31.77	0.8484	31.29	0.8244	31.55	0.8559	30.38	0.8333
Gu [15] (WNNM)	27.40	0.8248	32.42	0.8531	31.64	0.8303	31.93	0.8571	30.85	0.8413
Zhang [16]	**27.72**	0.8401	33.04	0.8609	**31.94**	0.8358	32.19	0.8661	30.12	**0.8774**
Dong [17] (ARCNN)	27.62	0.8389	32.53	0.8591	31.50	0.8434	32.05	0.8711	30.93	0.8531
Jiang [6]	27.44	**0.8448**	**33.25**	**0.8678**	31.83	0.8532	**33.11**	**0.8838**	**31.41**	0.8624
Proposed	27.33	0.8429	32.75	0.8623	31.63	**0.8567**	32.51	0.8802	31.06	0.8605

Table 5 PSNR (*d*B) and SSIM performance of the proposed method on standard image databases Set5, Set14, and Urban100. The bold face numbers signify the best observed performance

		JPEG only		Jiang [6]		Proposed	
Dataset	QF	PSNR	SSIM	PSNR	SSIM	PSNR	SSIM
Set5	5	26.20	0.7216	29.20	0.8387	**29.40**	**0.8397**
	10	29.06	0.8184	**31.40**	**0.8854**	31.33	0.8805
	15	30.55	0.8562	–	–	**32.05**	**0.8947**
Set14	5	24.92	0.6791	26.89	**0.7914**	**26.93**	0.7587
	10	27.52	0.7889	**28.91**	**0.8336**	28.34	0.8204
	15	28.91	0.8348	–	–	**28.97**	**0.8433**
Urban100	5	23.39	0.6988	–	–	**25.05**	**0.7759**
	10	25.65	0.7984	–	–	**26.51**	**0.8358**
	15	26.94	0.8378	–	–	**26.99**	**0.8449**

[11, 12, 14, 16] and denoising algorithms [13, 15, 17], as well as with a fully CNN-based sub-sampling/restoration system similar to ours [6, Fig. 2]. However, the restoration CNN in [6] is different from ours, and their training procedure involves retraining the sub-sampling and restoration CNNs to each other in an iterative manner. This is not only a complicated time-consuming procedure, but its convergence does not appear to be guaranteed, due to the presence of the JPEG codec (essentially a quantizer) placed between the two CNNs. It is noticeable that our method largely outperforms all denoising-only methods. Furthermore, our method performs practically very close to the fully CNN-based method of [6] in most cases. This is likely due to our CNN-2 architecture being better in super-resolution and denoising than the RecCNN used in [6]. In Table 5, we present the performance of our approach on image datasets Set5, Set14, and Urban100 from [24]. For comparison, this table also shows the performance of the fully CNN-based method from [6] where available. These results also support the previous observation that the two methods perform nearly identically.

Finally, we investigate the robustness of our trained compression system, against the choice of JPEG quality factor which determines the bit rate of the encoded images. This is important as the image restoration CNN (CNN-2) in our system is trained for a specific QF value. If the trained system shows high sensitivity to QF, then this type of systems will suffer considerable performance degradations in practice where QF is chosen to achieve some target bit rate which may differ from the QF value used during training. Figure 4 shows the PSNR and SSIM values obtained by testing the systems trained for QF setting 5, 10, and 15, at various QF values in the range of 5–30. Also shown is the performance of direct JPEG coding. It can be seen that a network trained for a given QF remains superior to JPEG-only coding for a good range of QF values. For example, the network trained for QF = 15 is superior to the JPEG encoder up to QF $\approx$ 25. Given that the method is most effective at low bit rates (low QF values), these results suggest that designing a few systems to cover a range of QF values (e.g., 0–50) will be all that is required.

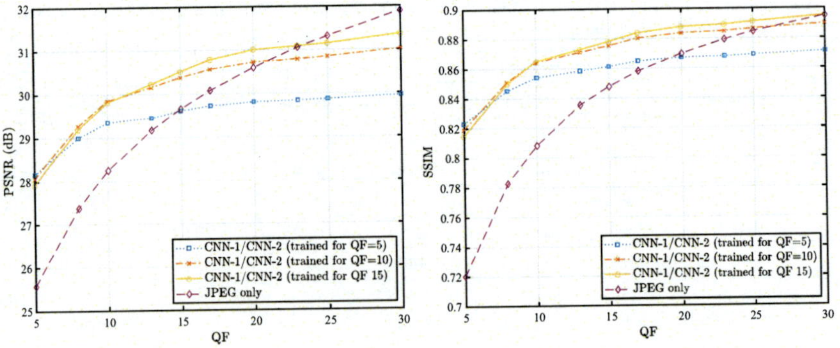

Fig. 4 PSNR (left) and SSIM (right) of the proposed system and the JPEG codec, when tested at QFs in the range of 5–30

4 Concluding Remarks

We have considered applying state-of-the-art CNN-based nonlinear image processing to extract additional coding gains from a standard image codec. Our approach uses image sub-sampling as a pre-processing step to reduce the coding bit rate and subsequent post-processing to restore the decoded image to improve the quality, thus achieving an overall coding gain. To this end, we investigated a novel CNN architecture for restoring the codec output, which has been derived from highly effective image super-resolution and denoising CNN architectures. Experimental results show that the system investigated in this chapter can not only outperform JPEG coding without sub-sampling and JPEG coding with sub-sampling enhanced by various coding artifact reduction methods but is also competitive with a recently reported fully CNN-based image coding system which requires more complex end-to-end training.

References

1. Ma, S., Zhang, X., Jia, C., Zhao, Z., Wang, S., Wang, S.: Image and video compression with neural networks: a review. IEEE Trans. Circuits Syst. Video Technol. **30**(6), 1683–1697 (2020)
2. Ballé, J., Laparra, V., Simoncelli, E.P.: End-to- end optimized image compression. In: 5th International Conference on Learning Representations (ICLR) (2017)
3. Toderici, G., Vincent, D., Johnston, N., Hwang, S.J., Minnen, D., Shor, J., Covell, M.: Full resolution image compression with recurrent neural networks. In: IEEE International Conference on Computer Vision and Pattern Recognition (CVPR) (2017)
4. Theis, L., Shi, W., Cunningham, A., Huszár, F.: Lossy image compression with compressive autoencoders. In: The International Conference on Learning Representations (2017)
5. Tao, W., Jiang, F., Zhang, S., Ren, J., Shi, W., Zuo, W., Guo, X., Zhao, S.: An end-to-end compression framework based om convolutional neural networks. In: Data Compression Conference (DCC), p. 463 (2017)

6. Jiang, F., Tao, W., Liu, S., Ren, J., Guo, X., Zhao, D.: An end-to-end compression framework based on convolutional neural networks. IEEE Trans. Circuits Syst. Video Technol. **28**(10), 3007–3018 (2018)
7. Ma, H., Liu, D., Xiong, R., Wu, F.: A CNN-based image compression scheme compatible with JPEG-2000. In: IEEE International Conference on Image Processing (ICIP), pp. 704–708 (2019)
8. Chang, I.-H., Chang, P.C., Liu, T.S.: Adaptive subsampling JPEG image coding. In: International Conference on Consumer Electronics, pp. 264–265 (1995)
9. Lin, W., Dong, L.: Adaptive downsampling to improve image compression at low bit rates. IEEE Trans. Image Process. **15**(9), 2513–2521 (2006)
10. Zhang, Y., Zhao, D., Zhang, J., Xiong, R., Gao, W.: Interpolation-dependent image downsampling. IEEE Trans. Image Process. **20**(11), 3291–3296 (2011)
11. Zhang, X., Xiong, R., Fan, X., Ma, S., Gao, W.: Compression artifact reduction by overlapped-block transform coefficient estimation with block similarity. IEEE Trans. Image Process. **22**(12), 4613–4626 (2013)
12. Ren, J., Liu, J., Li, M., Bai, W., Guo, Z.: Image blocking artifacts reduction via patch clustering and low-rank minimization. In: Data Compression Conference (DCC), p. 516 (2013)
13. Dabov, K., Foi, A., Katkovnik, V., Egiazarian, K.: Image denoising by sparse 3-d transform-domain collaborative filtering. IEEE Trans. Image Process. **16**(8), 2080–2095 (2007)
14. Chang, H., Ng, M.K., Zeng, T.: Reducing artifacts in JPEG decompression via a learned dictionary. IEEE Trans. Signal Process. **62**(3), 718–728 (2014)
15. Gu, S., Zhang, L., Zuo, W., Feng, X.: Weighted nuclear norm minimization with application to image denoising. In: IEEE Conference on Computer Vision and Pattern Recognition, p. 2862–2869 (2014)
16. Zhang, J., Xiong, R., Zhao, C., Zhang, Y., Ma, S., Gao, W." CONCOLOR: constrained non-convex low-rank model for image deblocking. IEEE Trans. Image Process. **25**(3), 1246–1259 (2016)
17. Dong, C., Deng, Y., Loy, C.C., Tang, X.: Compression artifacts reduction by a deep convolutional network. In: Proceedings of the IEEE International Conference on Computer Vision, pp. 576–584 (2015)
18. Dong, C., Loy, C.C., Tang, X.: Accelerating the super-resolution convolutional neural network. In: 14th European Conf. on Computer Vision (ECCV), pp. 391–407 (2016)
19. Zhang, K., Zuo, W., Chen, Y., Meng, D., Zhang, L.: Beyond a Gaussian denoiser: Residual learning of deep CNN for image denoising. IEEE Trans. Image Process. **26**(7), 3142–3155 (2017)
20. He, K., Zhang, X., Ren, S., Su, J.: Deep residual learning for image recognition. In: IEEE Conference on Computer Vision and Pattern Recognition (CVPR) (2016)
21. Keras. Online, https://keras.io
22. Kingma, D.P., Ba, J.L.: Adam: A method for stochastic optimization. In: International Conference on Learning Representations (ICLR) (2015)
23. He, K., Zhang, X., Ren, S., Sun, J.: Delving deep into rectifiers: Surpassing human-level performance on ImageNet classification. In: IEEE International Conference on Computer Vision, pp. 1026–1034 (2015)
24. Datsets for machine learning. aritifical intelligenece, and statistics. Online, https://deepai.org/datasets

Optical Wireless Satellite Networks versus Optical Fiber Terrestrial Networks: The Latency Perspective

Invited Chapter

Aizaz U. Chaudhry and Halim Yanikomeroglu

1 Introduction

Laser inter-satellite links (LISLs) [1] between satellites in upcoming low Earth orbit (LEO) or very low Earth orbit (VLEO) satellite constellations will enable the creation of optical wireless satellite networks (OWSNs) among satellites in these constellations. These satellite networks are also referred to as free-space optical satellite networks. The LISLs will be crucial in ensuring low-latency paths within the OWSN. In their absence, a long-distance inter-continental connection over the OWSN between two cities, such as New York and Dublin, will have to go back and forth between grounds stations and satellites, which will negatively impact network latency.

The primary use case of an OWSN that is formed by LISLs in upcoming LEO/VLEO satellite constellations, like Phase I of SpaceX's Starlink [2], could be the provision of long-distance low-latency communications. An OWSN can provide low-latency communications as a premium service to the financial hubs around the world, and this use case can easily recover the cost of deploying and sustaining such an OWSN. A one millisecond advantage can translate into $100 million a year in revenues for a major brokerage firm in trading stocks at the stock exchange [3], and an advantage of a few milliseconds in financial stock markets may mean billions of dollars of revenues for these financial firms. Technological solutions, such as low-latency communications networks, are being highly sought after by these firms, and a low-latency OWSN could be the perfect solution.

Unlike optical communications in optical fiber terrestrial networks (OFTNs) where data is sent using an optical carrier (i.e., a laser beam) over a guided medium

A. U. Chaudhry (✉) · H. Yanikomeroglu
Department of Systems and Computer Engineering, Carleton University, Ottawa, ON, Canada
e-mail: auhchaud@sce.carleton.ca; halim@sce.carleton.ca

H. Nguyen et al. (eds.), *30th Biennial Symposium on Communications 2021*,
Signals and Communication Technology, https://doi.org/10.1007/978-3-031-06947-5_17

like optical fiber, data in OWSNs is sent over LISLs by employing laser beams between satellites over the vacuum in space – an unguided medium. The ratio of the speed of light in vacuum to the speed of light in a medium is called the refractive index or index of refraction of that medium [4]. A higher refractive index for a medium translates into slower transmission of light through that medium. The light will travel through a medium at 1/2 the speed of light in vacuum if it has a refractive index of 2. The speed of light in vacuum is usually denoted by c [5].

Optical fibers are generally made of glass and have an index of refraction of approximately 1.5. This indicates that the speed of light in optical fibers is approximately $c/1.5$, which means that the speed of light in vacuum is approximately 50% higher than the speed of light in optical fiber. This has critical significance in OWSNs. The higher speed of light in vacuum gives OWSNs a critical advantage over OFTNs in terms of latency for long-distance data communications.

The end-to-end delay in the network from source to destination comprises transmission delay, processing delay, queueing delay, and propagation delay [6]. In OFTNs or OWSNs, propagation delay means the delay caused by the transmission of the optical signal along the medium, i.e., optical fiber or vacuum. The propagation delay is directly related to the end-to-end distance between the source and the destination and gets very significant for long-distance data communications [7]. In this chapter, we investigate the network latency of OWSNs versus OFTNs and we define latency (or network latency) as the propagation delay from the source to the destination. For this comparison, we employ Starlink's Phase I constellation and assume LISLs between satellites in this constellation to realize an OWSN. We consider three different scenarios for long-haul inter-continental data communications, including connections between New York and Dublin, Sao Paulo and London, and Toronto and Sydney. We find minimum-latency paths between cities over the two networks for each scenario. The results indicate that the OWSN outperforms the OFTN in terms of latency. The greater the inter-continental distance between cities, the higher the improvement in latency with OWSN compared to OFTN.

The rest of the chapter is organized as follows. Related work is briefly discussed in Sect. 2. The details of different steps of our methodology for calculating latency of an inter-continental connection between cities over the OFTN and the OWSN are given in Sect. 3. The results of the comparison of these two networks in terms of latency are presented in Sect. 4. Conclusions and future work are provided in Sect. 5.

2 Related Work

Creating an OWSN by using LISLs within a satellite constellation and the problem of routing over this network has been examined [8]. It is stated that an OWSN can provide lower latency communications than an OFTN for data communications over long distances of more than 3000 km.

Using a hypothetical constellation of 1600 LEO satellites at 550 km altitude at an inclination of 53° with respect to the Equator in 40 orbital planes (OPs) with 40 satellites in each OP, a median round trip time improvement of 70% with the satellite network was found when comparing with Internet latency [9]. However, this comparison was overly favorable to the satellite network as delays due to sub-optimal routing, congestion, queueing, and forward error correction were not accounted for in the satellite network, while such delays were considered in measuring Internet latency. In our work, we compare OWSN and OFTN in terms of propagation delay only. In fact, our comparison may be favorable to the OFTN as we consider the shortest distance between two cities over the OFTN along the surface of the Earth. This is not the case in reality, and the OFTN may not provide the shortest path along Earth's surface between cities in different continents. For example, instead of following the shortest path to connect two points on Earth's surface, long-haul submarine optical fiber cables are laid along paths that avoid earthquake-prone areas and difficult seabed terrains with high slopes [10]. Also, instead of a hypothetical constellation, we employ the satellite constellation for Phase I of Starlink for this comparison.

In an earlier work [11], we investigated a use case for OWSNs to check their suitability for providing low-latency communications over long distances as their primary service. It was shown that an OWSN operating at 550 km altitude outperforms an OFTN in terms of latency when communication distances are greater than 3000 km. In this chapter, we build on our work in [11] and investigate the network latency of OWSNs versus OFTNs under more realistic scenarios for long-distance inter-continental data communications.

3 Methodology

3.1 Optical Fiber Terrestrial Network

The refractive index of a single-mode optical fiber (manufactured by Corning®) that is suitable for long-distance communications is 1.4675 at 1310 nm operating wavelength [12]. Consequently, the speed of light in this optical fiber is $c/1.4675$. The exact value of the speed of light in vacuum or c is 299,792,458 m/s [13]. We consider the speed of light in optical fiber in the OFTN to be $c/1.4675$, i.e., 204,287,876 m/s.

The latency of an inter-continental long-distance connection between two cities over the OFTN along the surface of the Earth is calculated by dividing the shortest distance between the two cities with this speed of light in optical fiber. The shortest distance between two cities is calculated by using their coordinates (latitudes and longitudes) on the surface of the Earth. The radius of the Earth is considered as 6378 km for this calculation. For example, the shortest distance between New York–Dublin, Sao Paulo–London, and Toronto–Sydney inter-continental connections

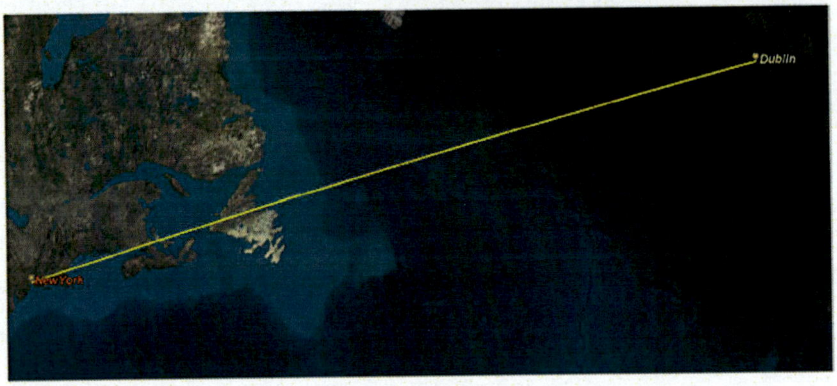

Fig. 1 Shortest distance path between New York and Dublin over the OFTN

along the surface of the Earth is calculated as 5121.30 km, 9514.30 km, and 15,584.58 km, respectively. For this purpose, we use the coordinates of the financial stock markets (i.e., New York Stock Exchange, Dublin Stock Exchange, Sao Paulo Stock Exchange, London Stock Exchange, Toronto Stock Exchange, and Sydney Stock Exchange) within these cities. Figure 1 illustrates the shortest distance path in yellow color between New York and Dublin over the OFTN along the Earth's surface.

3.2 *Optical Wireless Satellite Network*

To investigate the network latency of OWSNs versus OFTNs, we employ the satellite constellation for Phase I of Starlink. This constellation will have a total of 1584 LEO satellites that will be deployed at 550 km altitude at an inclination of 53° with respect to the Equator in 24 OPs and each OP will have 66 satellites [2]. When assuming this constellation to be uniform, the spacing between OPs is 15° and the spacing between satellites within an OP is 5.45°.

LISLs are assumed between satellites in this constellation to realize an OWSN. The LISL range for all satellites is assumed to be 1500 km, and the LISL range is defined as the distance over which a satellite can form a laser inter-satellite link with another satellite. The resulting connectivity for a satellite (i.e., satellite *x10101* in this example) in the OWSN is shown in Fig. 2. It consists of four intra-OP neighbors (i.e., four neighbors in the same OP within the LISL range including two in the front, i.e., *x10102* and *x10103*, and two at the rear, i.e., *x10165* and *x10166*), two adjacent OP neighbors (i.e., two nearest neighbors within the LISL range in right and left OPs, i.e., *x10265* and *x12454*), and six crossing OP neighbors within the LISL range in three different crossing OPs (i.e., *x11229* in crossing OP *x112*; *x11325*, *x11326*, *x11327*, and *x11328* in crossing OP *x113*; and *x11424* in crossing OP *x114*).

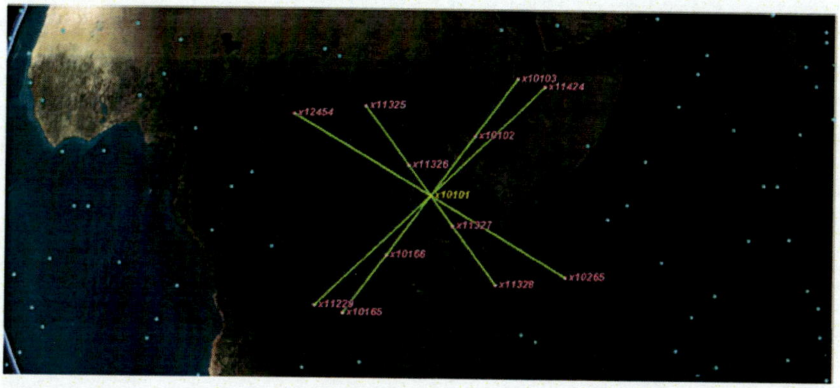

Fig. 2 Satellite connectivity in the OWSN

We consider the exact value of the speed of light in vacuum for calculating latency in the OWSN. The latency of an LISL in the OWSN is calculated by dividing the length of that link (i.e., the distance between satellites at the two ends of that link) by the speed of light in vacuum. Then, Dijkstra's shortest path algorithm [14] is employed to find the shortest path between two cities over the OWSN in terms of link latency, which is in fact the minimum-latency route between cities over the OWSN. It includes the latency of the optical link from the ground station (GS) in the source city on Earth to the ingress satellite of the OWSN in space, the latencies of the LISLs in this path, and the latency of the optical link from the egress satellite of the OWSN in space to the GS in the destination city on Earth. The time is divided into time slots of 1 s duration. A shortest path (i.e., a route with minimum latency) is calculated at each time slot where a time slot represents a snapshot of the OWSN at that second.

Compared to OFTN, an extra distance from a GS on Earth to ingress satellite and from egress satellite to a GS is involved when communicating over the OWSN. However, the extra latency due to this extra distance becomes insignificant for long-distance inter-continental communications over the OWSN.

4 Results

We compare OWSN and OFTN in terms of latency in three different inter-continental connection scenarios, including New York–Dublin, Sao Paulo–London, and Toronto–Sydney. To simulate the OWSN, we use the well-known satellite constellation simulator STK Version 12.1 [15]. We create the satellite constellation for Phase I of Starlink in STK using the parameters discussed in Sect. 3.

Distinct IDs are generated for the 1584 satellites within this constellation. The following IDs are created for the 24 OPs: {*x101*, *x102*, *x103*, ..., *x124*}. For the 66 satellites within each OP, distinct IDs are created, e.g., the following IDs are created for satellites in the first OP: {*x10101*, *x10102*, *x10103*, ..., *x10166*}.

The satellites in this constellation travel at speeds of approximately 7.6 km/s. This means that the GS to ingress satellite link, satellite to satellite links, and egress satellite to GS link or the latencies of these links are constantly changing. Consequently, the shortest path of the inter-continental connection over the OWSN between two cities (and/or its latency) also changes at every time slot.

We run the simulation for 1 h and find the shortest path in terms of latency between two cities over this OWSN at every second (or time slot). For example, the shortest paths calculated at the first twenty time slots for the New York–Dublin inter-continental connection and their latencies are given in Table 1. Furthermore, the shortest path for this inter-continental connection over the OWSN at first time slot is shown in Fig. 3 in yellow. This shortest path consists of the GS at New York (located at the New York Stock Exchange); satellites *x11503* (ingress), *x11504*, *x11506*, *x11508*, and *x11607* (egress) in the OWSN; and the GS at Dublin (located at the Dublin Stock Exchange). The latency of this shortest path decreases over the next three time slots as indicated in Table 1. Note that a shortest path with lower latency becomes available at the fifth time slot, and this shortest path is shown in Fig. 4. The latency of this path increases at succeeding time slots, and at eleventh time slot, a shortest path with lower latency is found as can be seen from Table 1. This shortest path is shown in Fig. 5; it continues to exist till eighteenth time slot after which it becomes unavailable and a new shortest path with a higher latency is established at the nineteenth time slot. Due to the high-speed movement of satellites along their OPs, the shortest path and/or its latency change at every time slot.

Table 2 shows the results of the comparison of OWSN and OFTN in terms of latency in milliseconds. The latency of an inter-continental connection over the OWSN shown in this table is the average of the latencies of all shortest paths that are calculated at all time slots over the simulation duration of 1 h.

Compared to OFTN, the latency improvement with the OWSN is 5.00 ms, 9.93 ms, and 17.95 ms for New York–Dublin, Sao Paulo–London, and Toronto–Sydney long-distance connections, respectively. These results clearly indicate that the OWSN outperforms the OFTN in all scenarios in terms of network latency. In fact, the longer the length of the inter-continental connection between a pair of cities, the higher the improvement in latency provided by the OWSN compared to OFTN.

5 Conclusions

We examine the network latency of OWSN versus OFTN under three different scenarios for long-distance inter-continental data communications. We simulate the OWSN by using LISLs among satellites in the constellation for Phase I of Starlink. We find the shortest path between two cities in terms of link latency (i.e., the minimum-latency route between cities) over the OWSN at every time slot. Then, we compare the average latency of shortest paths at all time slots in the OWSN

Table 1 Shortest paths at first twenty time slots over the OWSN for New York–Dublin inter-continental connection

Time slot (s)	Shortest path	Latency (ms)
1	GS at New York, satellites *x11503*, *x11504*, *x11506*, *x11508*, *x11607*, GS at Dublin	19.8152
2	GS at New York, satellites *x11503*, *x11504*, *x11506*, *x11508*, *x11607*, GS at Dublin	19.8148
3	GS at New York, satellites *x11503*, *x11504*, *x11506*, *x11508*, *x11607*, GS at Dublin	19.8145
4	GS at New York, satellites *x11503*, *x11504*, *x11506*, *x11508*, *x11607*, GS at Dublin	19.8144
5	GS at New York, satellites *x11503*, *x11505*, *x11507*, *x11607*, GS at Dublin	19.8042
6	GS at New York, satellites *x11503*, *x11505*, *x11507*, *x11607*, GS at Dublin	19.8042
7	GS at New York, satellites *x11503*, *x11505*, *x11507*, *x11607*, GS at Dublin	19.8044
8	GS at New York, satellites *x11503*, *x11505*, *x11507*, *x11607*, GS at Dublin	19.8047
9	GS at New York, satellites *x11503*, *x11505*, *x11507*, *x11607*, GS at Dublin	19.8052
10	GS at New York, satellites *x11503*, *x11505*, *x11507*, *x11607*, GS at Dublin	19.8058
11	GS at New York, satellites *x11503*, *x11505*, *x11507*, *x11509*, GS at Dublin	19.6848
12	GS at New York, satellites *x11503*, *x11505*, *x11507*, *x11509*, GS at Dublin	19.6851
13	GS at New York, satellites *x11503*, *x11505*, *x11507*, *x11509*, GS at Dublin	19.6854
14	GS at New York, satellites *x11503*, *x11505*, *x11507*, *x11509*, GS at Dublin	19.6859
15	GS at New York, satellites *x11503*, *x11505*, *x11507*, *x11509*, GS at Dublin	19.6865
16	GS at New York, satellites *x11503*, *x11505*, *x11507*, *x11509*, GS at Dublin	19.6872
17	GS at New York, satellites *x11503*, *x11505*, *x11507*, *x11509*, GS at Dublin	19.6880
18	GS at New York, satellites *x11503*, *x11505*, *x11507*, *x11509*, GS at Dublin	19.6890
19	GS at New York, satellites *x10918*, *x11114*, *x11311*, *x11508*, *x11509*, GS at Dublin	20.2957
20	GS at New York, satellites *x10918*, *x11114*, *x11311*, *x11508*, *x11509*, GS at Dublin	20.3069

with latency of the long-distance inter-continental connection between these cities over the OFTN along the Earth's surface. In all scenarios, the OWSN performs better than the OFTN in terms of latency and provides an improvement of 5.00 ms,

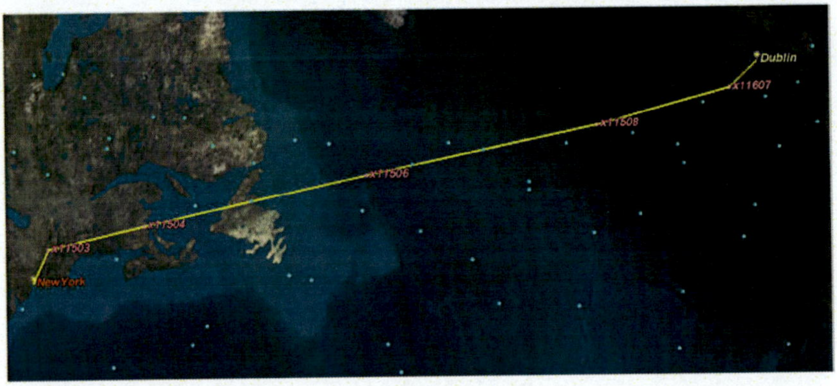

Fig. 3 Shortest path between New York and Dublin over the OWSN at first time slot

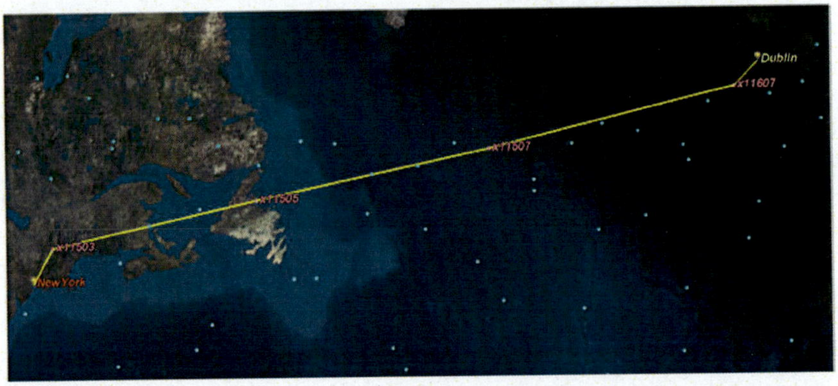

Fig. 4 Shortest path between New York and Dublin over the OWSN at fifth time slot

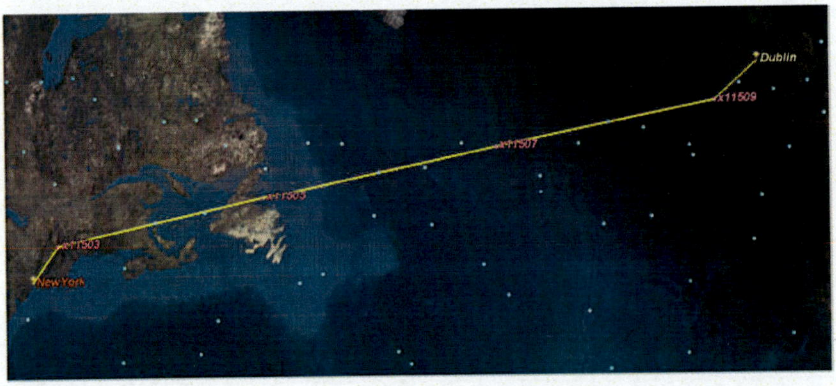

Fig. 5 Shortest path between New York and Dublin over the OWSN at eleventh time slot

Table 2 Latency – OWSN vs. OFTN

Inter-continental connection	Latency (ms)		Latency improvement	
	OFTN	OWSN	(ms)	(%)
New York–Dublin	25.07	20.07	5.00	19.94
Sao Paulo–London	46.57	36.64	9.93	21.32
Toronto–Sydney	76.29	58.34	17.95	23.53

9.93 ms, and 17.95 ms (or 19.94%, 21.32%, and 23.53%) for New York–Dublin, Sao Paulo–London, and Toronto–Sydney long-distance inter-continental connections, respectively.

We discover an interesting relationship between network latency over the OWSN and length of the inter-continental connection between cities. We find that the longer the inter-continental connection, the greater the gain due to the higher speed of light over free-space laser links in OWSN compared to the speed of light over optical fiber laser links in OFTN, and the better the improvement in network latency offered by the OWSN compared to OFTN. Based on these findings, we can say that OWSNs resulting from LISLs in upcoming LEO/VLEO satellite constellations, like Phase I of Starlink, can be the perfect solution for high-speed trading firms seeking low-latency inter-continental data communications among financial stock markets around the world.

In this work, we employed an LISL range of 1500 km for all satellites while calculating latency of an inter-continental connection for data communications over the OWSN. Different LISL ranges may impact latency differently, and as part of future work, we intend to investigate the effect of different LISL ranges on the latency of the OWSN.

Acknowledgments This work has been supported by the National Research Council Canada's (NRC) High Throughput Secure Networks program (CSTIP Grant #CH-HTSN-625) within the Optical Satellite Communications Consortium Canada (OSC) framework. The authors would like to thank AGI for the STK platform.

References

1. Chaudhry, A.U., Yanikomeroglu, H.: Laser intersatellite links in a Starlink constellation: a classification and analysis. IEEE Veh. Technol. Mag. **16**(2), 48–56 (2021)
2. SpaceX FCC update: SpaceX non-geostationary satellite system, Attachment A, Technical Information to Supplement Schedule S [Online]. Available: https://licensing.fcc.gov/myibfs/download.do?attachment_key=1569860 (2018)
3. Martin, R.: Wall Street's quest to process data at the speed of light. Information Week [Online]. Available: https://www.informationweek.com/wall-streets-quest-to-process-data-at-the-speed-of-light/d/d-id/1054287 (2007)
4. Gupta, S.C.: Textbook on Optical Fiber Communication and Its Applications. PHI Learning Private Limited, New Delhi (2018)
5. Hecht, J.: Understanding Lasers: An Entry-Level Guide. Wiley, Hoboken (2018)

6. Kurose, J.F., Ross, K.W.: Computer Networking – A Top Down Approach Featuring the Internet. Addison-Wesley, Boston (2009)
7. Corning Inc.: The super connected world: Optical Fiber Advances and Next Gen Backbone, Mobile Backhaul, and Access Networks [White Paper] [Online]. Available: https://www.corning.com/media/worldwide/coc/documents/Fiber/white-paper/wp6024.pdf (2012)
8. Handley, M.: Delay is not an option: low latency routing in space. In: Proceeding of the 17th ACM Workshop on Hot Topics in Networks, pp. 85–91, Redmond (2018)
9. Bhattacherjee, D., Singla, A.: Network topology design at 27,000 km/hour. In: Proceeding of the 15th International Conference on Emerging Networking Experiments and Technologies, pp. 341–354, Orlando (2019)
10. Wang, Z., Wang, Q., Moran, B., Zukerman, M.: Terrain constrained path planning for long-haul cables. Opt. Express. **27**(6), 8221–8235 (2019)
11. Chaudhry, A.U., Yanikomeroglu, H.: Free space optics for next-generation satellite networks. IEEE Consum. Electron. Mag. (Early Access). https://doi.org/10.1109/MCE.2020.3029772
12. Corning Inc.: Corning® Hermetic single-mode and multimode specialty optical fibers [Online]. Available: https://www.corning.com/media/worldwide/csm/documents/M0300013v00_Hermetic_Acrylate_March_2010.pdf (2010)
13. MacKay, R.J., Oldford, R.W.: Scientific method, statistical method and the speed of light. Stat. Sci. **15**(3), 254–278 (2000)
14. Dijkstra, E.W.: A note on two problems in connexion with graphs. Numer. Math. **1**, 269–271 (1959)
15. AGI: Systems Tool Kit (STK) [Online]. Available: https://www.agi.com/products/stk

Low-Complexity Framework for Performance Analysis of Energy Harvesting Communications

Invited Chapter

Dhawal Beohar, Leila Musavian, and Sonia Aïssa

1 Introduction

In this modern era, wireless technologies are emerging at an unprecedented rate in every sector of our society, including industrial communications, advance healthcare, smart cities, etc. [1–3]. With the significant increase in the next-generation wireless communication applications, it is imperative for the operating system to maintain the demands of the end users, such as higher data rates and reliable connectivity, while making efficient use of the available energy resources. In many applications, however, it is challenging to keep up with the growing energy demands, which can lead to serious problems of energy exhaustion in power-constrained networks, e.g., wireless sensor networks (WSNs) [4].

To address the challenge of power-constrained battery devices so as to meet the end users' quality of service (QoS) while maintaining the network lifetime, energy harvesting (EH) technology has come to the forefront. Communication systems empowered with EH can benefit to perpetuate the devices' batteries with alternative sources of energy via the implementation of EH batteries instead of the conventional non-rechargeable batteries or hardwired power [5]. As energy is directly scavenged from renewable sources available in the surrounding environment, for instance, from solar energy, the lifetime of these power-limited networks can be effectively expanded [6].

D. Beohar (✉) · L. Musavian
School of Computer Science and Electronic Engineering, University of Essex, Colchester, UK
e-mail: db18258@essex.ac.uk; leila.musavian@essex.ac.uk

S. Aïssa
Institut National de la Recherche Scientifique (INRS), University of Quebec, Montreal, QC, Canada
e-mail: sonia.aissa@inrs.ca

H. Nguyen et al. (eds.), *30th Biennial Symposium on Communications 2021*,
Signals and Communication Technology, https://doi.org/10.1007/978-3-031-06947-5_18

There exists a rich body of research work on EH communications that support the above-discussed advantages, e.g., on exploring transmission strategies to increase throughput and/or improve energy efficiency in EH-based wireless communication systems, on developing optimal power allocation strategies for such systems, or on analyzing their performances (see, e.g., [7–10] and the references therein).

In particular, several online transmission policies for EH-based wireless communication systems, defined as transmissions based on the assumption of knowledge of the arrival time and amount of the harvested energy at the transmitter, and not before or after transmission, have been introduced and also compared with offline transmission policies [11–15]. Specifically, in [11], the overall system throughput of an EH-based relaying system equipped with both, an EH battery and a fixed battery, was analyzed. Therein, closed-form expressions for the cumulative distribution function of individual links were derived to improve the performance of the system by using the harvest–use approach. Harvest–store–use and harvest–use EH communication protocols were presented in [12], wherein the corresponding system throughput and bit error rate were also analyzed. Furthermore, considering a multiuser network, the theoretical expressions for the outage probability, overall throughput, and ergodic capacity of a two-user EH communication system with full-duplex relaying and non-orthogonal multiple access were derived in [13]. Moreover, in [14], to analyze the throughput of EH-based communications, large deviation theory with Markov decision process was used to model the energy state while considering various energy constraints, such as energy overflow limits.

Several optimal transmission policies in EH communication systems with infinite battery capacity were also proposed, e.g., in [15–18]. Specifically, in [15], a throughput maximization problem in EH communications was studied and solved by proposing an online algorithm which is based on a developed comparison of transmissions over fading channels and static channels. In [16], minimization of the packet loss rate in EH communication systems was investigated. Therein, assuming that the accessible amount of harvested energy must support the transmission power of the EH communication to guarantee the real-time transmission success, an optimal energy allocation was proposed by using the Markov decision process and stochastic online algorithm. The online power scheduling problem was solved by using dynamic programming in a single-user EH communication system.

The aforementioned works, namely, [15–18], provided resource allocation solutions by adopting a specific underpinning assumption on the availability of sufficient amount of energy in the battery whenever the transmission is needed at a given (or required) transmission rate. This assumption cannot be satisfied in many real-time applications, since the amount of energy for transmission within a time-slot depends on the end user's required QoS and, in turn, on the channel state information in that time-slot. On the other hand, the source of the harvested energy is independent of the wireless fading channel conditions. Hence, one can confirm that the incoming energy into the battery and the outgoing energy from it are independent. This condition challenges the estimation of the exact energy level in the battery at any time instant. Besides, when the system suffers from severely weak fading channel conditions, guaranteeing a given QoS will require a significantly high transmit power for the data communication. Due to the variable status of the available energy in the EH battery, it is unavoidable for an energy-outage event to happen, which means that there will not be enough energy in the battery to perform the transmission. This challenge underpins the main motivation of the work in this chapter. Here, we assume that the EH-based communication system can tolerate some level of energy-outage and develop a low-complexity framework to analyze the performance of the system.

In particular, the main contributions of this chapter are threefold:

- We present a virtual battery queuing model for the EH battery and introduce an energy-outage probability constraint in the EH communication system.
- A low-complexity mathematical framework is introduced to analyze the performance of the EH communication system. In particular, a relationship between the fading channel condition and the incoming energy level is put forward through a unique framework while adhering to the energy-outage limit.
- The performance of the EH communication system is investigated in terms of energy-outage probability, and insightful discussions are provided.

2 System Modelling

2.1 System and Channel Models

The system model is sketched in Fig. 1. Point-to-point wireless EH communication over the Rayleigh block-fading channel is considered. The transmitter is equipped with a finite EH battery, modelled as an energy queue. The transmitter harvests solar energy and stores it in the energy queue of the EH battery.

At the receiver end, the received signal, $y(t)$, can be expressed as

$$y(t) = h(t)x(t) + n(t), \tag{1}$$

where $h(t)$ is the coefficient of the Rayleigh block-fading channel, $x(t)$ is the transmitted signal, $n(t)$ is a complex Gaussian random noise with zero mean and unit variance, and t refers to the time index.

The channel varies independently during transmission from one fading block to another but remains constant during a fading block [19]. The block duration is denoted by T. The fading channel coefficient $h(t)$ represents variations in different time-slots.

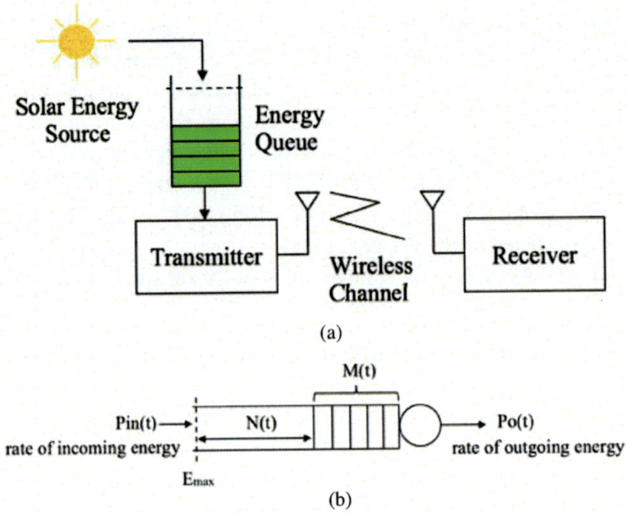

Fig. 1 The energy harvesting-based communication system. (**a**) System model. (**b**) Energy queue within the system model illustrated in (**a**)

2.2 *Physical Battery Model and Energy-Outage*

As shown in Fig. 1b, we consider a physical battery model that consists of a finite-capacity EH battery.

The EH battery is capable of storing the arriving energy in an energy queue up to a certain threshold level $E_{\max}$ (Joules). The accumulated energy in the battery and the empty portion of the battery are represented by two random variables, namely, $M(t)$ and $N(t)$, respectively.[1] The rate of the incoming energy into the battery from the EH process is denoted by $Pin(t)$, and the outgoing rate of energy from the battery is denoted by $Po(t)$; both rates are in the unit of joules/sec.

The instantaneous communication rate of the channel, $R(t)$ in nats/s/Hz, is expressed as

$$R(t) = \log\left(1 + \frac{h^2(t)Po(t)}{N_0 B}\right), \tag{2}$$

where $N_0 B$ represents the noise density per unit bandwidth and B is the system bandwidth.

[1] The time duration for the channel fading and the energy arrival epoch are considered to be the same.

Furthermore, we assume a minimum throughput as the QoS requirement which should be provided for the end user. Guaranteeing such QoS requirement in the EH communication system at all time is, however, not possible. In fact, the solar energy level that is harvested from the environment is not related to the condition of the communication channel or the user's required QoS. Actually, the outgoing energy from the battery is the quantity that depends on the channel conditions and the target QoS. As such, the status of the accumulated energy in the battery is difficult to know due to the non-dependency between the incoming rate of the energy and its outgoing rate. Therefore, ensuring the required QoS in the EH-based communication system all the time is a challenging task.

Recall that the channel fading is a random process and that it changes according to the unpredictable nature of the propagation conditions. When the channel is severely weak, the transmission is required to be done with a high power to deliver the target QoS. In order to carry out transmission with such high power, a sufficient amount of energy should be available in the battery. No transmission can take place if the amount of energy in the battery is less than the required outgoing energy needed to deliver the target QoS, i.e., $Po(t)T$, where T is the time-slot duration.

In this work, we propose a performance indicator, namely, energy-outage, which is defined as the event for not having enough energy in the transmitter's battery to send the data to its end destination. Also, we assume that the amount of energy that can be spent from the EH battery, at each time-slot, is limited by a maximum value of $P_{\max}T$.

Going back to the goal of this chapter, we aim to implement a probabilistic policy to assess the performance of the EH communication system by considering a threshold on the probability of energy-outage in the battery. By allowing a probabilistic energy-outage in the EH communication system, we can use the well-known mathematical theorem of LDP. To analyze the performance of the system, we note that the theory of large deviation models the probabilistic behavior of the tail events that are exponentially small [20].

Let us assume the fixed set of sequences of independent and identically distributed (i.i.d.) random variables $X_1, X_2, X_3, \ldots$, having mean μ and variance σ^2, and denote the sum of sequences by $S(n)$. With the help of the central limit theorem, the tail probability of $S(n)$ can be approximated, for any $e > 0$, as $\lim_{n\to\infty} \Pr\{S(n) > e\} = 0$, for any fixed value of n. The convergence of $S(n)$ to the expected value of the random variable is seen as n grows large. However, the central limit theory does not provide an answer to the convergence rate. On the other hand, the theory of large deviation is able to give answers regarding the convergence of $S(n)$ to a steady-state queue length $S(\infty)$ when $n \to \infty$, leading to

$$\lim_{n\to\infty} \frac{\log\big(\Pr\{S(\infty) \geq e\}\big)}{e} = -I, \tag{3}$$

where I is termed the rate function, which is used to measure the probabilities of rare events [20]. The estimation using the assumptions of the LDP theorem, as shown in (3), is only valid when e holds a fairly large value, in which case we have

$$\Pr\{S(\infty) \geq e\} \approx \exp(-Ie). \tag{4}$$

Now, we invoke tools from the LDP theory to find a statistical expression for the energy-outage, as the theory deals with the probabilities of events that are exponentially small. These tools will allow us to develop our mathematical framework by using the randomness property of energy-outage occurrence under QoS constraints.

For this purpose, we now define and formulate the concept of energy-outage as the event when the energy queue length is less than a required outgoing energy needed to deliver the target QoS, which can be formulated as

$$\Pr\{M(\infty) \leq Po(t)T\}, \tag{5}$$

where $M(\infty)$ indicates that the accumulated portion of queue length process converges in distribution to the steady-state queue length.

We need to enforce that the energy-outage probability is limited to a manageable value. For such, we introduce the condition

$$\Pr\{M(\infty) \leq Po(t)T\} \leq \text{EOP}, \tag{6}$$

where EOP is the energy-outage probability defined as the probability to measure the energy-outage event occurrence.

In stark contrast, the inequality shown in (5) is inverse of what makes the basis of the LDP theorem in (4) and, hence, cannot be implemented in the physical battery model. Therefore, in order to use the LDP assumptions and the discussed inequality in our proposal to formulate the energy-outage statistically, we introduce a new random variable which represents the empty side of the battery. Next, with the help of a virtual battery queuing model, the system performance will be studied by analyzing the energy-outage probability.

3 The Low-Complexity Performance Evaluation

3.1 Virtual Battery Queuing Model

We propose a virtual battery queuing model, as shown in Fig. 2. The proposed model is based on the discussed physical battery model but considers the empty side of the battery rather than the accumulated energy part as the queue model. Specifically, the role of the incoming random energy epoch into the battery and the random transmission power from the battery are exchanged. Hence, the virtual battery queuing model will functionally act as an opposite to the physical battery model. In fact, the virtual battery queuing model becomes competent with the assumptions considered in LDP, which was not the case with the physical battery model.

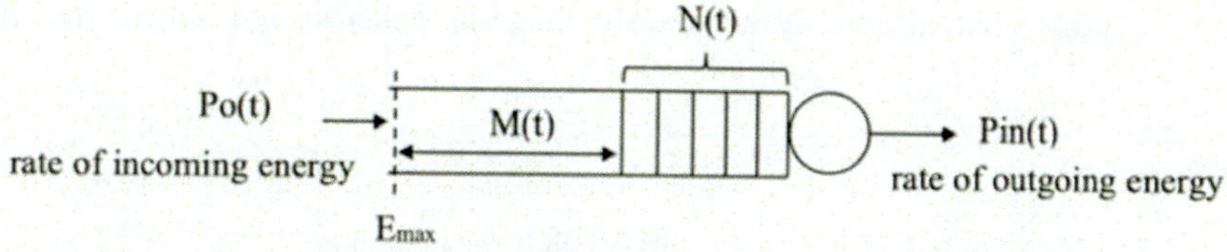

Fig. 2 The virtual battery queuing model

The proposed model is implemented so that a novel mathematical model can be established using the LDP theorem.

3.2 Energy-Outage Probability Analysis

To measure the performance of the EH-based communications, we will analyze the energy-outage probability.

In the proposed system model, transmissions will be possible only when the available energy in the battery is greater than $Po(t)T$. Violation of this condition is termed as energy-outage in our work. The probability of occurrence of this condition should be limited.

This condition is explained statistically in (6) pertaining to the physical battery model. But, when comparing the condition of energy-outage probability with the LDP assumption shown in (4), one can see that the inequalities in the expressions are inverse of each other, hence limiting the implementation of the LDP assumption. In order to fulfill the requirement of the LDP assumption in our work with regard to the energy-outage probability constraint, we have considered and proposed the virtual battery queuing model.

In the proposed virtual battery queuing model, a relationship between an energy-outage and the probability of the empty portion of the battery exceeding $E_{\max}$ and $P\mathrm{o}(t)T$ is obtained by considering the empty portion of the battery. Therefore, the energy-outage probability can now be formulated as

$$\Pr\big\{N(t) \geq (E_{\max} - P\mathrm{o}(t)T)\big\} \leq \mathrm{EOP}. \tag{7}$$

3.3 Statistical QoS Guarantees

According to the assumptions of the LDP theorem detailed in [21], an empty portion of energy queue $N(t)$, $t \geq 0$, converges in allocation to a finite random variable $N(\infty)$ for dynamic energy queue with constant and random arrivals. By using the assumption that the maximum power that can be used from battery at each time-slot is $P_{\max}$, we provide a simpler and tighter expression for the energy-outage probability which is mathematically tractable. Hence, by replacing $Po(t)$ with $P_{\max}$ in (7), we get a tighter limit as

$$\Pr\{N(\infty) \geq (E_{\max} - P_{\max}T)\} \leq \text{EOP}. \tag{8}$$

Using (4), the left-hand side of (8) can be estimated as

$$\begin{aligned} &\Pr\{N(\infty) \geq (E_{\max} - P_{\max}T)\} \\ &\approx \exp(-\theta(E_{\max} - P_{\max}T)), \end{aligned} \tag{9}$$

which holds only for the large values of $(E_{\max} - P_{\max}T)$.

For small values of $(E_{\max} - P_{\max}T)$, the energy-outage probability is calculated as

$$\begin{aligned} &\Pr\{N(\infty) \geq (E_{\max} - P_{\max}T)\} \\ &\approx \epsilon \times \exp(-\theta(E_{\max} - P_{\max}T)), \end{aligned} \tag{10}$$

where ϵ depicts the probability that the virtual battery buffer is in non-empty state or, in other words, the probability that the EH battery is not full.

The QoS exponent θ, $\theta \geq 0$, plays an important role as a measurement to analyze the overall system performance of the EH-based communication. As the QoS becomes more stringent, θ increases with a decrease in the source rate accommodation in a given wireless channel. Smaller values of θ, on the other hand, lead to slower decay rate [21].

3.4 *Effective Energy Harvesting*

We now introduce the concept of effective energy harvesting by taking motivations from the assumptions of LDP.

We assume the empty side of the energy queue in the virtual battery queuing model to have stationary and ergodic incoming energy rate at $t \geq 0$ as $Po(t)$. This incoming rate of energy can be modelled by its effective EH, asymptotically. Using the Gartner-Ellis theorem, the asymptotic log-MGF (moment generating function) of $Po(t)$ is defined as

$$\Lambda_{\text{EH}}(\theta) = \lim_{t \to \infty} \frac{1}{t} \log\left(E\left[\exp(\theta Po(t))\right]\right), \; \theta \geq 0, \tag{11}$$

where $E[\cdot]$ represents the expectation operation.

The rate of outgoing energy from the virtual battery, i.e., $Pin(t)$, is assumed to be fixed. Therefore, it is denoted by P_{fix}. In the following, the effective EH function of $Po(t)$ is defined as

$$\alpha(\theta) = \frac{\Lambda_{\mathrm{EH}}(\theta)}{\theta}. \tag{12}$$

By substituting (11) into (12), we get

$$\frac{\Lambda_{\mathrm{EH}}(\theta)}{\theta} = \frac{1}{\theta} \lim_{t\to\infty} \frac{1}{t} \log\Big(E\big[\exp(\theta Po(t))\big]\Big). \tag{13}$$

Then, by simplifying further, we obtain a low-complexity expression for the effective EH function, as follows:

$$\frac{\Lambda_{\mathrm{EH}}(\theta)}{\theta} = \frac{1}{\theta} \log\Big(E\big[\exp(\theta Po(t))\big]\Big). \tag{14}$$

3.5 *Formulation of an Upper Bound for P_{fix}*

Next, an upper bound for P_{fix} is formulated by considering the instantaneous rate of the channel (2) and the incoming energy process $Pin(t)$, which is assumed to be fixed and expressed as P_{fix}.

A QoS measure as a minimum required rate R_{min}, in the unit of nats/s/Hz, is considered in the system, which can be expressed as

$$R(t) = \log\left(1 + \frac{h^2(t) Po(t)}{N_0 B}\right) \geq R_{\mathrm{min}}. \tag{15}$$

Using (15) and substituting the value of $Po(t)$ in (14), we can derive an expression for the required arrival energy rate P_{fix}, which provides a relationship between the channel fading coefficient $h(t)$ and the user's minimum required rate R_{min}.

Finally, combining the derivations in (16) and (9) jointly gives us a low-complexity mathematical framework which is based on LDP assumptions under energy-outage constraint to analyze the performance of the EH communication system. Using both expressions, we can investigate the performance as well as the energy-outage probability in the system according to

$$\frac{\Lambda_{\mathrm{EH}}(\theta)}{\theta} \leq \frac{1}{\theta} \log\left(E\left[\exp\left(\theta \frac{(\exp(R_{\mathrm{min}}) - 1) N_0 B}{h^2(t)}\right)\right]\right), \tag{16}$$

with

$$P_{\mathrm{fix}} \leq \frac{\Lambda_{\mathrm{EH}}(\theta)}{\theta}.$$

4 Numerical Results and Validation

In this section, numerical results are presented to examine the energy-outage probability, EOP, and the fixed amount of incoming energy, P_{fix}, as a function of the QoS component, θ, for the EH communication system when using the developed framework.

In all simulations, and unless otherwise indicated, we assume the noise density is $N_0 B = 1$, the maximum level of energy that the battery can store is $E_{\max} = 40$ Joules, the minimum rate requirement $R_{\min}$ is set to 3.0 nats/s/Hz, and the duration of one time-slot is $T = 1$ sec. The Monte Carlo number of iterations is 10^5.

Firstly, in Fig. 3, the energy-outage probability, EOP, is plotted against the QoS component, θ, for various values of the transmission power, $\frac{P_{\max}}{N_0 B}$.[2] As observed from the plots, an increase in the QoS component leads to a decrease in the energy-outage probability. This verifies that for larger values of θ, the QoS is more rigid. Also, if we consider the QoS component to be 10^{-1}, the energy-outage probability increases with the increase in $\frac{P_{\max}}{N_0 B}$. For example, for a maximum transmission power of 5 dB, the energy-outage probability is approximately 2%, while for a transmission power of 14 dB, the energy-outage probability is 20%. This shows that the system will face increase in energy-outage with the increase in the maximum transmission power limit.

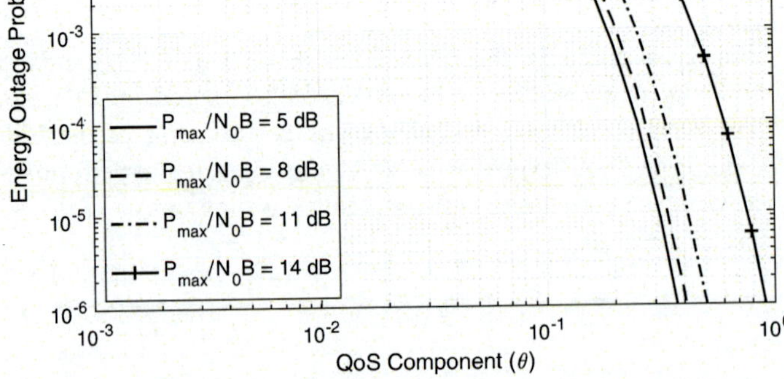

Fig. 3 Energy-outage probability versus the QoS component θ for various values of $P_{\max}/N_0 B$

[2] Hereafter, we use $P_{\max}$ instead of $\frac{P_{\max}}{N_0 B}$ for ease of reference, since $N_0 B$ is set to unity.

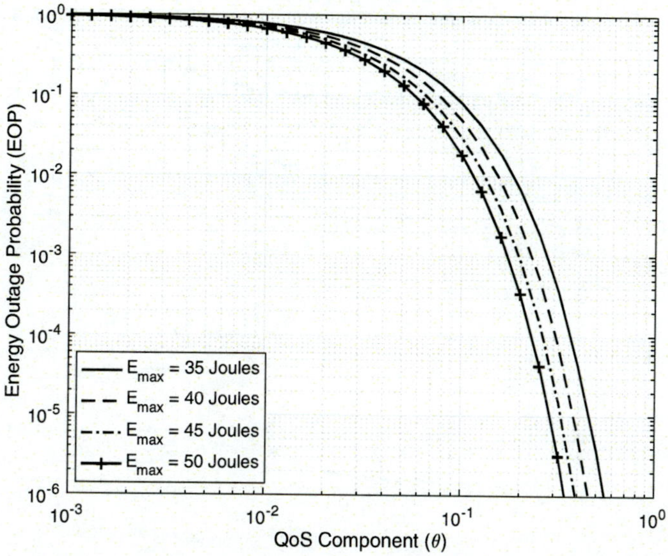

Fig. 4 Energy-outage probability versus the QoS component θ for various values of $E_{\max}$

In Fig. 4, we plot the energy-outage probability, EOP, versus the QoS component, θ, for various values of $E_{\max}$. In this experiment, the maximum transmission power ($P_{\max}$) is set to 10 dB. As it can be seen from the figure, when the QoS component increases, the energy-outage probability decreases. This verifies that for more rigid QoS exponent θ, the system will face less energy-outage. Fig. 4 also demonstrates that if we consider θ to be 10^{-1}, then EOP decreases for increased values of the battery capacity $E_{\max}$.

Finally, we analyze the variations of the rate of incoming energy, P_{fix}, as a function of the QoS component, θ, for various values of the minimum required rate, $R_{\min}$. The simulation results are shown in Fig. 5. Here, the maximum transmission value $P_{\max}$ is set to 10 dB. As observed from the figure, as the minimum required rate gets increased from 0.5 nats/s/Hz to 2.0 nats/s/Hz, P_{fix} also increases as θ increases. This can be attributed to the fact that in order to guarantee a higher QoS $R_{\min}$, higher transmit powers are required and, in turn, higher P_{fix} is needed to maintain the energy-outage probability.

5 Conclusion

In this chapter, we studied and analyzed the probability of energy-outage occurrence to measure the performance of energy harvesting (EH) communication system equipped with finite-sized EH battery. The problem of unknown status of the

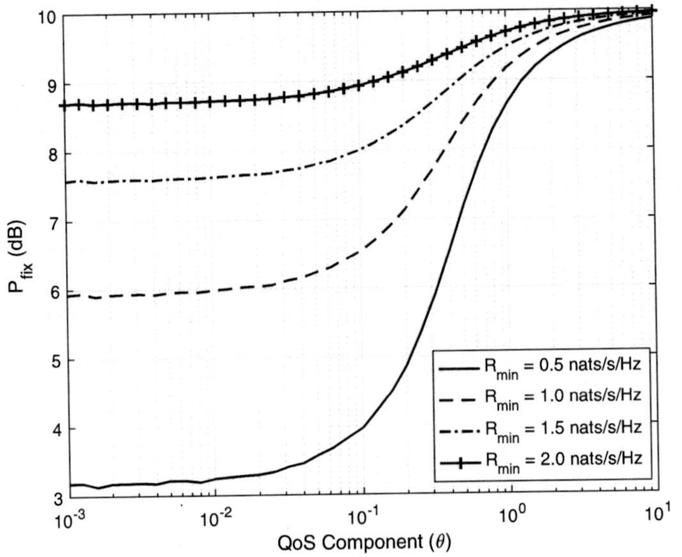

Fig. 5 Rate of incoming energy P_{fix} versus the QoS component θ for various values of the minimum required rate R_{min}

available energy in the battery was addressed through a probabilistic approach. Furthermore, a low-complexity mathematical framework was developed based on the theory of large deviation principle (LDP). Due to the discussed limitations of the physical battery model, we presented a virtual battery queuing model such that the assumptions of LDP can be satisfied, for subsequent use. Also, the energy-outage probability was calculated using novel statistical expressions while taking into account energy constraints of the EH-based communication system.

References

1. Seijo, O., Val, I., Lopez-Fernandez, J.A.: w-SHARP: Implementation of a high-performance wireless time-sensitive network for low latency and ultra-low cycle time industrial applications. IEEE Trans. Ind. Inf. **17**(5), 3651–3662 (2020)
2. Kong, P.: Cellular-assisted device-to-device communications for healthcare monitoring wireless body area networks. IEEE Sen. J. **20**(21), 13139–13149 (2020)
3. Ram, S.K., Chourasia, S., Das, B.B., Swain, A.K., Mahapatra, K., Mohanty, S.: A solar based power module for battery-less IoT sensors towards sustainable smart cities. In: Proceedings of IEEE Computer Society Annual Symposium on VLSI (ISVLSI), Cyprus, pp. 458–463 (2020)
4. Inamdar, M.A., Kumara, H.: Energy efficient 5G networks: Techniques and challenges. In: Proceedings, International Conference on Smart Electronics and Communication (ICOSEC), Trichy, India, pp. 1317–1322 (2020)
5. Li, Z., Saadatnia, Z., Yang, Z., Naguib, H.: A hybrid piezoelectric-triboelectric generator for low-frequency and broad-bandwidth energy harvesting. Energy Convers. Manag. **174**, 188–197 (2018)

6. Nundrakwang, S., Yingyong, P., Isarakorn, D.: Energy harvesting for self-powered systems. In: Proceedings of the International Conference on Applied Science and Technology (ICEAST), Chiang Mai, Thailand, pp. 1–4 (2020)
7. Gu, Y., Aissa, S.: RF-based energy harvesting in decode-and-forward relaying systems: Ergodic and outage capacities. IEEE Trans. Wirel. Commun. **14**(11), 6425–6434 (2015)
8. Gul, O.M., Demirekler, M.: Asymptotically throughput optimal scheduling for energy harvesting wireless sensor networks. IEEE Access **6**, 45004–45020 (2018)
9. Azarhava, H., Niya, J.M.: Energy efficient resource allocation in wireless energy harvesting sensor networks. IEEE Wirel. Commun. Lett. **9**(7), 1000–1003 (2020)
10. Zlatanov, N., Schober, R., Hadzi-Velkov, Z.: Asymptotically optimal power allocation for energy harvesting communication networks. IEEE Trans. Veh. Technol. **66**(8), 7286–7301 (2017)
11. Siddiqui, A.M., Musavian, L., Aissa, S., Ni, Q.: Performance analysis of relaying systems with fixed and energy harvesting batteries. IEEE Trans. Commun. **66**(4), 1386–1398 (2018)
12. Long, M., Chen, Y.: Performance analysis of energy harvesting communications using multiple time slots. IET Commun. **13**(3), 289–296 (2018)
13. Nguyen, B.C., Hoang, T.M., Pham, X.N., Tran, P.T.: Performance analysis of energy harvesting-based full-duplex decode-and-forward vehicle-to-vehicle relay networks with nonorthogonal multiple access. In: Wireless Commun. and Mobile Computing, Hindawi, vol. 2019 (2019)
14. Akin, S., Gursoy, M.C.: On the energy and data storage management in energy harvesting wireless communications. IEEE Trans. Commun. **67**(11), 8056–8071 (2019)
15. Wu, W., Wang, J., Wang, X., Shan, F., Luo, J.: Online throughput maximization for energy harvesting communication systems with battery overflow. IEEE Trans. Mobile Comput. **16**(1), 185–197 (2016)
16. Lei, L., Kuang, Y., Shen, X.S., Yang, K., Qiao, J., Zhong, Z.: Optimal reliability in energy harvesting industrial wireless sensor networks. IEEE Trans. Wirel. Commun. **15**(8), 5399–5413 (2016)
17. Zlatanov, N., Hadzi-Velkov, Z., Schober, R.: Asymptotically optimal power allocation for point-to-point energy harvesting communication systems. In: Proceedings of the IEEE Global Communications Conference (GLOBECOM), Atlanta, GA, pp. 2502–2507 (2013)
18. Wu, Q., Tao, M., Chen, W., Wu, J.: Optimal energy-efficient transmission for fading channels with an energy harvesting transmitter. In: Proceedings of the IEEE Global Communications Conference (GLOBECOM), Austin, TX, pp. 4294–4299 (2014)
19. Fan, R., Cui, J., Jiang, H., Gu, Q., An, J.: Power allocation for energy harvesting wireless communications with energy state information. IEEE Wirel. Commun. Lett. **8**(1), 201–204 (2018)
20. Weiss, A., Shwartz, A.: Large Deviations For Performance Analysis: Queues, Communication and Computing. Routledge, Milton Park (2019)
21. Akin, S.: Energy harvesting communications under energy underflow constraints. IEEE Commun. Lett. **21**(12), 2646–2649 (2017)

PAPR Analysis of Non-coherent OFDM-IM

Ali Fazeli and Ha H. Nguyen

1 Introduction

Orthogonal division multiplexing with index modulation (OFDM-IM) utilizes the indices (i.e., locations) of subcarriers as an additional information-bearing domain and leads to certain advantages over the conventional OFDM system [1]. One way to realize this technique is to divide the input bits into *index-selector* bits and *mapper* bits. In the first step, the index-selector bits activate a specific combination of subcarriers. Then, the activated subcarriers are modulated by constellation symbols to carry the mapper bits. Therefore, the set of subcarrier activation patterns forms a code book in the index domain, where "extra" bits (index selector) can be mapped onto. In general, active subcarriers can be modulated with symbols (to transmit mapper bits), a procedure similar to the conventional OFDM. Yet, an OFDM-IM system can be designed so that information is only transferred in the index domain [2]. A chief advantage of this approach is the ability to carry out non-coherent communications in fast fading scenarios [3].

Although OFDM-IM has certain advantages over the conventional OFDM, it inherits several drawbacks of the conventional OFDM, including the high peak-to-average power ratio (PAPR). High PAPR of OFDM signals is considered as one of the main challenges in using this technique in many real-world applications, such as low-power Internet of Things (IoT) devices [4]. The problem with high PAPR begins with amplification of OFDM signals before transmission. A characteristic of

A. Fazeli (✉)
Department of Electrical & Computer Engineering, University of Saskatchewan, Saskatoon, SK, Canada
e-mail: ali.fazeli@usask.ca

H. H. Nguyen *(deceased)*
Saskatoon, SK, Canada

H. Nguyen et al. (eds.), *30th Biennial Symposium on Communications 2021*,
Signals and Communication Technology, https://doi.org/10.1007/978-3-031-06947-5_19

every amplifier is its maximum input power threshold. As long as the input power is below the threshold, the output is a linearly amplified version of the input, a desirable outcome. When the input power exceeds the amplifier's threshold, the device saturates and the input signal suffers distortion caused by the nonlinearity of the saturated amplifier. Clipping, as an example of distortion in such situations, causes out-of-band power leakage and consequently degrades the performance. High PAPR in an OFDM system can easily force the amplifier to operate in its nonlinear region, even at low values of the signal average power. Consequently, extensive research has long been conducted to reduce the PAPR of an OFDM system. Block coding [5], interleaving [6], and tone reservation [7] are some of the most effective methods on PAPR reduction.

As a step toward practical realization of OFDM-IM, in this chapter, we investigate the PAPR in an OFDM system based on index modulation. Specifically, a non-coherent OFDM-IM system, where information is transmitted only in the index domain, is considered. We show that the PAPR of such a system reduces as the variance of Hamming weights of the activation code gets smaller. In the ideal case, provided their existence for specific code rates, low-weight constant-weight codes are the best choice in terms of minimizing the PAPR. However, as will be shown, such a choice may not yield the best error performance. To elaborate on this fact, we show how the inherent properties of a block code, such as the minimum Hamming distance and Hamming weight spectrum of the code, set a trade-off between spectral efficiency, error performance, and the PAPR.

Notations Throughout this chapter, boldface letters are used to denote column vectors; for a vector $\mathbf{v}$, the ith element is represented by $[\mathbf{v}]_i$; $\mathcal{CN}(\mathbf{0}, \mathbf{I}_V)$ represents a circularly symmetric complex Gaussian vector with mean $\mathbf{0}$ and identity covariance matrix; diag($\mathbf{v}$) denotes a diagonal matrix with elements of $\mathbf{v}$ on the main diagonal; for two sets ψ_1 and ψ_2, set difference is represented by $\psi_1 \setminus \psi_2$; $\|\cdot\|_0$ denotes the cardinality of a set; and finally, $\binom{V}{k} = \frac{V!}{k!(V-k)!}$.

2 System Model and PAPR Analysis

The block diagram of an OFDM-IM modulator that transmits information in the index domain is depicted in Fig. 1. The *Index Encoder* maps (uniquely) the input information sequence $\mathbf{u}$ of length $\log_2 N$ to a codeword (activation pattern) $\mathbf{I}^{[k]}$, $1 \leq k \leq N$, of length V in a $(V, \log_2 N)$ binary block code $\mathcal{B}$. The selected codeword enters the *OFDM Modulator* to activate subcarriers with power $P_{\text{activ}} > 0$ in accordance with indices of the 1 entries in the selected codeword. The nth time sample of the baseband equivalent to the output of the OFDM modulator can be written as

$$[\mathbf{i}^{[k]}]_n = [\text{IFFT}\{\mathbf{I}^{[k]}\}]_n = \frac{\sqrt{P_{\text{activ}}}}{V} \sum_{l=0}^{V-1} [\mathbf{I}^{[k]}]_\ell e^{j\frac{2\pi \ell n}{V}}, \tag{1}$$

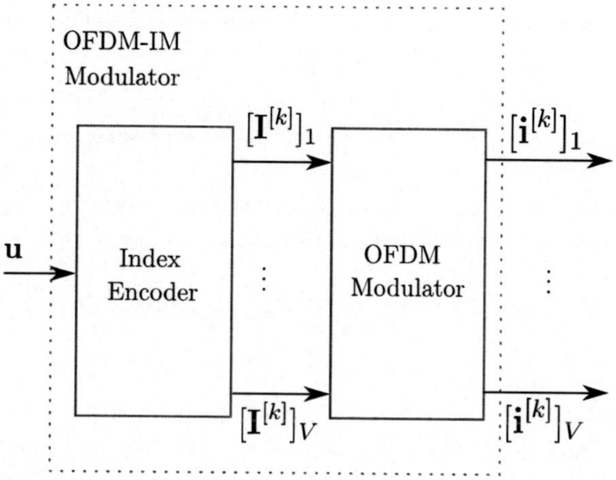

Fig. 1 Block diagram of a non-coherent OFDM-IM modulator

where we have ignored the cyclic prefix portion of the signal as it does not have any effects on our analysis.

For $1 \leq k \leq N$, denote the Hamming weight of $\mathbf{I}^{[k]}$ by Q_k. Then, the power of $\mathbf{I}^{[k]}$ can be computed as

$$(\mathbf{I}^{[k]})^\top \mathbf{I}^{[k]} = P_{\text{activ}} Q_k. \tag{2}$$

From Parseval's theorem, the signal power in time domain is

$$(\mathbf{i}^{[k]})^\top \mathbf{i}^{[k]} = \frac{P_{\text{activ}} Q_k}{V}. \tag{3}$$

Since the codebook is composed of N equiprobable codewords, the average power of the transmitted signal (or equivalently, the codebook power) can be written as

$$P_{\text{avg}} = \frac{P_{\text{activ}}}{VN} \sum_{k=1}^{N} Q_k. \tag{4}$$

It follows from (1) that the maximum instantaneous power (peak power) of the kth transmitted signal can be computed as

$$\begin{aligned} \max_n \{|[\mathbf{i}^{[k]}]_n|^2\} &= \frac{P_{\text{activ}}}{V^2} \sum_{l=0}^{V-1} \sum_{u=0}^{V-1} [\mathbf{I}^{[k]}]_\ell [\mathbf{I}^{[k]}]_u e^{j\frac{2\pi(\ell-u)n}{V}} \\ &= \frac{P_{\text{activ}} Q_k^2}{V^2}, \end{aligned} \tag{5}$$

where the second equality holds because the argument is maximized at $n = 0$. Therefore, the absolute peak power of the code is given by

$$P_{\text{peak}} = \frac{P_{\text{activ}}}{V^2} \max_k Q_k^2. \tag{6}$$

The PAPR is defined as the ratio of the peak power to the average power of the codebook, and given as and (6), we have

$$\text{PAPR} = \frac{P_{\text{peak}}}{P_{\text{avg}}} = \frac{N}{V} \frac{\max_k Q_k^2}{\sum_{k=1}^{N} Q_k}. \tag{7}$$

Now, consider a $(V, \log_2 N)$ binary code $\mathcal{B}$ with the maximum Hamming weight $1 \leq Q_{\max} \leq V$ and the minimum weight $Q_{\min} \leq Q_{\max}$. We define N_i, $i = Q_{\min}, \ldots, Q_{\max}$, as the number of codewords in $\mathcal{B}$ with Hamming weight i. Then, (7) can be rewritten as

$$\begin{aligned} \text{PAPR} &= \frac{N}{V} \frac{Q_{\max}^2}{\sum_{k=Q_{\min}}^{Q_{\max}} k \cdot N_k} \\ &= \frac{N}{V} \frac{Q_{\max}^2}{N_{Q_{\max}} Q_{\max} + \sum_{k=Q_{\min}}^{Q_{\max}-1} k \cdot N_k} \\ &= \frac{N}{V} \frac{Q_{\max}}{N_{Q_{\max}} + \sum_{k=Q_{\min}}^{Q_{\max}-1} \frac{k}{Q_{\max}} \cdot N_k}, \end{aligned} \tag{8}$$

where

$$\sum_{k=Q_{\min}}^{Q_{\max}} N_k = N, \tag{9}$$

and

$$0 \leq N_k \leq \binom{V}{k}, \quad Q_{\min} \leq k \leq Q_{\max}. \tag{10}$$

To establish a connection between the behavior of the PAPR and properties of the codebook, consider the following simple upper bound on the PAPR:

$$\begin{aligned} \text{PAPR} &\leq \text{PAPR}_{\text{u}} \\ &= \frac{N}{V} \frac{Q_{\max}}{N_{Q_{\max}} + \frac{Q_{\min}}{Q_{\max}} \cdot (N - N_{Q_{\max}})}, \end{aligned} \tag{11}$$

which is simply resulted from (8) by replacing $\frac{k}{Q_{\max}}$ (in the denominator) with $\frac{Q_{\min}}{Q_{\max}}$, for every $Q_{\min} \leq k \leq Q_{\max} - 1$. Based on the above upper bound, the following observations are made with respect to the PAPR of the non-coherent OFDM-IM system:

- PAPR_{u} is a strictly *increasing* function of $Q_{\max}$, strictly *decreasing* function of $Q_{\min}$, and strictly *decreasing* function of $\frac{Q_{\min}}{Q_{\max}}$.
- It follows from the above observation that the PAPR of the system decreases when the ratio $\frac{Q_{\min}}{Q_{\max}}$ approaches to 1. Hence, for a fixed $Q_{\max}$, using constant-weight codes (for which $Q_{\max} = Q_{\min}$) gives the minimum PAPR. On the other hand, when $\frac{Q_{\min}}{Q_{\max}}$ is fixed, the PAPR increases with $Q_{\max}$.
- It is concluded from the above second observation that, among all binary codes of the same rate, using the constant-weight code having the minimum Hamming weight results in the minimum PAPR.
- For a fixed code size N and a fixed Hamming weight distribution, the PAPR decreases with the length V of the code. This means that the PAPR can be reduced at the cost of reduced spectral efficiency.

According to the above analysis, although constant-weight codes with smaller Hamming weights are preferred in terms of PAPR, in general, the use of such codes may cost the system a lower spectral efficiency and/or a higher error probability. This is elaborated in more detail in the following remarks.

Remark 1 The maximum size of a constant-weight code of weight Q is $N = \binom{V}{Q}$. It is easy to see that for $Q \leq \lfloor \frac{V}{2} \rfloor$, N increases with Q. Therefore, to provide a higher spectral efficiency, one would increase Q up to $\lfloor \frac{V}{2} \rfloor$, but doing so also increases the PAPR.

Now, suppose the spectral efficiency (rate) $r = \frac{\log_2 N}{V}$ is fixed, and there exist two constant-weight codes, $\mathcal{B}_1$ and $\mathcal{B}_2$, with weights Q_1 and Q_2, respectively, where $Q_2 > Q_1$, which can achieve the desired spectral efficiency. According to the previous analysis, the PAPR of $\mathcal{B}_1$ is smaller than that of $\mathcal{B}_2$. Comparing the asymptotic error performance of the two codes is more involved. According to (4), for a constant-weight code of weight Q, the power of an active index is equal to $P_{\text{activ}} = \frac{P_{\text{avg}} V}{Q}$. Therefore, by increasing the weight Q of the code, the power allocated to each active index decreases, which in turn causes an increase in the probability of error (a negative coding gain). At a quick glance, one may expect that by increasing the weight of a constant-weight code, both PAPR and error performance become worst. However, this is not always true for the error performance.

It is shown in [8] that the non-coherent OFDM-IM system as considered in this paper achieves a diversity order

$$L_{M\backslash m} = \min_{k,j}\{\max\{L_{k\backslash j}, L_{j\backslash k}\}\}, \tag{12}$$

where $L_{k\setminus j} = \| \psi_k \setminus \psi_j \|_0$ ($L_{j\setminus k} = \| \psi_j \setminus \psi_k \|_0$), with ψ_i denoting the set of active indices of an arbitrary codeword $\mathbf{I}^{[i]}$. In the above expression, $\max\{L_{k\setminus j}, L_{j\setminus k}\}$ is interpreted as the *partial Hamming distance* of $\mathbf{I}^{[k]}$ and $\mathbf{I}^{[j]}$. The result in (12) simply states that a non-coherent OFDM-IM system achieves a diversity order equal to the minimum partial Hamming distance among all pairs of codewords in the codebook. It can be easily shown that, for a constant-weight code, the partial Hamming distance between any pair of codewords is exactly half of their Hamming distance. Consequently, in the context of non-coherent OFDM-IM, the diversity order achieved by a constant-weight code is half of the *minimum Hamming distance of the code*.

As noted in Remark 1, by increasing the weight Q up to $\lfloor \frac{V}{2} \rfloor$, the maximum size of the code increases. An implication of this fact is that by fixing the transmission rate and increasing the weight of a constant-weight code, the minimum Hamming distance of the code can be increased, and hence the higher diversity order of the non-coherent OFDM-IM system. This stems from the fact that when the maximum size of a code increases, there will be more freedom in expurgating the code (through coding techniques) to produce a code having a lower rate, but a larger minimum Hamming distance. The above discussion is summarized in the following remarks.

Remark 2 For a given spectral efficiency, although the PAPR and coding gain of a constant-weight code degrade by increasing its Hamming weight, the error performance of the corresponding non-coherent OFDM-IM system can improve, thanks to the higher diversity order that can be achieved by proper coding techniques. This presents a trade-off between PAPR and error performance in using constant-weight codes for non-coherent OFDM-IM systems.

Remark 3 It was shown in [3] that in a non-coherent OFDM-IM system, where the index domain is the sole medium to carry information, using a variable-weight activation code can yield a diversity order higher than the maximum diversity order achieved by using a constant-weight code. As such, the analysis in this paper clearly shows that the worse error performance resulted from using a constant-weight code is the price one has to pay for minimizing the PAPR.

In the next section, we present some examples to corroborate the analysis performed in this section.

3 Examples

Example 1 Consider an OFDM system with $V = 8$ subcarriers and three codebooks $\mathcal{B}_1$, $\mathcal{B}_2$, and $\mathcal{B}_3$ listed in Table 1. All codes are (8, 3) constant-weight codes, with $\mathcal{B}_1$ having a Hamming weight $Q = 1$, whereas the other two codes have a weight of 4. Also, since PAPR is not a function of P_{activ}, we normalize $P_{\text{activ}} = 1$. Figures 2 and 3 plot the envelopes of instantaneous power P_{inst} of all codewords

Table 1 An example of three (8, 3) constant-weight codes

$\mathcal{B}_1$	$\mathcal{B}_2$	$\mathcal{B}_3$
10000000	11110000	11110000
01000000	01111000	00001111
00100000	00111100	11001100
00010000	00011110	11000011
00001000	00001111	00111100
00000100	11011000	00110011
00000010	01101100	10101010
00000001	00110110	01010101

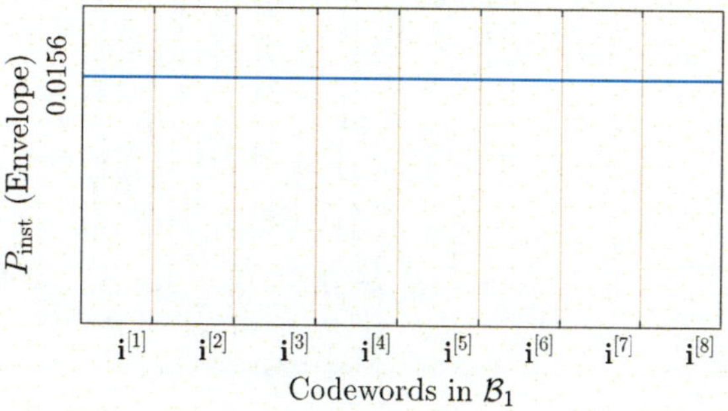

Fig. 2 Envelope of the instantaneous power of all codewords in codebook $\mathcal{B}_1$ in Example 1

in the three codebooks. As can be seen from Fig. 2, all codewords in $\mathcal{B}_1$ have a peak power of 0.0156, which can also be obtained from (5). From (4), the code has an average power of 0.125. So, the PAPR of $\mathcal{B}_1$ is $\text{PAPR}_{\mathcal{B}_1} = 0.1248$. Similarly, from Fig. 3, we observe a peak power of 0.25 for both weight-4 codes, $\mathcal{B}_2$ and $\mathcal{B}_3$. The average power for both $\mathcal{B}_2$ and $\mathcal{B}_3$ is equal to 0.5, which results in $\text{PAPR}_{\mathcal{B}_2} = \text{PAPR}_{\mathcal{B}_3} = 0.5$. Consequently, the PAPR of $\mathcal{B}_1$ is approximately $\frac{1}{4}$ of the PAPR of the other two codes, which is as expected. It should be stressed that, although in this example all codes have the same rate, constant-weight codes of weight 4 and length $V = 8$ can transmit a maximum number of $\lfloor \log_2 \binom{8}{4} \rfloor = 6$ bits, twice as much when compared to the constant-weight code of weight 1 and length 8. This is consistent with the discussion following Remark 1 that the cost of enjoying a smaller PAPR is a lower spectral efficiency.

Now, let us compare error performance in using these three codes in a non-coherent OFDM-IM system, under the scenario that every index (subcarrier) is independently affected by frequency non-selective Rayleigh fading and additive white Gaussian noise [2]. The received signal after cyclic prefix removal and FFT operation can be written as

$$\mathbf{y} = \underline{\mathbf{h}}_D \cdot \mathbf{I}^{[k]} + \mathbf{n}, \tag{13}$$

where $\underline{\mathbf{h}}_D$ =diag($\mathbf{h}$), $\mathbf{h} \sim \mathcal{CN}(\mathbf{0}, \mathbf{I}_V)$ is the frequency domain channel vector, and $\mathbf{n} \sim \mathcal{CN}(\mathbf{0}, N_0\mathbf{I}_V)$ is the vector of noise samples.

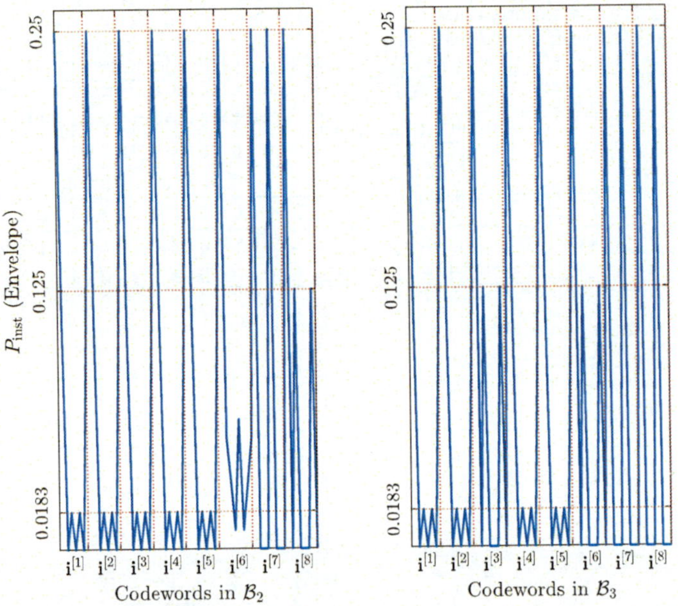

Fig. 3 Envelope of the instantaneous power of all codewords in codebooks $\mathcal{B}_2$ and $\mathcal{B}_3$ in Example 1

Figure 4 depicts the codeword error probability (CEP) for different values of SNR when the maximum likelihood (ML) receiver is employed. Note that for the fair comparison of different codes, we should ensure that the average power per transmitted information bit is the same for all codes. To this end, we plot CEP versus SNR per bit, defined as

$$\rho_b = \frac{W}{N \log_2 N} P_{\text{activ}}, \tag{14}$$

where $\sum_{i=1}^{N} Q_i$ is the sum weight of all codewords in the codebook. Therefore, (14) ensures that a larger power P_{activ} is allocated to the active index of the code that has a smaller weight (i.e., $\mathcal{B}_1$). The behavior of the CEP curves in Fig. 4 is in complete agreement with Remark 2. To see this, let us first compare the performance of codes $\mathcal{B}_1$ and $\mathcal{B}_2$. Code $\mathcal{B}_2$ is an example of an inefficient code in terms of the PAPR and error performance. Specifically, with its larger Hamming weight, its PAPR is larger than the PAPR of code $\mathcal{B}_1$. Also, the larger Hamming weight has caused a coding loss for $\mathcal{B}_2$ as compared with $\mathcal{B}_1$. Finally, the code is not wisely designed to maximize the minimum Hamming distance, which makes the diversity order of $\mathcal{B}_2$ the same as that of $\mathcal{B}_1$ (equal to 1). Thus, not only $\mathcal{B}_2$ has a larger PAPR than $\mathcal{B}_1$, but its error performance is also worst. Codes $\mathcal{B}_1$ and $\mathcal{B}_3$ can be seen as good codes in terms of the PAPR and diversity order, respectively. Since $\mathcal{B}_1$ is a constant-weight

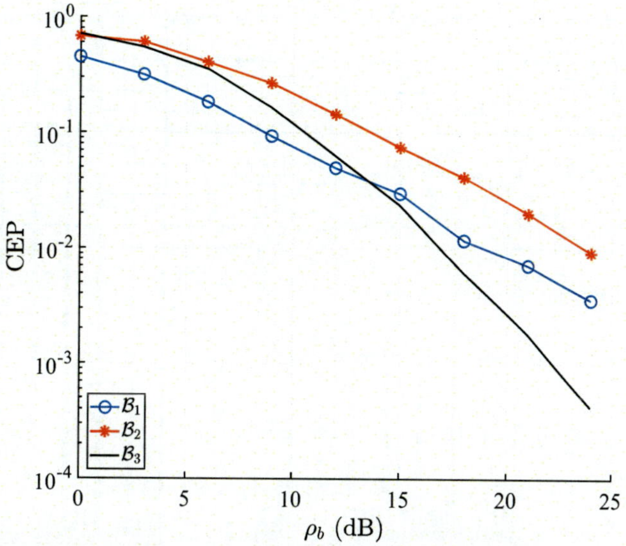

Fig. 4 Codeword error probabilities of three codes considered in Example 1

code of the lowest possible weight, it has the smallest PAPR among all codes of length 8. On the other hand, code $\mathcal{B}_3$ is better designed in the sense that its minimum Hamming distance is twice that of $\mathcal{B}_1$ and $\mathcal{B}_2$. Hence, $\mathcal{B}_3$ exploits a diversity order twice that of $\mathcal{B}_1$ and $\mathcal{B}_2$, which can be clearly seen by comparing slopes of the CEP curves (it is noted that the minimum Hamming distance of $\mathcal{B}_3$ is 4). Therefore, although having a higher PAPR, $\mathcal{B}_3$ significantly outperforms $\mathcal{B}_1$ and $\mathcal{B}_2$ at high SNRs. ♦

Example 2 This example serves to corroborate Remark 3 in the previous section regarding the trade-off between PAPR and error rate for constant-weight and variable-weight codes. Consider the following two (16, 8) codes.

Constant-Weight Code A group of order 21 with two generators $(0, 2, 1, 4, 5, 3, 6)$ $(7, 9, 8, 11, 12, 10, 13)$ and $(0, 2, 4)(1, 3, 5)(7, 9, 11)(8, 10, 12)$, acting on 20 base codewords of length 16, generates 322 codewords of length 16 [9]. In [3], it is shown that this code can be used in a non-coherent OFDM-IM system to carry 8 information bits per codeword, while exploiting a diversity order of 2, which is the maximum achievable diversity order among all constant-weight activation patterns of the same rate. This code has a constant weight of 5, and hence according to (6), the peak power is equal to $P_{\text{peak}} = 0.0977$ (again we assume $P_{\text{activ}} = 1$ as it does not affect the PAPR). Figure 5 plots the envelope of the instantaneous power P_{inst} of 8 codewords in the codebook. According to (4), the average power of the code is $\frac{5}{16}$. As such, we have PAPR$=\frac{16\times.0977}{5} = 0.3126$.

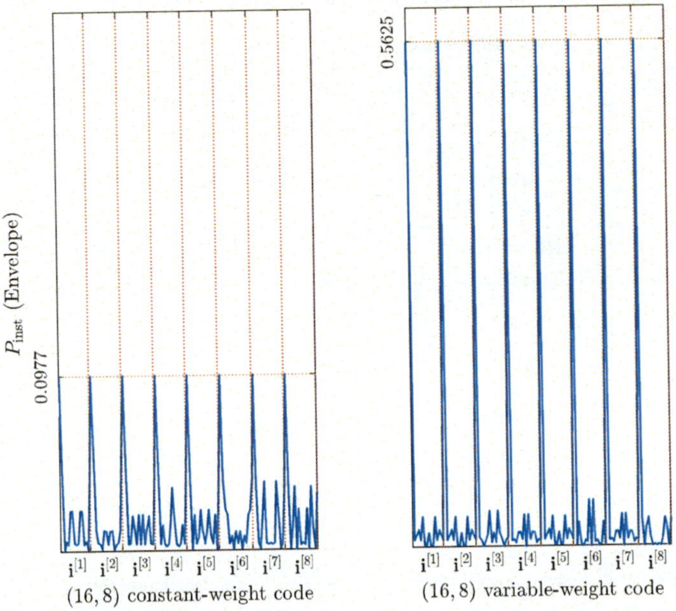

Fig. 5 Envelope of the instantaneous power of 8 codewords in constant-weight and variable-weight codes of Example 2

Variable-Weight Code A (16, 8) variable-weight code can be obtained by puncturing the 16th dimension of the (17,8) cyclic code, whose generating idempotent is $X^3 + X^5 + X^6 + X^7 + X^{10} + X^{11} + X^{12} + X^{14}$ [10]. This code, also capable of carrying 8 bits per codeword, is proved to extract a diversity order of 3 in a non-coherent OFDM-IM system [3]. The maximum Hamming weight of this code is 12, which means $P_{\text{peak}} = 0.5625$. This can be verified in Fig. 5, where the envelope of the instantaneous power of 8 codewords in the codebook having the largest peak power is plotted. On the other hand, the sum of Hamming weights of all codewords equals 2048. Then, $P\text{avg} = 0.5$ is directly resulted from (4). In conclusion, for this code, we have $\text{PAPR} = \frac{0.5625}{0.5} = 1.125$.

Comparing the above results, it can be seen that the PAPR of the variable-weight code is approximately 3.6 times larger than that of the constant-weight code, which means that the constant-weight code is a better choice in terms of the PAPR. However, regarding error performance, the variable-weight code is a better code, specially at high SNRs, as it enjoys a diversity order 1.5 times higher than that of its constant-weight counterpart. This fact can be verified in Fig. 6, which plots the exact and asymptotic CEP curves (versus SNR per bit) for the above two codes. As can be seen, the asymptotic curve has a slope 3 for the variable-weight code, while the high-SNR slope of the constant-weight code is 2. ♦

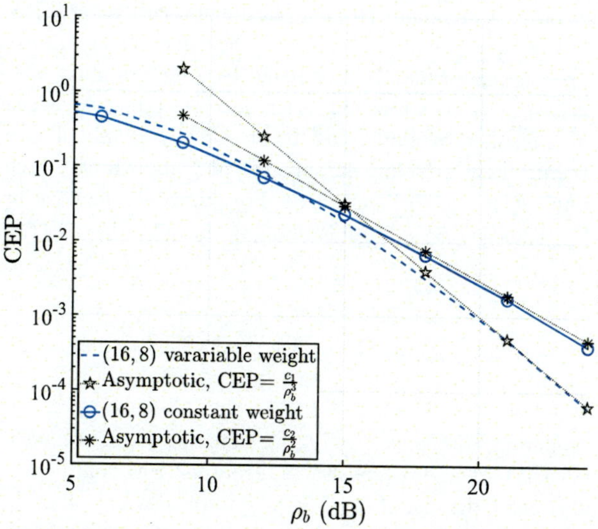

Fig. 6 Exact and asymptotic CEPs for the (16, 8) variable-weight and constant-weight codes

Remark 4 Obviously, PAPR is the property of the transmitted signal and does not depend on statistical properties of the channel. However, the probability of error is affected by the channel, and the CEP results presented in the previous examples are for the case of uncorrelated OFDM subcarriers. The assumption of uncorrelated subcarriers can be realized in practice by subcarrier interleaving of a proper depth. When interleaving cannot be applied, the subcarriers are generally correlated. It is shown in [8] that although the error performance of non-coherent OFDM-IM degrades under correlated subcarriers, one can still design codes that harvest the maximum diversity order equal to the number of independent channel taps in the time domain.

Example 2 is specifically provided to show the trade-off between achievable diversity and PAPR of constant-weight and variable-weight codes. It was concluded that for a fixed rate, a constant-weight code with the smallest weight results in the lowest PAPR. However, it is noted that finding constant-weight codes having good distance properties is difficult, especially for long codes. It appears that the best practical codes designed in the literature are variable-weight codes. In the following, we study properties of a practical code (which is a variable-weight code) when used to generate activation patterns of the proposed IM system.

Example 3 Consider the Reed–Muller code RM(3,7), which is a (128, 64) binary code. The maximum weight of the code is $Q_{\max} = 128$, and the weight distribution of this code is given in [11]. Substituting these values into (8) results in PAPR= 2. ♦

Example 4 (Conventional OFDM) In the non-coherent OFDM-IM system studied in this chapter, patterns of active subcarriers carry information. When every subcarrier is independently modulated by L information bits, a conventional 2^L-ary OFDM system is realized. As a special case, a non-coherent binary OFDM system is resulted by employing the on–off keying modulation in every subcarrier. This is equivalent to having the space of all V-tuple binary vectors as our codebook, i.e., a universal (V, V) binary code. For this code, we have $N = 2^V$, $Q_{\max} = V$, $Q_{\min} = 0$, and $N_k = \binom{V}{k}$. Substituting these values into (8) results

$$\text{PAPR} = 2^V \frac{V}{\sum_{k=0}^{V} k \cdot \binom{V}{k}} = 2. \tag{15}$$

♦

Comparing Examples 3 and 4, we note that (conventional) non-coherent binary OFDM and OFDM-IM (constructed from the RM code) have the same PAPR. On the other hand, the rate of the OFDM system is twice that of the OFDM-IM system. However, regarding the achievable diversity, the OFDM-IM scheme is superior. Indeed, the code in Example 3 has a minimum Hamming distance of 16 and a minimum partial Hamming distance of 8. This means that it can exploit a diversity order of 8, which is eight times higher than the unit diversity of the conventional OFDM.

4 Conclusion

This chapter examined the PAPR in connection with the design of activation patterns for non-coherent OFDM-IM systems. We showed that the PAPR is proportional to the square of the maximum Hamming weight and reciprocal of the average Hamming weight of the activation patterns. Hence, a smaller variance of the distribution of Hamming weights leads to a lower PAPR. For a fixed rate, the lowest PAPR is achieved by using a constant-weight code having the lowest weight. On the other hand, we showed that although constant-weight codes are good choices in terms of the PAPR, they may have lower diversity orders and hence poorer error performance compared to variable-weight codes.

References

1. Basar, E., Aygolu, U., Panayirci, E., Poor, H.V.: Orthogonal frequency division multiplexing with index modulation. IEEE Trans. Signal Process. **61**(22), 5536–5549 (2013)
2. Choi, J.: Noncoherent OFDM-IM and its performance analysis. IEEE Trans. Wirel. Commun. **17**(1), 352–360 (2018)

3. Fazeli, A., Nguyen, H.H.: Code design for non-coherent index modulation. IEEE Commun. Lett. **24**, 477–481 (2020)
4. Schulze, H., Luders, C.: Theory and Applications of OFDM and CDMA: Wideband Wireless Communications. Wiley, Hoboken (2005)
5. Ahn, H.-J., Shin, Y., Im, S.: A block coding scheme for peak-to-average power ratio reduction in an orthogonal frequency division multiplexing system. In: Proceedings of the IEEE Vehicular Technology Conference, Tokyo, pp. 56–60 (2000)
6. Jayalath, A.D.S., Tellambur, C.: The use of interleaving to reduce the peak-to-average power ratio of an OFDM signal. In: Proceedings of the IEEE Global Telecommunications Conference. Conference Record, San Francisco, CA, pp. 82–86 (2000)
7. Slimane, S.B.: Peak-to-average power ratio reduction of OFDM signals using broadband pulse shaping. In: Proceedings of the IEEE Vehicular Technology Conference, Vancouver, BC, pp. 889–893 (2002)
8. Fazeli, A., Nguyen, H.H., Hanif, M.: Generalized OFDM-IM with noncoherent detection. IEEE Trans. Wirel. Commun. **19**(7), 4464–4479 (2020)
9. Brouwer, A.E., Etzion, T.: Some new distance-4 constant weight codes. Adv. Math. Commun. **5**(3), 417–424 (2011)
10. Delsarte, P., Piret, P.: Bounds and constructions for binary asymmetric error-correcting codes. IEEE Trans. Inform. Theory **27**(1), 125–128 (1981)
11. Sugino, M., Ienaga, Y., Tokura, N., Kasami, T.: Weight distribution of (128, 64) Reed-Muller code (Corresp.). IEEE Trans. Inform. Theory **17**(5), 627–628 (1971)

Maximum Multipath Diversity in Coded Cooperative Relay Networks

Alireza Rahmati and Alagan Anpalagan

1 Introduction

Relaying techniques have garnered a significant interest, since the communication range can be significantly extended especially whenever multiple antennas may not be practical for the cellular mobile devices due to size and power limitations. However, for wideband channels, it would still suffer from the frequency selectivity arising from multipath scattering which causes the so-called intersymbol interference (ISI), leading to an inevitable performance degradation. An efficient approach to mitigate ISI is the use of orthogonal frequency division multiplexing (OFDM). In [1], the decoding of OFDM systems with linear constellation pre-coding is done using the maximum-likelihood (ML) detector, and such an approach has been shown to achieve the maximum possible diversity even when full-rate pre-coders are employed. Although significant diversity gains were demonstrated, the complexity of the ML equalizers increases with the channel memory, signal constellation size, and the number of transmit and receive antennas.

The primary work in [2] introduced a variety of cooperative relaying protocols using repetition coding. The most important among them are the amplify-and-forward (AF) protocol and the decode-and-forward (DF) protocol. In [3], the diversity performance of coded beamforming schemes in AF multiple antenna relay systems for frequency-selective channels is investigated, and a code construction criterion to achieve full diversity is obtained. Even though the relay operation for the AF relay is simple, their transceivers require expensive radio frequency amplifiers. For the DF relay network, a cooperative MRC (C-MRC) was proposed in [4], and it was proved that complicated C-MRC, for the uncoded single-antenna DF

A. Rahmati and A. Anpalagan (✉)
ELCE Department, Ryerson University, Toronto, ON, Canada
e-mail: alagan@ryerson.ca

H. Nguyen et al. (eds.), *30th Biennial Symposium on Communications 2021*,
Signals and Communication Technology, https://doi.org/10.1007/978-3-031-06947-5_20

relay network, can achieve full diversity. This, in turn, places significant additional computational and power consumption loads on the receiver side. In a multiple relay networks, there are several methods for utilizing the spatial diversity of the system. Best relay selection [5] and distributed space-time coding [6] are other ways of obtaining diversity. Cooperative beamforming [7] is a useful method for achieving full diversity order and uses bandwidth and power more efficiently. The extra bandwidth and energy for sending back this information from destination to relays in a distributed network is one of the drawbacks of cooperative distributed beamforming.

Coded cooperation uses the available bandwidth efficiently and adjusts its coding rate to adapt to the varying channel conditions. The key characteristic of coded cooperation is integrating channel coding and cooperation, which is achieved by partitioning. In other words, coded cooperation allows the user to transmit part of the codeword, while the other part is transmitted by a partner user through partial or complete decoding. In the research literature, several works have been proposed to investigate code partitioning, e.g., using convolutional codes [8], using distributed space-time coding [9], using distributed low density parity check codes [10], as well as, using distributed turbo coding schemes [11]. The merits of such schemes come at the cost of high computational complexity; however, in applications where the receiver has limited processing power, it is important to investigate equalization techniques that strike a balance between computational complexity and performance. In this work, we propose a low-complexity iterative receiver for modified coded cooperation transmission in relay networks. We further show that the proposed receiver can collect full multipath and spatial diversity orders at low-complexity implementation. In our earlier paper [12], an iterative receiver method has been proposed to recovery symbols on spectral nulls and erasures in Rayleigh fading and erasure channels, respectively.

The rest of this chapter is organized as follows. In Sect. 2 modified coded cooperative diversity for OFDM system is reviewed. The performance analysis for our proposed algorithm is provided in Sect. 3, and 4 concludes this chapter.

2 Modified Coded Cooperative Diversity for OFDM Systems in Relay Networks

Two source nodes, A and B, are paired as partners for transmission of their data to a common destination node D. In a relay network, all terminals are equipped with a single transmit and receive antenna. It is assumed that the half-duplex constraint is imposed on the cooperating nodes. The channel impulse responses (CIRs) for A $\rightarrow$ B, B $\rightarrow$ A, A $\rightarrow$ D, and B $\rightarrow$ D links for each transmission block are given by $\mathbf{h_{AB}} = \mathbf{h_{BA}} = [h_{AB}(0), h_{AB}(1), \ldots, h_{AB}(L_{AB})]^T$, $\mathbf{h_{AD}} = [h_{AD}(0), h_{AD}(1), \ldots, h_{AD}(L_{AD})]^T$, and $\mathbf{h_{BD}} = [h_{BD}(0), h_{BD}(1), \ldots, h_{BD}(L_{BD})]^T$ respectively, where L_{AB}, L_{AD}, and L_{BD} denote the corresponding channel memory lengths. Reciprocal channel is assumed between two given nodes of A and B. Node A in OFDM modified coded cooperative diversity transmits a part of its data in cooperation step k, and node B transmits other part of node A's data in cooperation step $k+1$.

Note that when a user acts as relay for its partner, it simultaneously transmits its own data packet and the packet for which it acts as relay, using a form of superposition modulation in modified coded cooperation [13]. For transmitting a part of node A's data, a diagonal puncturing matrix of $\mathbf{G}_t = \text{diag}(g_{t1}, \ldots, g_{tN})$, $g_{tn} \in \{0, 1\}$, $n \in \{1, 2, ..., N\}$, is inserted after the IFFT transform at the transmit side of node A in step k, see Fig. 1a, where $\mathbf{c'}^{\mathbf{k}}_{\mathbf{A}} = [c'^{k}_{A}(1), c'^{k}_{A}(2), \ldots, c'^{k}_{A}(N)]^T$ and $c'^{k}_{A}(n)$ as a 2^Q−PSK or 2^Q−QAM symbol sequence, where Q is the number of bits per symbol, and are segmented into blocks with size N to provide node A's total data $\mathbf{c'}^{\mathbf{k}}_{\mathbf{A}}$. Then, $\mathbf{c'}^{\mathbf{k}}_{\mathbf{A}}$ is punctured with block of $\mathbf{G}_t$ to provide node A's data in cooperation step k, $\mathbf{s}^{\mathbf{k}}_{\mathbf{A}}$. The remaining data of node A is transmitted by the puncturing matrix of $\mathbf{I}_N - \mathbf{G}_t$ at node B in step $k+1$ where $\mathbf{I}_N$ is identity matrix of size N. Because of the application of the matrix $\mathbf{G}_t$ and $\mathbf{I}_N - \mathbf{G}_t$ in time domain, we have post-coded OFDM [14] in both nodes of A and B. This post-coding scheme can be made equivalent to a pre-coding scheme by selecting,

$$\left.\begin{aligned} \mathbf{s}^{\mathbf{k}}_{\mathbf{A}} &= \mathbf{G}_t \mathbf{F}^H_N \mathbf{c}'^{\mathbf{k}}_{\mathbf{A}} \\ \mathbf{s}^{\mathbf{k}}_{\mathbf{A}} &= \mathbf{F}^H_N \mathbf{G}_f \mathbf{c}'^{\mathbf{k}}_{\mathbf{A}} \end{aligned}\right\} \Rightarrow \mathbf{G}_f = \mathbf{F}_N \mathbf{G}_t \mathbf{F}^H_N, \tag{1}$$

where $\mathbf{G}_f$ is a frequency domain pre-coding matrix. Note that $\mathbf{G}_f$ is a circulant matrix due to the FFT processing on the diagonal matrix $\mathbf{G}_t$. Also, $\mathbf{G}_f$ is a singular matrix due to the special structure of the diagonal matrix $\mathbf{G}_t$ with entries $g_{ti} \in \{0, 1\}$. In modeling the erasure channel, $\mathbf{G}_t$ is a realization of spreading codes $\mathbf{G}_t = \text{diag}(\mathbf{F}^H_N \mathbf{cf})$ (please see equation (17) in [14]) with bandwidth efficiency, constant Euclidean distance, and low computational complexity. This allows us to model OFDM symbols in erasure channel as post-coded OFDM with $\uparrow L = 1$. In other words, for the first time, a simple post-coded OFDM in erasure channel with up-sampling $\uparrow L = 1$.

In Table 1, we compare the complexity of our system and the proposed system in [14]. The data symbols are spread across Q distinct groups of M subcarriers in pre-coded OFDM partial spreading. Note that our proposed OFDM system is capable of lowering the implementation cost of OFDM system. Figure 1a and b shows the post-coded and its equivalent pre-coded OFDM system for node A. To remove inter-block interference and make the channel matrix circulant, a cyclic prefix (CP) with length $l = \max(L_{AB}, L_{AD})$ and $l = \max(L_{BA}, L_{BD})$ is added between adjacent information blocks at node A and node B, respectively. Conventional and the iterative reconstruction of transmitted data in step k, after removing CP, are summarized as follows.

In conventional OFDM systems, node B employs the receiver Fig. 1c and the received signal after the FFT transform can be written as

Table 1 Comparison of computation cost of different operations in pre-coded and post-coded OFDM systems

	Post-coded OFDM [14] (full spreading)	Post-coded OFDM [14] (partial spreading $N = MQ$)	Proposed system (with $\uparrow L = 1$)	Pre-coded OFDM with Zero forcing
FFT	$\mathcal{O}(N\,logN)$	$\mathcal{O}(N\,logN)$	$\mathcal{O}(N\,logN)$	$\mathcal{O}(NL\,logNL)$
IFFT	$\mathcal{O}(NL\,logN)$	$\mathcal{O}(NL\,logN)$	$\mathcal{O}(N\,logN)$	$\mathcal{O}(NL\,logNL)$
Encoding	$\mathcal{O}(N)$	$\mathcal{O}(N)$	$\mathcal{O}(N)$	$\mathcal{O}(N^2L)$
Detection	$\mathcal{O}(N^{3.5})$	$\mathcal{O}(QM^{3.5})$	$\mathcal{O}(N^2)$	$\mathcal{O}(N^3)$

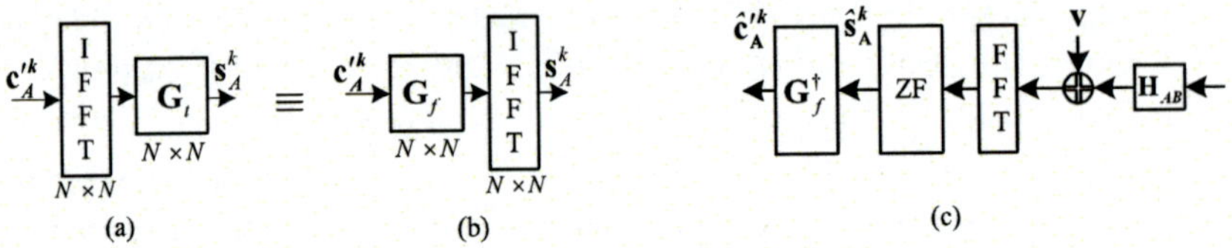

Fig. 1 Node A in cooperation step k as (**a**) post-coded OFDM and (**b**) its equivalent pre-coded OFDM system. (**c**) Node B in cooperation step $k+1$ to decode $\mathbf{c'^k_A}$ from part of received signal $\mathbf{s^k_A}$

$$\mathbf{r^k_B} = \mathbf{F_N H_{AB} s^k_A} + \mathbf{F_N \nu_{AB}} = \mathbf{D_{AB} G_f c'^k_A} + \mathbf{F_N \nu_{AB}}, \tag{2}$$

in which

$$\mathbf{D_{AB}} = \mathrm{diag}(H_{AB}(0), H_{AB}(1), \ldots, H_{AB}(N-1)) = \mathbf{F_N H_{AB} F^H_N}, \tag{3}$$

where $H_{AB}(n) = \sum_{l=0}^{L} h_{AB}(l)\exp(-j2\pi nl/N)$ is the frequency response of the underlying FIR channel evaluated at the FFT grid, and $\nu_{\mathbf{AB}}$ denotes the white Gaussian noise with zero mean and variance N_0. Hence, after zero forcing (ZF) equalization block, we obtain $\hat{\mathbf{s}}^{\mathbf{k}}_{\mathbf{A}} = \mathbf{G_f c'^k_A} + \mathbf{D^{-1}_{AB} F_N \nu_{AB}}$, where F_N is the FFT matrix. The statement of the problem is that given the output $\mathbf{G_f c'^k_A}$ vector, we wish to find the input vector $\mathbf{c'^k_A}$, i.e., to find the inverse of the system (see Fig. 2a). An iterative receiver was proposed in [12], which can calculate $\mathbf{G}^{\dagger}_{f}$ indirectly with the finite number of iterations. The overall processing in uniform sampling stage can be concisely written as $\mathbf{G}_t\mathbf{F^H_N}$. Note that at the output of this stage, node B has the uniform sampled version of node A's data because of the puncturing matrix $\mathbf{G}_t$ that discards some symbols from each OFDM symbol in time domain. For reconstructing lost symbols from the remaining ones, $\boldsymbol{\Psi}_{\mathbf{A}} = \mathbf{G}_t\mathbf{F^H_N}\hat{\mathbf{s}}^{\mathbf{k}}_{\mathbf{A}}$, the iterative reconstruction block in Fig. 2b which uses the puncturing matrix $\mathbf{G}_t$ is used. This iterative reconstruction block can be modelled as

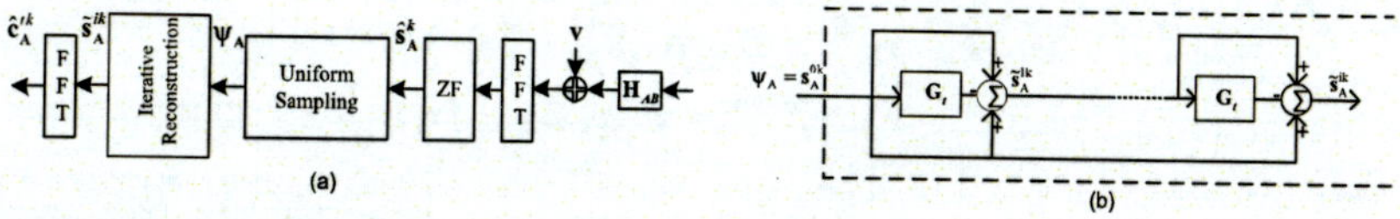

Fig. 2 (a) Node B in cooperation step $k+1$ to iteratively decode $\mathbf{c}'^{\mathbf{k}}_{\mathbf{A}}$ from part of received signal $\mathbf{s}^{\mathbf{k}}_{\mathbf{A}}$ and (b) iterative reconstruction block

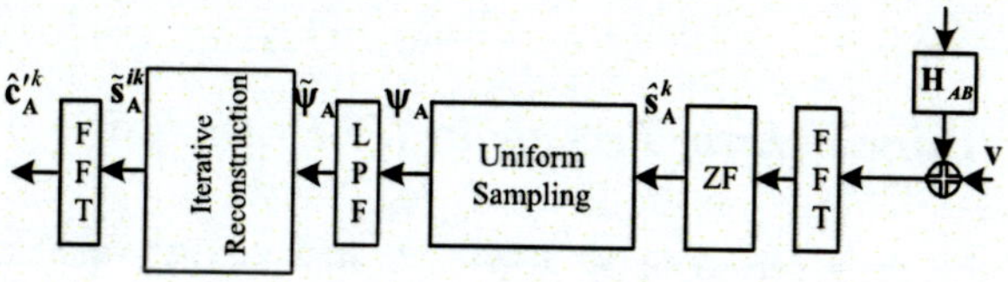

Fig. 3 Convergent iterative receiver for node B in cooperation step $k+1$

$$\begin{cases} \tilde{\mathbf{s}}^{ik}_{\mathbf{A}} = \boldsymbol{\Psi}_{\mathbf{A}} + (\mathbf{I}_N - \mathbf{G}_t)\,\tilde{\mathbf{s}}^{(i-1)k}_{\mathbf{A}} & i > 0, \\ \tilde{\mathbf{s}}^{0k}_{\mathbf{A}} = \boldsymbol{\Psi}_{\mathbf{A}} & i = 0, \end{cases} \tag{4}$$

where $\tilde{\mathbf{s}}^{ik}_{\mathbf{A}}$ and $\tilde{\mathbf{s}}^{(i-1)k}_{\mathbf{A}}$ are the reconstructed vector of node A's data in ith and $(i-1)$th iteration for kth step of cooperation, respectively. The use of iterative equalizer and decoding is common in communication systems [12]. It has been shown in [15] that $\mathbf{c}'^{\mathbf{k}}_{\mathbf{A}}$ can be completely obtained from a part of transmitted data, $\mathbf{s}^{\mathbf{k}}_{\mathbf{A}}$, in node B at the infinite iteration as follows:

$$\hat{\mathbf{c}}'^{\mathbf{k}}_{\mathbf{A}} = \mathbf{c}'^{\mathbf{k}}_{\mathbf{A}} + \mathbf{D}^{-1}_{\mathbf{AB}}\mathbf{F}_{\mathbf{N}}\boldsymbol{\nu}_{\mathbf{AB}}. \tag{5}$$

It can also be shown that the $\hat{\mathbf{c}}'^{\mathbf{k}}_{\mathbf{A}}$ for arbitrary ith iteration can be obtained from

$$\hat{\mathbf{c}}^{\mathbf{k}}_{\mathbf{A}} = \mathbf{F}_{\mathbf{N}}\left(\mathbf{I}_{\mathbf{N}} - (\mathbf{I}_{\mathbf{N}} - \mathbf{G}_t)^{i+1}\right)\left(\mathbf{G}_t\mathbf{F}^{\mathbf{H}}_{\mathbf{N}}\mathbf{c}'^{\mathbf{k}}_{\mathbf{A}} + \mathbf{F}^{\mathbf{H}}_{\mathbf{N}}\mathbf{D}^{-1}_{\mathbf{AB}}\mathbf{F}_{\mathbf{N}}\boldsymbol{\nu}_{\mathbf{AB}}\right). \tag{6}$$

To have a convergent iterative receiver, oversampled OFDM system is defined in [15]. The receive side for oversampled OFDM system which uses the convergent iterative reconstruction block is depicted in Fig. 3. Based on this figure, we can write

$$\hat{\mathbf{c}}'^{\mathbf{k}}_{\mathbf{A}} = \mathbf{F}_N\left(\mathbf{I}_N - (\mathbf{I}_N - \mathbf{F}^H_N\mathbf{E}\mathbf{F}_N\mathbf{G}_t)^{i+1}\right)\left(\mathbf{G}_t\mathbf{F}^{\mathbf{H}}_{\mathbf{N}}\mathbf{c}'^{\mathbf{k}}_{\mathbf{A}} + \mathbf{F}^{\mathbf{H}}_{\mathbf{N}}\mathbf{D}^{-1}_{\mathbf{AB}}\mathbf{F}_{\mathbf{N}}\boldsymbol{\nu}_{\mathbf{AB}}\right). \tag{7}$$

Hence, the received signals after N−point FFT at destination node D in cooperation steps k and $k+1$ can be summarized as follows:

$$\begin{aligned}&\text{step } k: \ \mathbf{r_D}^{\mathbf{k}} = \mathbf{D_{AD}G_f c'^{k}_{A}} + \mathbf{F_N \nu_{AD}}\\&\text{step } k+1: \mathbf{r_D}^{\mathbf{k+1}} = \mathbf{D_{BD}}\left(\mathbf{I}_N - \mathbf{G_f}\right)\hat{\mathbf{c}}'^{\mathbf{k}}_{\mathbf{A}} + \mathbf{F_N \nu_{BD}},\end{aligned} \tag{8}$$

in which node A transmits a fraction of its own data, $\mathbf{G_f c'^{k}_{A}}$, in cooperation step k, and node B transmits the remaining data of node A, $\left(\mathbf{I}_N - \mathbf{G_f}\right)\hat{\mathbf{c}}'^{\mathbf{k}}_{\mathbf{A}}$, in cooperation step $k+1$.

2.1 Pairwise Error Probability (PEP) Analysis

Note that the received signal at node B after the FFT transform can be written as $\mathbf{r^k_B} = \mathbf{F_N H_{AB} s^k_A} + \mathbf{F_N \nu_{AB}}$. In [16], it was shown that if the channel taps $h_{AB}(l)$ are Gaussian and independent, then it is possible to obtain maximal (i.e., $(L+1)$-order) multipath diversity using maximum-likelihood (ML) decoding, which estimates $\mathbf{c'^k_A}$ from $\mathbf{c'^k_A} = \arg\min \left\| \mathbf{r^k_B} - \mathbf{D_{AB}G_f c'^k_A} \right\|$, where $\|.\|$ denotes the Frobenius matrix norm. Here, we will prove that maximum diversity is possible with proposed iterative block equalization at node B for a specific equalizer $\mathbf{T} = \mathbf{F_N}\left(\mathbf{I_N} - (\mathbf{I_N} - \mathbf{G}_t)^{i+1}\right)\mathbf{F^H_N D^{\dagger}_{AB}}$, see (6), that satisfies $\mathbf{T}\left(\mathbf{D_{AB}G_f}\right) = \mathbf{I_N}$.

Proof First, $\mathbf{T}\left(\mathbf{D_{AB}G_f}\right) = \mathbf{I_N}$ can be derived as follows:

$$\begin{aligned}\mathbf{T}(\mathbf{D_{AB}G_f}) &= \mathbf{F_N}\left(\underbrace{\mathbf{I_N} - \overbrace{(\mathbf{I_N} - \mathbf{G}_t)}^{\mathbf{I_N}-\mathbf{G}_t}{}^{i+1}}_{\mathbf{G}_t}\right)\mathbf{F^H_N D^{\dagger}_{AB} D_{AB} G_f}\\&= \mathbf{F_N G}_t\mathbf{F^H_N}(\mathbf{G_f}) = \mathbf{G^{\dagger}_f}(\mathbf{G_f}) = \mathbf{I_N}.\end{aligned} \tag{9}$$

Note that the same approach can be considered for equalizer designing in the oversampled OFDM system which practically converges, Appendix II.C in [15].

If we use $\mathbf{T} = \mathbf{F_N}\left(\mathbf{I_N} - (\mathbf{I_N} - \mathbf{G}_t)^{i+1}\right)\mathbf{F^H_N D^{\dagger}_{AB}}$ equalizer at node B, we have

$$\hat{\mathbf{c}}'^{\mathbf{k}}_{\mathbf{A}} = \mathbf{c'^k_A} + \mathbf{F_N}\left(\mathbf{I_N} - (\mathbf{I_N} - \mathbf{G}_t)^{i+1}\right)\mathbf{F^H_N D^{\dagger}_{AB} F_N \nu_{AB}} = \mathbf{c'^k_A} + \mathbf{n}, \tag{10}$$

where $\mathbf{n} = \mathbf{F_N}\left(\mathbf{I_N} - (\mathbf{I_N} - \mathbf{G}_t)^{i+1}\right)\mathbf{F^H_N D^{\dagger}_{AB} F_N \nu_{AB}}$. The estimate of the transmitted vector is then given by finding $\arg\min \left\| \hat{\mathbf{c}}'^{\mathbf{k}}_{\mathbf{A}} - \mathbf{c'^k_A} \right\|$ for all possible values of $\mathbf{c'^k_A}$. This is a suboptimal decision rule, since the noise vector $\mathbf{n}$ is no longer white and its covariance depends on the equalizer matrix.

In what follows, for simplicity, we drop the k and $\mathbf{A}$ in (10). For the PEP analysis, we will be looking at the probability that a symbol $\mathbf{c'}_i$ is sent, but another block $\mathbf{c'}_j \neq \mathbf{c'}_i$ is received at node B. It is straightforward to show that

$$P\left[\mathbf{c'}_i \rightarrow \mathbf{c'}_j|\mathbf{h}_{AB}\right] = Q\left(\frac{d_{ij}}{\sqrt{2N_0\left\|\left(\mathbf{D_{AB}F_N}(\mathbf{I_N}-(\mathbf{I_N}-\mathbf{G}_t)^{i+1})\mathbf{F_N^H}\right)^{\dagger H}\mathbf{e}_{ij}\right\|^2}}\right), \tag{11}$$

where $d_{ij} = \left\|\mathbf{c'}_j - \mathbf{c'}_i\right\|$, $\mathbf{e}_{ij} = {(\mathbf{c'}_i - \mathbf{c'}_j)}/{\left\|\mathbf{c'}_j - \mathbf{c'}_i\right\|}$, and $Q(.)$ is the error function. Applying the Chernoff bound to (11), we get

$$Q\left(\frac{d_{ij}}{\sqrt{2N_0\left\|\left(\mathbf{D_{AB}F_N}(\mathbf{I_N}-(\mathbf{I_N}-\mathbf{G}_t)^{i+1})\mathbf{F_N^H}\right)^{\dagger H}\mathbf{e}_{ij}\right\|^2}}\right) \leq \frac{1}{2}\exp\left(\frac{-d_{ij}^2}{4N_0\|\mathbf{T}\|^2}\right) \tag{12}$$

where the following property of Frobenius norm is used:

$$\begin{aligned}
&\left\|\left(\mathbf{D_{AB}F_N}\left(\mathbf{I_N}-(\mathbf{I_N}-\mathbf{G}_t)^{i+1}\right)\mathbf{F_N^H}\right)^{\dagger H}\mathbf{e}_{ij}\right\| \\
&\leq \left\|\left(\mathbf{D_{AB}F_N}\left(\mathbf{I_N}-(\mathbf{I_N}-\mathbf{G}_t)^{i+1}\right)\mathbf{F_N^H}\right)^{\dagger H}\right\|\left\|\mathbf{e}_{ij}\right\| \\
&= \left\|\left(\mathbf{D_{AB}F_N}\left(\mathbf{I_N}-(\mathbf{I_N}-\mathbf{G}_t)^{i+1}\right)\mathbf{F_N^H}\right)^{\dagger H}\right\| = \|\mathbf{T}\|.
\end{aligned} \tag{13}$$

Henceforth, we show if there exists some constant $C > 0$ independent of $\mathbf{h}_{AB}$ such that $\|\mathbf{T}\|^{-2} \leq C\|\mathbf{h}_{AB}\|^2$, then the iterative equalizer $\mathbf{T} = \mathbf{F_N}\left(\mathbf{I_N}-(\mathbf{I_N}-\mathbf{G}_t)^{i+1}\right)\mathbf{F_N^H}\mathbf{D}^{\dagger}_{\mathbf{AB}}$ provides maximum multipath diversity.

By the premise, the PEP can be upper-bounded by $\exp\left(\frac{-d_{ij}^2C\|\mathbf{h}_{AB}\|^2}{4N_0}\right)$. Averaging this upper bound with respect to the distribution of the Gaussian channel vector $\mathbf{h}_{AB}$, we obtain $E_{\mathbf{h}}\left[P\left[\mathbf{c'}_i \rightarrow \mathbf{c'}_j|\mathbf{h}_{AB}\right]\right] \leq \left(\frac{-d_{ij}^2C}{4N_0}\right)^{-r}\left(\prod_{l=0}^{r-1}\lambda_l\right)^{-l}$, where λ_l are the eigenvalues of $R_{\mathbf{h}_{AB}} = E\left[\mathbf{h}_{AB}\mathbf{h}_{AB}^H\right]$, r is the rank of $R_{\mathbf{h}_{AB}}$, and $\frac{-d_{ij}^2C}{4N_0} = K \times SNR$, where K is a constant. We observe that r is an exponent of the SNR, which means that there is rth-order diversity. Since the maximum diversity offered by the channel is given by r, it proves that the iterative equalizer $\mathbf{T} = \mathbf{F_N}\left(\mathbf{I_N}-(\mathbf{I_N}-\mathbf{G}_t)^{i+1}\right)\mathbf{F_N^H}\mathbf{D}^{\dagger}_{\mathbf{AB}}$ provides maximum multipath diversity.

□

3 Results and Discussion

In this section, we present simulation results to support the analytical results derived in this chapter. For OFDM modulation, $N = 128$ subcarriers are used with 16-QAM. Figure 4 compares the performance of the modified coded cooperation system using the iterative receiver at the destination with the space-time coded OFDM system with the same bandwidth. We examine the following cases: (i) nTx = 2, nRx = 1, (ii) nTx = 1, nRx = 1 for iterative receiver, and (iii) nTx = 2, nRx = 2. We observe from the figure that the slopes of the performance curves corresponding to cases (i) and (ii) are the same, which agrees with the analytical results derived for diversity order in full cooperation between two nodes. In other words, the diversity order that is linearly proportional to maximum number of diversity paths (nTx×nRx) for case (i) is the same as case (ii). As we used nTx = 1, nRx = 1 for proposed iterative receiver, it confirms that the maximum diversity is possible with the proposed iterative block at the partner node for the specific equalizer at the partner node. A glance at Fig. 5 provided reveals that better PEP performance can be achieved with the employment of the iterative receiver, rather than the conventional one in the cooperation scheme in relay networks. The performance of DF relaying scheme for OFDM system with the same bandwidth and cooperation level is also shown in Fig. 5, where (511, 130) BCH codes are applied at the cooperative nodes. Due to the error propagation, the performance of DF relaying system is poor. PEP of the space time coded cooperation in OFDM system is also depicted in Fig. 5, which is very similar to that the coded cooperation has benefited from the iterative receiver.

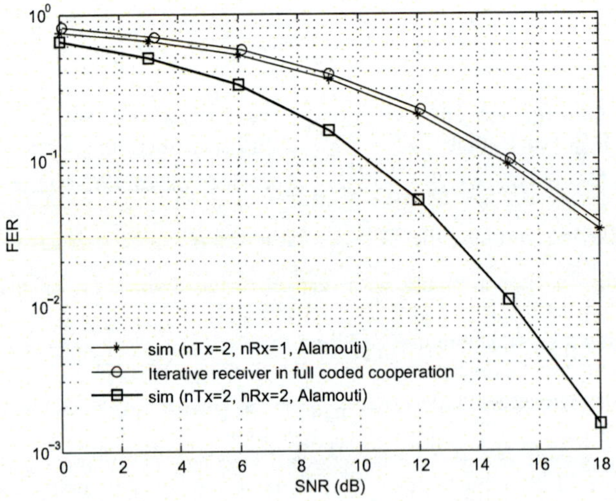

Fig. 4 FER vs. SNR at the destination node for the iterative receiver and the one using Alamouti decoding

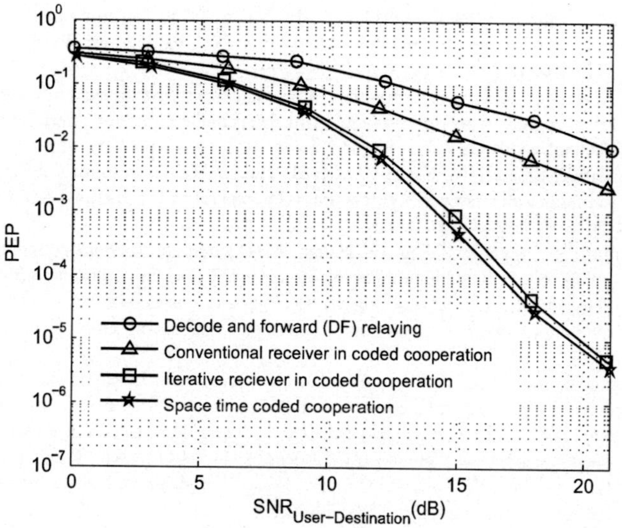

Fig. 5 User node PEP versus the user destination channel SNR for the coded cooperation using the iterative receiver, space-time coded cooperation, and DF relaying scheme in the OFDM system

4 Conclusion

We proposed a novel reduced-complexity iterative-based receiver for modified coded cooperative transmissions in relay networks. We showed that, by incorporating simple puncturing technique, our receiver is able to collect full antenna and multipath diversity gains. The performance of the proposed scheme in full-duplex will be derived in our future works.

References

1. Venturino, L., et al.: Design of linear dispersion codes for practical MIMO-OFDM systems. IEEE J. Sel. Top. Signal Process. **1**(1), 178–188 (2007)
2. Laneman, J.N., et al.: Cooperative diversity in wireless networks: efficient protocols and outage behavior. IEEE Trans. Inform. Theory **50**(12), 3062–3080 (2004)
3. Song, C., Lee, I.: Diversity analysis of coded beamforming in MIMO-OFDM amplify-and-forward relaying systems. IEEE Trans. Wirel. Commun. **10**(8), 2445–2450 (2011)
4. Wang, T., Cano, A., Giannakis, G., Laneman, J.: High-performance cooperative demodulation with decode-and-forward relays. IEEE Trans. Commun. **55**(7), 1427–1438 (2007)
5. Silva, S., Amarasuriya, G., Tellambura, C., Ardakani, M.: Relay selection strategies for MIMO two-way relay networks with spatial multiplexing. IEEE Trans. Commun. **63**(12), 4694–4710 (2015)

6. Yang, K., Yang, N., Xing, C., Wu, J., Zhang, Z.: Space-time network coding with transmit antenna selection and maximal-ratio combining. IEEE Trans. Wirel. Commun. **14**(4), 2106–2117 (2015)
7. Afana, A., Ghrayeb, A., Asghari, V.R., Affes, S.: Distributed beamforming for two-way DF relay cognitive networks under primary-secondary mutual interference. IEEE Trans. Veh. Technol. **64**(9), 3918–3930 (2015)
8. Karbalayghareh, A., Nasiri-Kenari, M., Hejazi, M.: Convolutional network-coded cooperation in multi-source networks with a multi-antenna relay. IEEE Trans. Commun. **13**(8), 4323–4333 (2014)
9. Sun, H., Ng, S.X., Dong, C., Hanzo, L.: Decode-and-forward cooperation-aided triple-layer turbo-trellis-coded hierarchical modulation. IEEE Trans. Commun. **63**(4), 1136–1148 (2015)
10. Schwandter, S., Amat, A.G.i, Matz, G.: Spatially-coupled LDPC codes for decode-and-forward relaying of two correlated sources over the BEC. IEEE Trans. Commun. **62**(4), 1324–1337 (2014)
11. Lu, X., Li, J., Liu, Y.: New soft-encoding relay (SoER) mechanisms for wireless relay systems: convolutional and turbo constructions. IEEE Trans. Wirel. Commun. **15**(4), 2443–2457 (2016)
12. Rahmati, A., Raahemifar, K., Tsiftsis, T., Anpalagan, A., Azmi, P.: OFDM signal recovery in deep faded erasure channel. IEEE Access **7**, 1–1, 10 (2018)
13. Larsson, E.G., Vojcic, B.R.: Cooperative transmit diversity based on superposition modulation. IEEE Commun. Lett. **9**(9), 778–780 (2005)
14. Shah, S.F.A., Tewfik, A.H.: Design and analysis of post-coded OFDM systems. IEEE Trans. Wirel. Commun. **7**(12), 4907–4918 (2008)
15. Rahmati, A., Raahemifar, K., Anpalagan, A., Tsiftsis, T.A., Azmi, P., Miridakis, N.I.: Superposition modulation based cooperation for oversampled OFDM signals. IEEE Trans. Commun. **65**(11), 4791–4802 (2017)
16. Chung, C.-D.: Spectral precoding for constant-envelope OFDM. IEEE Trans. Commun. **58**(2), 555–567 (2010)

Index

H. Nguyen et al. (eds.), *30th Biennial Symposium on Communications 2021*, Signals and Communication Technology, https://doi.org/10.1007/978-3-031-06947-5